Food Addiction and Eating Addiction

Food Addiction and Eating Addiction: Scientific Advances and Their Clinical, Social and Policy Implications

Special Issue Editors

Adrian Carter
Tracy Burrows
Charlotte Hardman

MDPI • Basel • Beijing • Wuhan • Barcelona • Belgrade • Manchester • Tokyo • Cluj • Tianjin

Special Issue Editors

Adrian Carter	Tracy Burrows	Charlotte Hardman
Monash University	University of Newcastle	University of Liverpool
Australia	Australia	UK

Editorial Office
MDPI
St. Alban-Anlage 66
4052 Basel, Switzerland

This is a reprint of articles from the Special Issue published online in the open access journal *Nutrients* (ISSN 2072-6643) (available at: https://www.mdpi.com/journal/nutrients/special_issues/Food_Addiction_Eating_Addiction).

For citation purposes, cite each article independently as indicated on the article page online and as indicated below:

LastName, A.A.; LastName, B.B.; LastName, C.C. Article Title. *Journal Name* **Year**, *Article Number*, Page Range.

ISBN 978-3-03936-358-2 (Pbk)
ISBN 978-3-03936-359-9 (PDF)

© 2020 by the authors. Articles in this book are Open Access and distributed under the Creative Commons Attribution (CC BY) license, which allows users to download, copy and build upon published articles, as long as the author and publisher are properly credited, which ensures maximum dissemination and a wider impact of our publications.

The book as a whole is distributed by MDPI under the terms and conditions of the Creative Commons license CC BY-NC-ND.

Contents

About the Special Issue Editors . . . . **vii**

Adrian Carter, Charlotte A. Hardman and Tracy Burrows
Food Addiction and Eating Addiction: Scientific Advances and Their Clinical, Social and Policy Implications
Reprinted from: *Nutrients* **2020**, *12*, 1485, doi:10.3390/nu12051485 . . . . **1**

May Thet Khine, Atsuhiko Ota, Ashley N. Gearhardt, Akiko Fujisawa, Mamiko Morita, Atsuko Minagawa, Yuanying Li, Hisao Naito and Hiroshi Yatsuya
Validation of the Japanese Version of the Yale Food Addiction Scale 2.0 (J-YFAS 2.0)
Reprinted from: *Nutrients* **2019**, *11*, 687, doi:10.3390/nu11030687 . . . . **5**

Kirrilly M. Pursey, Oren Contreras-Rodriguez, Clare E. Collins, Peter Stanwell and Tracy L. Burrows
Food Addiction Symptoms and Amygdala Response in Fasted and Fed States
Reprinted from: *Nutrients* **2019**, *11*, 1285, doi:10.3390/nu11061285 . . . . **19**

Shanon L. Casperson, Lisa Lanza, Eram Albajri and Jennifer A. Nasser
Increasing Chocolate's Sugar Content Enhances Its Psychoactive Effects and Intake
Reprinted from: *Nutrients* **2019**, *11*, 596, doi:10.3390/nu11030596 . . . . **29**

Siddharth Sarkar, Kanwal Preet Kochhar and Naim Akhtar Khan
Fat Addiction: Psychological and Physiological Trajectory
Reprinted from: *Nutrients* **2019**, *11*, 2785, doi:10.3390/nu11112785 . . . . **41**

Rachel C. Adams, Jemma Sedgmond, Leah Maizey, Christopher D. Chambers and Natalia S. Lawrence
Food Addiction: Implications for the Diagnosis and Treatment of Overeating
Reprinted from: *Nutrients* **2019**, *11*, 2086, doi:10.3390/nu11092086 . . . . **57**

Susana Jiménez-Murcia, Zaida Agüera, Georgios Paslakis, Lucero Munguia, Roser Granero, Jéssica Sánchez-González, Isabel Sánchez, Nadine Riesco, Ashley N Gearhardt, Carlos Dieguez, et al.
Food Addiction in Eating Disorders and Obesity: Analysis of Clusters and Implications for Treatment
Reprinted from: *Nutrients* **2019**, *11*, 2633, doi:10.3390/nu11112633 . . . . **93**

Laurence J. Nolan and Steve M. Jenkins
Food Addiction Is Associated with Irrational Beliefs via Trait Anxiety and Emotional Eating
Reprinted from: *Nutrients* **2019**, *11*, 1711, doi:10.3390/nu11081711 . . . . **109**

Stephanie E. Cassin, Daniel Z. Buchman, Samantha E. Leung, Karin Kantarovich, Aceel Hawa, Adrian Carter and Sanjeev Sockalingam
Ethical, Stigma, and Policy Implications of Food Addiction: A Scoping Review
Reprinted from: *Nutrients* **2019**, *11*, 710, doi:10.3390/nu11040710 . . . . **123**

Helen K. Ruddock, Michael Orwin, Emma J. Boyland, Elizabeth H. Evans and Charlotte A. Hardman
Obesity Stigma: Is the 'Food Addiction' Label Feeding the Problem?
Reprinted from: *Nutrients* **2019**, *11*, 2100, doi:10.3390/nu11092100 . . . . **141**

Kerry S. O'Brien, Rebecca M. Puhl, Janet D. Latner, Dermot Lynott, Jessica D. Reid, Zarina Vakhitova, John A. Hunter, Damian Scarf, Ruth Jeanes, Ayoub Bouguettaya and Adrian Carter
The Effect of a Food Addiction Explanation Model for Weight Control and Obesity on Weight Stigma
Reprinted from: *Nutrients* **2020**, *12*, 294, doi:10.3390/nu12020294 . . . . **159**

About the Special Issue Editors

Adrian Carter (Ph.D.): Associate Professor Adrian Carter is an NHMRC Research Fellow and Director, Community Engagement and Neuroethics, Turner Institute for Brain and Mental Health, Monash University. He is also: Director, Neuroethics Program, ARC Centre of Excellence for Integrative Brain Function; Co-Chair, Neuroethics and Responsible Research and Innovation Committee, Australian Brain Alliance; Co-Editor-in-Chief, Neuroethics (Springer); and sits on the Board of Directors, International Neuroethics Society. His research examines the impact of neuroscience on our understanding and treatment of mental and neurological disorders. Dr Carter has been an advisor to the WHO, OECD, European Monitoring Centre for Drugs and Drug Addiction, and United Nations Office on Drugs and Crime.

Tracy Burrows (Ph.D.): Associate Professor Tracy Burrows is a NHMRC Research Fellow at The University of Newcastle and a researcher at the Hunter Medical Research Institute. She is recognized as a Fellow of the Dietitians Association of Australia. Her research focuses on understanding eating behavior with an interest in mental health populations, dietary assessment and weight management.

Charlotte Hardman (Ph.D.): Dr Charlotte Hardman is a Senior Lecturer in the University of Liverpool. She is also: Fellow of the Higher Education Academy (FHEA); Co-ordinator of the North West Network of the UK Association for the Study of Obesity; and Advisory Editor for the journal Appetite. Her research focuses on the psychology of food-related behaviour and the application of this knowledge to interventions for behaviour change. She has received research funding from UK Research and Innovation, the European Commission, and the Wellcome Trust. She has published over 60 peer-reviewed articles in high-impact journals such as JAMA Psychiatry, Nature Reviews Endocrinology and the International Journal of Obesity.

Editorial

Food Addiction and Eating Addiction: Scientific Advances and Their Clinical, Social and Policy Implications

Adrian Carter [1,*], Charlotte A. Hardman [2] and Tracy Burrows [3]

1. School of Psychological Sciences and the Turner Institute for Brain and Mental Health, Monash University, Clayton, VIC 3800, Australia
2. Department of Psychology, University of Liverpool, Liverpool L69 7ZA, UK; charlotte.hardman@liverpool.ac.uk
3. School of Health Sciences, Faculty of Health, University of Newcastle, Newcastle, NSW 2308, Australia; tracy.burrows@newcastle.edu.au
* Correspondence: adrian.carter@monash.edu; Tel.: +61-3-9902-9431

Received: 1 May 2020; Accepted: 14 May 2020; Published: 20 May 2020

There is a growing understanding within the literature that certain foods, particularly those high in refined sugars and fats, may have addictive potential for some individuals. Moreover, individuals who are overweight and have obesity display dietary intake patterns that resemble the ways in which individuals with substance use disorders consume addictive drugs. While food addiction is not yet recognized in the Diagnostic and Statistical Manual of Mental Disorders, Fifth Edition (DSM-5), there are many similarities with substance use disorders, and a growing acceptance that some forms of obesity should be treated as a food addiction. Despite growing research in this area, there remain many unresolved questions about the science of food addiction and its potential impact upon: how we treat overweight and overeating; stigmatization and discrimination of people who are overweight; internalized weight bias and treatment seeking; as well as policies to reduce excess weight and overeating.

This interdisciplinary special issue collects 10 articles, including reviews and original research, that further our understanding and application of the science of addictive eating. These papers span a broad range of areas, including basic science, clinical assessment tools, neural responses to addictive foods, as well as insights into future treatments and public health policies, and the possible stigma associated with food addiction.

1. Validation of Food Addiction Scales

Validation of the Japanese Version of the Yale Food Addiction Scale 2.0 (J-YFAS 2.0)

The Yale Food Addiction Scale (YFAS) is the most widely used diagnostic tool for food addiction, and has been translated into numerous languages, including Italian, French, German, Spanish, Arabic Chinese and Turkish. In this special issue, Khine and colleagues [1] describe the translation and validation of the Japanese version of the Yale Food Addiction Scale 2.0 (J-YFAS 2.0), carried out in 731 undergraduate students. The J-YFAS 2.0 has a one-factor structure and adequate convergent validity and reliability, similar to the YFAS 2.0 in other languages. Prevalence of J-YFAS 2.0-diagnosed mild, moderate, and severe food addiction was 1.1%, 1.2%, and 1.0% respectively.

2. Neural Responses Underlying Food Addiction

2.1. Food Addiction Symptoms and Amygdala Response in Fasted and Fed States

Pursey et al. [2] conducted a small pilot study to explore the association between food addiction symptoms and activation in the basolateral amygdala and central amygdala. 12 females, aged 24.1 ± 2.6 years, completed two functional magnetic resonance imaging (fMRI) scans (fasted and fed) while viewing high-calorie food images and low-calorie food images. Food addiction symptoms were assessed using the Yale Food Addiction Scale. Participants had a mean BMI of 27.4 ± 5.0 kg/m^2, and food addiction symptom score of 4.1 ± 2.2. The results found a significant positive association, between food addiction symptoms, and higher activation of the left basolateral amygdala to high-calorie versus low-calorie foods in the fasted session, but not the fed session. There were no significant associations with the central amygdala in either session.

2.2. Increasing Chocolate's Sugar Content Enhances Its Psychoactive Effects and Intake

This study by Caperson and colleagues explored the potential psychoactive effect of chocolate [3]. Participants consumed 5 g of a commercially available chocolate with increasing amounts of sugar (90% cocoa, 85% cocoa, 70% cocoa, and milk chocolate). After each chocolate sample, participants completed the Psychoactive Effects Questionnaire (PEQ) and the Binge Eating Scale (BES). Participants were also allowed to eat as much as they wanted of each of the different chocolates. Casperson et al. [3] found that the excitement subscale of the PEQ increased (relative to baseline) after the 90% cocoa. The Morphine–Benzedrine Group subscale (containing questions about wellbeing and euphoria) and the Morphine subscale (focusing on attitudes and physical sensations) increased after the 85th cocoa sample. This suggests incremental increases in the sugar content of chocolate has a psychoactive effect which enhances the addictive-like eating response.

3. Implications for Treatment

3.1. Food Addiction: Implications for the Diagnosis and Treatment of Overeating

The validity of a food addiction diagnosis remains controversial, despite a growing body of preclinical, neurobiological and clinical evidence supporting it. This literature review discusses the DSM-5 diagnostic criteria for substance use disorders, to summarize evidence for food addiction. Adams and colleagues [4] concluded that there is evidence to suggest that, for some individuals, food can induce addictive-type behaviors similar to those seen with other addictive substances. However, as several DSM-5 criteria have limited application to overeating, they argue that the term 'food addiction' is likely to apply only in a minority of cases. Research investigating the underlying psychological causes of overeating within the context of food addiction has also led to some novel treatment approaches, such as cognitive training tasks and neuro-modulation interventions.

3.2. Food Addiction in Eating Disorders and Obesity: Analysis of Clusters and Implications for Treatment

This study by Jimenez-Murcia et al. [5] identified three distinct clusters of food addiction in those with eating disorders and obesity. The study was conducted in 234 participants who scored positive on the Yale Food Addiction Scale 2.0. Cluster 1, classified as "dysfunctional", was associated with the highest prevalence of other specified feeding or eating disorders and bulimia nervosa, as well as the highest eating disorder severity and psychopathology, and more dysfunctional personality traits. Cluster 2, classified as "moderate", was associated with high prevalence of bulimia nervosa and binge eating disorders, and moderate levels of eating disorder psychopathology. Cluster 3, classified as "adaptive", was characterized by high prevalence of obesity and binge eating disorders, low levels of eating disorder psychopathology and more functional personality traits. The authors suggest that the identification of types of food addiction traits may allow for more personalized treatment to improve outcomes.

3.3. Food Addiction Is Associated with Irrational Beliefs via Trait Anxiety and Emotional Eating

Irrational beliefs are believed to be one of the prime causes of psychopathologies, including anxiety and depression. A study of 239 adults by Nolan and colleagues [6] investigated whether food addiction and emotional eating are associated with irrational beliefs. Questionnaires measuring food addiction, irrational beliefs, emotional eating, depression, trait anxiety, and anthropometry were assessed and reported. They found that irrational beliefs were significantly positively correlated with food addiction, emotional eating, depression and trait anxiety. Results also showed that irrational beliefs were associated with higher food addiction via higher trait anxiety and emotional eating acting in a serial pathway. As such, targeting irrational beliefs as a treatment in individuals who experience food addiction and emotional eating may be a reasonable approach for clinicians.

3.4. Fat Addiction: Psychological and Physiological Trajectory

A number of recent studies have attempted to parse out the psychological and physiological etiology of food addiction. This review article by Sarkar et al. [7] examines the specific role of dietary fats in compulsive overeating. They review preclinical, psychological and clinical evidence to argue for the addiction to fat rich diets as a prominent subset of food addiction. They then discuss the clinical implications of "fat addiction" for society.

4. Associations between Food Addiction, Stigma and Public Policy

4.1. Ethical, Stigma, and Policy Implications of Food Addiction: A Scoping Review

This scoping review by Cassin and colleagues [8] examines the potential ethical, stigma and health policy implications of food addiction described in the current literature. Their findings suggest that the literature on potential ethical implications was mostly focused on debates regarding individualized responsibility and sources for blame. Potential stigma focused on evidence of internalized and externalized stigma when food addiction is used as the explanation for obesity. The policy implications of food addiction largely drew on comparisons with the historic regulation of the tobacco industry to manage food addiction in policy, and the current challenges in classifying foods in terms of their addictive potential.

4.2. Obesity Stigma: Is the 'Food Addiction' Label Feeding the Problem?

There is significant debate around whether describing someone as addicted to food would increase or decrease weight-based stigma. Ruddock et al. [9] examined the effect of the food addiction label on stigmatizing attitudes towards an individual with obesity, and towards people with obesity more generally (i.e., general stigma). They presented the results of two online studies, where participants (n = 439, n = 523) read a short description about a woman described as 'very overweight'. They found that a food addiction label may exacerbate stigmatizing attitudes towards an individual with obesity. However, the label appears to have no effect on general weight-based stigma. Stigmatizing attitudes towards people with obesity also appeared to be more pronounced in individuals with low levels of addiction-like eating behaviors, compared to high levels of addiction-like eating.

4.3. The Effect of a Food Addiction Explanation Model for Weight Control and Obesity on Weight Stigma

In the final paper of this special issue, O'Brien and colleagues [10] reported on two experimental studies examining the impact of a food addiction model of obesity and weight control on weight stigma. In both experiments, participants were randomized to receive one of two newspaper articles: one describing obesity as the result of a brain-based food addiction, and the other describing obesity as the result of diet and exercise. The food addiction explanation for weight control and obesity did not increase weight stigma, and resulted in lower stigma than the diet and exercise explanation, which attributes obesity to personal control. Their findings highlight the need for evidence-based health messaging about the causes of obesity, and the need for communications that do not exacerbate weight stigma.

Author Contributions: T.B. developed an initial draft of the manuscript that was updated by A.C. All authors revised the final manuscript. All authors have read and agreed to the published version of the manuscript.

Funding: This research received no external funding.

Conflicts of Interest: The authors declare no conflict of interest.

References

1. Khine, M.T.; Ota, A.; Gearhardt, A.N.; Fujisawa, A.; Morita, M.; Minagawa, A.; Li, Y.; Naito, H.; Yatsuya, H. Validation of the Japanese Version of the Yale Food Addiction Scale 2.0 (J-YFAS 2.0). *Nutrients* **2019**, *11*, 687. [CrossRef] [PubMed]
2. Pursey, K.M.; Contreras-Rodriguez, O.; Collins, C.E.; Stanwell, P.; Burrows, T.L. Food Addiction Symptoms and Amygdala Response in Fasted and Fed States. *Nutrients* **2019**, *11*, 1285. [CrossRef]
3. Casperson, S.L.; Lanza, L.; Albajri, E.; Nasser, J.A. Increasing Chocolate's Sugar Content Enhances Its Psychoactive Effects and Intake. *Nutrients* **2019**, *11*, 596. [CrossRef]
4. Adams, R.C.; Sedgmond, J.; Maizey, L.; Chambers, C.D.; Lawrence, N.S. Food Addiction: Implications for the Diagnosis and Treatment of Overeating. *Nutrients* **2019**, *11*, 2086. [CrossRef] [PubMed]
5. Jiménez-Murcia, S.; Agüera, Z.; Paslakis, G.; Munguia, L.; Granero, R.; Sánchez-González, J.; Sánchez, I.; Riesco, N.; Gearhardt, A.N.; Dieguez, C.; et al. Food Addiction in Eating Disorders and Obesity: Analysis of Clusters and Implications for Treatment. *Nutrients* **2019**, *11*, 2633. [CrossRef] [PubMed]
6. Nolan, L.; Jenkins, S. Food Addiction Is Associated with Irrational Beliefs via Trait Anxiety and Emotional Eating. *Nutrients* **2019**, *11*, 1711. [CrossRef] [PubMed]
7. Sarkar, S.; Kochhar, K.P.; Khan, N.A. Fat Addiction: Psychological and Physiological Trajectory. *Nutrients* **2019**, *11*, 2785. [CrossRef] [PubMed]
8. Cassin, S.E.; Buchman, D.Z.; Leung, S.E.; Kantarovich, K.; Hawa, A.; Carter, A.; Sockalingam, S. Ethical, Stigma, and Policy Implications of Food Addiction: A Scoping Review. *Nutrients* **2019**, *11*, 710. [CrossRef] [PubMed]
9. Ruddock, H.K.; Orwin, M.; Boyland, E.J.; Evans, E.H.; Hardman, C.A. Obesity Stigma: Is the 'Food Addiction' Label Feeding the Problem? *Nutrients* **2019**, *11*, 2100. [CrossRef] [PubMed]
10. O'Brien, K.S.; Puhl, R.M.; Latner, J.D.; Lynott, D.; Reid, J.D.; Vakhitova, Z.; Bouguettaya, A. The Effect of a Food Addiction Explanation Model for Weight Control and Obesity on Weight Stigma. *Nutrients* **2020**, *12*, 294. [CrossRef] [PubMed]

© 2020 by the authors. Licensee MDPI, Basel, Switzerland. This article is an open access article distributed under the terms and conditions of the Creative Commons Attribution (CC BY) license (http://creativecommons.org/licenses/by/4.0/).

Article

Validation of the Japanese Version of the Yale Food Addiction Scale 2.0 (J-YFAS 2.0)

May Thet Khine [1], Atsuhiko Ota [1,*], Ashley N. Gearhardt [2], Akiko Fujisawa [1], Mamiko Morita [3], Atsuko Minagawa [3], Yuanying Li [1], Hisao Naito [1] and Hiroshi Yatsuya [1]

1. Department of Public Health, Fujita Health University School of Medicine, 1-98 Dengakugakubo, Kutsukake-cho, Toyoake, Aichi 470-1192, Japan; maythet7@gmail.com (M.T.K.); akifuji@fujita-hu.ac.jp (A.F.); liyy@fujita-hu.ac.jp (Y.L.); naitoh@fujita-hu.ac.jp (H.N.); yatsuya@fujita-hu.ac.jp (H.Y.)
2. Department of Psychology, University of Michigan, 2268 East Hall, 530 Church Street, Ann Arbor, MI 48109, USA; agearhar@umich.edu
3. Faculty of Nursing, Fujita Health University School of Health Sciences, 1-98 Dengakugakubo, Kutsukake-cho, Toyoake, Aichi 470-1192, Japan; mamorita@fujita-hu.ac.jp (M.M.); mina@fujita-hu.ac.jp (A.M.)
* Correspondence: ohtaa@fujita-hu.ac.jp; Tel.: +81-562-93-2453; Fax: +81-562-93-3079

Received: 9 February 2019; Accepted: 19 March 2019; Published: 22 March 2019

Abstract: The Yale Food Addiction Scale 2.0 (YFAS 2.0) is used for assessing food addiction (FA). Our study aimed at validating its Japanese version (J-YFAS 2.0). The subjects included 731 undergraduate students. Confirmatory factor analysis indicated the root-mean-square error of approximation, comparative fit index, Tucker–Lewis index, and standardized root-mean-square residual were 0.065, 0.904, 0.880, and 0.048, respectively, for a one-factor structure model. Kuder–Richardson α was 0.78. Prevalence of the J-YFAS 2.0-diagnosed mild, moderate, and severe FA was 1.1%, 1.2%, and 1.0%, respectively. High uncontrolled eating and emotional eating scores of the 18-item Three-Factor Eating Questionnaire (TFEQ R-18) ($p < 0.001$), a high Kessler Psychological Distress Scale score ($p < 0.001$), frequent desire to overeat ($p = 0.007$), and frequent snacking ($p = 0.003$) were associated with the J-YFAS 2.0-diagnosed FA presence. The scores demonstrated significant correlations with the J-YFAS 2.0-diagnosed FA symptom count ($p < 0.01$). The highest attained body mass index was associated with the J-YFAS 2.0-diagnosed FA symptom count ($p = 0.026$). The TFEQ R-18 cognitive restraint score was associated with the J-YFAS 2.0-diagnosed FA presence ($p < 0.05$) and symptom count ($p < 0.001$), but not with the J-YFAS 2.0-diagnosed FA severity. Like the YFAS 2.0 in other languages, the J-YFAS 2.0 has a one-factor structure and adequate convergent validity and reliability.

Keywords: food addiction; Japan; validation; Yale Food Addiction Scale 2.0

1. Introduction

The idea of food addiction (FA) is receiving increased interest [1]. Evidence is emerging that certain types of foods (e.g., highly processed foods with high levels of refined carbohydrates and/or added fat) may be capable of triggering addictive-like eating behaviors (e.g., loss of control, withdrawal, and cravings) in some individuals, which can lead to significant impairment or distress [2,3]. Obesity and eating disorders such as bulimia nervosa (BN), binge eating disorders (BED), along with psychiatric disorders such as depression, posttraumatic stress disorder, attention-deficit hyperactivity disorder, have been reported as potential correlates with FA [4–6]. Relevant pharmacological findings have been reported. Highly processed sweetened and fatty foods trigger a rewarding effect through the release of dopamine [7]. Repeated eating of hyper-palatable food down-regulates the dopaminergic response, resulting in impulsive and compulsive responses to food cues [8]. Food craving—an intense desire to eat a specific food—activates the hippocampus, insula, and caudate nucleus, similar to drug craving [9]. On the other hand, there has been a lot of debate regarding the extent to which food can be addictive in the same way as drugs. Controversies exist, for instance, as to whether FA represents a specific construct as addiction that is distinct from other eating disorders, such as BED, and whether neurobiological changes underlying FA behaviors are sufficiently ascertained in humans [10,11].

The Yale Food Addiction Scale (YFAS) is the most commonly used measure to assess FA, although FA is not included in the Diagnostic and Statistical Manual of Mental Disorders, 5th edition (DSM-5) [12] and controversy exists regarding its definition [11]. The original YFAS applies the DSM 4th edition (DSM-IV) diagnostic criteria for substance dependence to the consumption of highly palatable foods (e.g., chocolate, ice cream, and pizza) [13,14]. Later, the scale was replaced with the Yale Food Addiction Scale 2.0 (YFAS 2.0) in response to the revision of the Substance-Related and Addictive Disorders criteria in the DSM-5 [15]. The YFAS 2.0 additionally introduced the following four diagnostic criteria: craving, use despite interpersonal or social consequences, failure in role obligations, and use in physically hazardous situations. It also introduced a severity classification. The YFAS 2.0 is available not only in English but also in German, French, Italian, Spanish, and Arabic [16–20]. The YFAS 2.0 exhibits good internal consistency, as well as convergent, discriminant, and incremental validity [15–20]. Associations of the YFAS-diagnosed FA with obesity, eating disorders, and psychiatric disorders have been accumulated. The YFAS 2.0-defined FA prevalence is supposed to draw a J-shape curve according to body mass index (BMI): 3.3–15.8% in healthy general populations [15–20], 17.2–47.4% in obese population [16,21], and 15.0% in underweight population [21]. Women and patients with eating disorders (BN and BED) and mental disorders (depression, sleep disturbance, and general psychiatric status) were more likely to have FA diagnosed with the YFAS 2.0 [15,17–19].

The current study aimed to validate the Japanese version of YFAS 2.0 (J-YFAS 2.0). Scant evidence regarding FA is available in Asia. The previous version of YFAS was translated into Chinese [22,23] and Malay [24]. Using these questionnaires, researchers reported that a FA diagnosis was assigned to 6.9–9.2% of Chinese teenage students [22,23] and 10.4% of Malay obese adults [24]. FA prevalence in Japan has not been reported so far, to the best of our knowledge. The YFAS 2.0 has not yet been translated into Asian languages. Development of the J-YFAS 2.0 enables examining the FA prevalence in Japan, comparing it with other countries and regions, and exploring the mechanism of FA. Referring to previous research [15–19], we hypothesized that (1) the J-YFAS has a one-factorial structure for the 11 J-YFAS 2.0 diagnostic criteria (structural validity); (2) underweight, overweight, obesity, uncontrolled and emotional eating, frequent desire to overeat, frequent snacking, and mood and anxiety disorders are associated with the J-YFAS 2.0-diagnosed FA (convergent validity); (3) cognitive restraint in eating is not associated with the J-YFAS 2.0-diagnosed FA (discriminant validity); and (4) the internal consistency is good for the 11 J-YFAS 2.0 diagnostic criteria (reliability).

2. Subjects and Methods

2.1. Study Design

We employed a cross-sectional design. All data were collected from a questionnaire survey. The present study was completed in accordance with the Declaration of Helsinki and the Ethical Guidelines for Medical and Health Research Involving Human Subjects established by the Ministry of Education, Culture, Sports, Science and Technology and the Ministry of Health, Labour and Welfare, Japan. We obtained the approval by the Ethics Review Committee of Fujita Health University, Japan (HM17-110 and HM18-155). All subjects provided their informed written consent for participation in the present study.

2.2. Subjects

This study was conducted with a convenience sample of undergraduate students from a private medical and health science university in Japan. The authors (A.O., M.M., and A.M.) explained the study purpose and methods to the students in the classes. Paper-based questionnaires were then distributed. Of the 759 students to whom the questionnaires were distributed, 752 (99%) were returned. Those who did not provide informed consent ($n = 2$) and who did not fully complete the J-YFAS 2.0 ($n = 18$) were excluded from the analysis. One student who replied to experience desire to overeat 50 times per week was excluded as this reply was a significant outlier. Consequently, we retained the remaining 731 students (96%) as the subjects.

2.3. J-YFAS 2.0

As with the YFAS 2.0 [15], the J-YFAS 2.0 is a 35-item self-administered questionnaire (Table S1). It assesses food consumption during the past 12 months. A Likert-scale ranging from 0 (never) through 7 (every day) is employed as a response option for each item. The items assess clinical impairment/distress and the following 11 diagnostic criteria: (1) eating larger amounts for a longer period than intended (consumed more than intended); (2) persistent desire or repeated unsuccessful attempts to quit eating (unable to cut down or stop); (3) spending considerable time or activity obtaining or eating food or recovering from eating (great deal of time spent); (4) giving up or reducing important social, occupational, or recreational activities due to eating (important activities given up); (5) continued eating despite knowledge of adverse consequences (use despite physical/emotional consequences); (6) development of tolerance (tolerance); (7) characteristic withdrawal symptoms (withdrawal); (8) continued eating despite interpersonal or social problems (use despite interpersonal/social problems); (9) failure to fulfil major role obligation at work, school, and home due to eating (failure in role obligation); (10) eating even in physically hazardous situations (use in physically hazardous situations); and (11) craving, strong desire, or urge for certain foods (craving). Each item is scored dichotomously based on the threshold determined by the YFAS 2.0 validation paper [15]. If any item that corresponds to the diagnostic criteria or clinical severity meets the clinical threshold, this criterion is endorsed. There are two scoring methods: the symptom count and the diagnostic threshold. For the symptom count scoring method, the diagnostic criteria for which the subjects meet are summed together. For the diagnostic threshold, the clinically significant impairment/distress criterion has to be met and two or more diagnostic criteria have to be met. The J-YFAS 2.0 FA diagnostic severity is classified as mild (2–3 criteria met plus impairment/distress), moderate (4–5 criteria met plus impairment/distress), and severe (6–11 criteria met plus impairment/distress).

For the development of the J-YFAS 2.0, the original English YFAS 2.0 [15] was translated into Japanese by the three Japanese authors (A.O., A.F., and H.Y.) and back-translated into English by an external professional translator who had no previous knowledge of the YFAS 2.0. Discrepancies between the back-translation and the original were resolved by consensus amongst the three Japanese authors and an American author (A.N.G.), who developed the original YFAS 2.0. We added two food examples, wagashi (Japanese traditional confectionery) and instant noodles (as salty snacks), in the introductory part, considering that food preference in Japan differs from western countries.

2.4. Variables for Convergent and Discriminant Validity

2.4.1. Body Mass Index (BMI)

Each subject self-reported their current and highest attained BMI. The questionnaire included a table indicating BMI from the weights and heights so that the subjects could choose their BMI from the following options: <16.0, 16.0–16.9, 17.0–18.4, 18.5–22.9, 23.0–24.9, 25.0–29.9, and $\geq$30.0 kg/m^2. No one chose <16.0 kg/m^2 for their current or highest attained BMI.

It was reported that Japanese tended to under-report their body weights and the tendency was more prominent among those with high BMI than those with low BMI [25]. We arranged the categorical response options for BMI to minimize the shame that subjects may feel for self-reporting their actual BMI.

2.4.2. Three-Factor Eating Questionnaire Revised 18-Item Version (TFEQ R-18)

The TFEQ R-18 is a self-assessment tool used to measure the following three types of eating behaviors: Cognitive restraint, uncontrolled eating, and emotional eating [26]. Cognitive restraint is a control over food intake in order to influence body weight and body shape [26]. Uncontrolled eating is a tendency to overeat food with the feeling of being out of control [27]. Emotional eating is a tendency to eat in response to negative emotions [27]. The higher the score is, the greater the levels of cognitive restraint, uncontrolled eating, and emotional eating are. We chose the corresponding items for the current study from the Japanese version of the original 51-item TFEQ [28].

2.4.3. Desire to Overeat

No validated questionnaire was available in Japanese to evaluate binge eating frequency. Thus, we asked the frequency of desiring to overeat with a single question, "How many times per week did you feel you wanted to eat more even after eating quite a lot of food during the last two hours?" The subjects filled in the number of the times.

2.4.4. Snacking Frequency

A frequency of snacking (eating and drinking outside of breakfast, lunch, or dinner) was self-reported. No validated questionnaire was available in Japanese to evaluate the frequency of snacking. Thus, we developed a single question, "How many days per week are you snacking?" for this evaluation. The subjects chose one of the following options: none, 2–3 days, 4–5 days, and almost every day. The snack included foods and drinks that contained any calories. Zero-calorie drinks, such as coffee and tea without milk and sugar, and vitamin and mineral supplements were excluded from the snack.

2.4.5. Kessler Psychological Distress Scale (K6)

The Japanese version of K6 was used as an indicator of mood and anxiety disorders [29]. A K6 score of 13 or greater was regarded as having such disorders.

2.5. Statistical Analyses

Confirmatory factor analysis (CFA) was conducted to assess the one-factor structure for the 11 J-YFAS 2.0 diagnostic criteria. Clinically significant impairment/distress was not included in this CFA analysis. The model fit was evaluated with the root-mean-square error of approximation (RMSEA), comparative fit index (CFI), Tucker–Lewis index (TLI), and standardized root-mean-square residual (SRMR). For assessing the reliability, internal consistency was calculated for the 11 J-YFAS 2.0 diagnostic criteria with Kuder–Richardson's α (KR-20). Convergent and discriminant validity was examined with chi-square test, t-test, analysis of variance (ANOVA), and Spearman's rank correlation. We examined whether the current and highest attained BMI, TFEQ R-18 cognitive restraint, uncontrolled eating, and emotional eating scores, frequency of desire to overeat, snacking frequency, and K6 score were associated with the J-YFAS 2.0-diagnosed FA. Not only the presence and severity (mild, moderate, and severe) but also the symptom count was used as the J-YFAS 2.0-diagnosed FA index, given the small numbers of subjects diagnosed as having FA. We could not apply the chi-square test to examine the associations of BMI, high K6 score, and the snacking frequency with the J-YFAS 2.0-diagnosed FA severity, since more than 20% of all cells had an expected frequency of less than five. Effect size indices were calculated [30–32]. Subjects with missing responses were excluded from the corresponding analyses. SPSS version 23.0 (IBM, Armonk, NY, USA) and Amos Version 23.0 (IBM, Chicago, IL, USA) were used for statistical calculations.

3. Results

3.1. Subjects' Characteristics

Most subjects were women (78.5%, n = 574) (Table 1). The mean (standard deviation) age was 20.8 (1.8) years. The years and majors included fourth-year medical technology students, first- to fourth-year nursing students, and third-year medical students. Around 80% of the subjects reported normal-weight BMI, 18.5–24.9 kg/m^2.

Table 1. Subject characteristics (n = 731).

Characteristics	Frequency (%) or Mean (SD)
Sex	
Men	156 (21.3%)
Women	574 (78.5%)
Age (year)	20.8 (1.8)
Years and Majors	
Fourth-year medical technology students	149 (20.4%)
First-year nursing students	142 (19.4%)
Second-year nursing students	132 (18.1%)
Third-year medical students	111 (15.2%)
Fourth-year nursing students	99 (13.5%)
Third-year nursing students	98 (13.4%)
Current body mass index (BMI) (kg/m^2)	
16.0–16.9	17 (2.3%)
17.0–18.4	108 (14.8%)
18.5–22.9	521 (71.3%)
23.0–24.9	57 (7.8%)
25.0–29.9	21 (2.9%)
30 and above	6 (0.8%)
Highest attained BMI (kg/m^2) *	
16.0–16.9	3 (0.4%)
17.0–18.4	60 (8.2%)
18.5–22.9	493 (67.4%)
23.0–24.9	117 (16.0%)
25.0–29.9	51 (7.0%)
30 and above	7 (1.0%)
Kessler Psychological Distress Scale (K6) score	4.6 (4.5)
13 or greater	45 (6.2%)
Three-factor Eating Questionnaire-R 18 (TFEQ R-18) score	
Cognitive restraint	37.0 (20.2)
Uncontrolled eating	35.5 (19.9)
Emotional eating	29.7 (27.5)
Desire to overeat	0.5 (1.0) (Range: 0–7)
Snacking frequency per week	
None	89 (12.2%)
2–3 days	276 (37.8%)
4–5 days	157 (21.5%)
Almost every day	208 (28.5%)
J-YFAS 2.0-diagnosed food addiction (FA)	
No FA	707 (96.7%)
Mild FA	8 (1.1%)
Moderate FA	9 (1.2%)
Severe FA	7 (1.0%)

SD: standard deviation. There were missing responses for sex (n = 1), age (n = 1), current BMI (n = 1), K6 (n = 4), the TFEQ R-18 cognitive restraint (n = 8), uncontrolled eating (n = 12), and emotional eating (n = 2), desire to overeat (n = 1), and snacking frequency (n = 1). * Highest attained BMI means the highest weight ever (when not pregnant) during the lifetime.

3.2. CFA and Internal Consistency

The RMSEA, CFI, TLI, and SRMR were 0.065, 0.904, 0.880, and 0.048, respectively. One diagnostic criterion (failure in role obligation) indicated a factor loading of 0.31 (Table 2). The other diagnostic criteria had factor loadings of 0.41 or higher. The KR-20 was 0.78 for the 11 diagnostic criteria.

Table 2. Diagnostic criteria of the Japanese version of the Yale Food Addiction Scale 2.0 ($n = 731$).

Diagnostic Criteria	Met Criteria	Did Not Meet Criteria	Factor Loading
Consumed more than intended	82 (11.2%)	649 (88.8%)	0.57 ***
Unable to cut down or stop	124 (17.0%)	607 (83.0%)	0.52 ***
Great deal of time spent	30 (4.1%)	701 (95.9%)	0.45 ***
Important activities given up	25 (3.4%)	706 (96.6%)	0.41 ***
Use despite physical/emotional consequences	45 (6.2%)	686 (93.8%)	0.55 ***
Tolerance	31 (4.2%)	700 (95.8%)	0.50 ***
Withdrawal	90 (12.3%)	641 (87.7%)	0.62 ***
Use despite interpersonal/social problems	98 (13.4%)	633 (86.6%)	0.54 ***
Failure in role obligation	29 (4.0%)	702 (96.0%)	0.31 ***
Use in physically hazardous situations	42 (5.7%)	689 (94.3%)	0.56 ***
Craving	21 (2.9%)	710 (97.1%)	0.50 ***
Impairment/distress	29 (4.0%)	702 (96.0%)	

*** $p < 0.001$, calculated with confirmatory factor analysis.

3.3. J-YFAS 2.0-Diagnosed FA Prevalence

The mean J-YFAS 2.0-diagnosed FA symptom count was 0.84 (SD = 1.61; range = 0–11). The proportions of the subjects who met the threshold for each diagnostic criterion ranged from 2.9–17.0% (Table 2). A total of 24 (3.3%) subjects were regarded as having FA: 8 (1.1%) received a mild, 9 (1.2%) a moderate, and 7 (1.0%) a severe FA diagnosis using the J-YFAS 2.0 (Table 1). All subjects who were diagnosed as FA with the J-YFAS 2.0 were women. Sex was significantly associated with the J-YFAS 2.0-diagnosed FA ($p = 0.004$, Fisher's Exact Test).

3.4. Convergent and Discriminant Validity

For convergent validity, neither the current nor the highest attained BMI was associated with the presence of J-YFAS 2.0-diagnosed FA (Table 3). The highest attained BMI was associated with the J-YFAS 2.0-diagnosed FA symptom count, while the current BMI was not (Table 4). The effect size was small for the association between the highest attained BMI and the J-YFAS 2.0-diagnosed FA symptom count—the η^2 was 0.02. A high K6 score and snacking frequency were significantly associated with the J-YFAS 2.0-diagnosed FA presence and the J-YFAS 2.0-diagnosed FA symptom count (Tables 3 and 4, respectively). TFEQ R-18 uncontrolled eating and emotional eating scores and desire to overeat were significantly associated with the J-YFAS 2.0-diagnosed FA presence, severity, and symptom count (Tables 5–7, respectively).

For discriminant validity, there was a significant association between the TFEQ R-18 cognitive restraint score and the J-YFAS 2.0-diagnosed FA presence (Table 5). Its effect size was small—the Cohen's d was 0.44. There was no significant association between the cognitive restraint score and the J-YFAS 2.0-diagnosed FA severity (Table 6). We found a significant correlation between the cognitive restraint score and the J-YFAS 2.0-diagnosed FA symptom count (Table 7). Its Spearman's rank correlation coefficient was 0.143.

Table 3. Associations of body mass index (BMI), the Kessler Psychological Distress Scale (K6) score, and snacking frequency with the J-YFAS 2.0-diagnosed food addiction (FA) absence/presence.

	FA Absent (n = 707)	FA Present (n = 24)	Chi-Square	p Value	Effect Size (V)
Current BMI (kg/m^2)					
16.0–16.9	16 (94.1%)	1 (5.9%)			
17.0–18.4	105 (97.2%)	3 (2.8%)			
18.5–22.9	503 (96.5%)	18 (3.5%)	1.421	0.922	0.04
23.0–24.9	55 (96.5%)	2 (3.5%)			
25.0–29.9	21 (100%)	0 (0%)			
30 and above	6 (100%)	0 (0%)			
Highest attained BMI (kg/m^2) *					
16.0–16.9	3 (100%)	0 (0%)			
17.0–18.4	58 (96.7%)	2 (3.3%)			
18.5–22.9	478 (97.0%)	15 (3.0%)	1.522	0.911	0.05
23.0–24.9	113 (96.6%)	4 (3.4%)			
25.0–29.9	48 (94.1%)	3 (5.9%)			
30 and above	7 (100%)	0 (0%)			
K6 score					
12 or less	665 (97.5%)	17 (2.5%)	22.565	<0.001	0.18
13 or greater	38 (84.4%)	7 (15.6%)			
Snacking Frequency Per Week					
None	89 (100%)	0 (0%)			
2–3 days	272 (98.6%)	4 (1.4%)	13.855	0.003	0.14
4–5 days	151 (96.2%)	6 (3.8%)			
Almost every day	194 (93.3%)	14 (6.7%)			

Chi-square test was used. The numbers of missing responses were as follows: current BMI (n = 1), K6 score (n = 4), and snacking frequency (n = 1). * Highest attained BMI means the highest weight ever (when not pregnant) during the lifetime.

Table 4. Associations of body mass index (BMI), the Kessler Psychological Distress Scale (K6) score, and snacking frequency with the J-YFAS 2.0-diagnosed food addiction (FA) symptom count (n = 731).

	FA Symptom Count	F/t Value	p Value	Pairwise Difference a	Effect Size (η^2)/(d)
Current BMI (kg/m^2)					
16.0–16.9	1.0 (2.7)				
17.0–18.4	0.6 (1.2)				
18.5–22.9	0.8 (1.6)	1.375	0.231		0.01
23.0–24.9	1.1 (1.8)				
25.0–29.9	1.4 (2.1)				
30 and above	1.3 (1.5)				
Highest attained BMI (kg/m^2) *					
16.0–16.9 (1)	0.7 (0.6)				
17.0–18.4 (2)	0.6 (1.7)				
18.5–22.9 (3)	0.8 (1.6)	2.555	0.026	(2), (3), (4) < (5)	0.02
23.0–24.9 (4)	0.7 (1.1)				
25.0–29.9 (5)	1.5 (2.3)				
30 and above (6)	1.1 (1.5)				
K6 score					
12 or less	0.8 (1.4)	−3.060	0.004		0.95
13 or greater	2.2 (3.2)				
Snacking frequency per week					
None (1)	0.4 (0.7)				
2–3 days (2)	0.5 (1.0)	15.986	<0.001	(1), (2) < (3), (4)	0.06
4–5 days (3)	1.0 (1.9)				
Almost every day (4)	1.4 (2.1)				

Analysis of variance (for BMI and snacking frequency) and t-test (for K6 score) were used. FA symptom counts are shown as mean (standard deviation). The numbers of missing responses were as follows: Current BMI (n = 1), K6 score (n = 4), and snacking frequency (n = 1). a Pairwise differences were of p < 0.05 (Bonferroni corrected). * Highest attained BMI means the highest weight ever (when not pregnant) during the lifetime.

Table 5. Associations of the 18-item Three-Factor Eating Questionnaire (TFEQ R-18) scores and frequency of desiring to overeat with the J-YFAS 2.0-diagnosed food addiction (FA) absence/presence.

	FA Absent (n = 707)	FA Present (n = 24)	t Value	p Value	Effect Size (d)
Cognitive restraint	36.7 (20.1)	45.7 (21.7)	−2.097	0.036	0.44
Uncontrolled eating	34.5 (19.1)	64.1 (23.1)	−7.246	<0.001	1.54
Emotional eating	28.6 (26.4)	63.9 (34.6)	−4.959	<0.001	1.32
Desire to overeat	0.41 (0.91)	1.7 (2.0)	−2.965	0.007	1.28

t-test was used. TFEQ R-18 scores and frequency of desiring to overeat are shown as mean (standard deviation). The numbers of missing responses were as follows: TFEQ R-18 cognitive restraint, n = 8 (7 from FA absent, 1 from FA present); uncontrolled eating, n = 12 (11 from FA absent, 1 from FA present); emotional eating, n = 2 (all from FA absent); and desire to overeat, n = 1 (from FA absent).

Table 6. Associations of the 18-item Three-Factor Eating Questionnaire (TFEQ R-18) scores and frequency of desiring to overeat with the J-YFAS 2.0-diagnosed food addiction (FA) severity.

	FA Absent (n = 707)	Mild FA (n = 8)	Moderate FA (n = 9)	Severe FA (n = 7)	F Value	p Value	Pairwise Difference [a]	Effect Size (η^2)
Cognitive restraint	36.7 (20.1)	43.1 (17.8)	48.6 (24.3)	45.2 (25.6)	1.56	0.197		0.01
Uncontrolled eating	34.5 (19.1)	56.0 (24.5)	62.5 (20.5)	75.1 (23.2)	18.80	<0.001	1 < 2,3,4	0.07
Emotional eating	28.6 (26.4)	43.1 (37.3)	74.1 (26.1)	74.6 (34.4)	16.06	<0.001	1 < 3,4	0.06
Desire to overeat	0.4 (0.9)	0.6 (1.2)	1.8 (1.5)	2.7 (2.9)	19.53	<0.001	1 < 3, 4; 2 < 4	0.07

Analysis of variance was used. TFEQ R-18 scores and frequency of desiring to overeat are shown as mean (standard deviation). The numbers of missing responses were as follows: TFEQ R-18 cognitive restraint, n = 8 (7 from FA absent, 1 from Moderate FA); uncontrolled eating, n = 12 (11 from FA absent, 1 from Moderate FA); emotional eating, n = 2 (all from FA absent); and desire to overeat, n = 1 (from FA absent). [a] Pairwise differences were of p < 0.05 (Bonferroni corrected). 1 = No FA, 2 = Mild FA, 3 = Moderate FA, 4 = Severe FA.

Table 7. Spearman's rank correlation coefficients among the J-YFAS 2.0-diagnosed food addiction (FA) symptom count, the 18-item Three-Factor Eating Questionnaire (TFEQ R-18) scores, and frequency of desiring to overeat (n = 731).

	FA Symptom Count	Cognitive Restraint	Uncontrolled Eating	Emotional Eating	Desire to Overeat
FA symptom count					
Cognitive restraint	0.143 ***				
Uncontrolled eating	0.403 ***	0.248 ***			
Emotional eating	0.296 ***	0.258 ***	0.619 ***		
Desire to overeat	0.277 ***	0.039	0.449 ***	0.361 ***	

*** p < 0.001. The numbers of missing responses were as follows: TFEQ R-18 cognitive restraint (n = 8), uncontrolled eating (n = 12), emotional eating (n = 2), and desire to overeat, (n = 1).

4. Discussion

We examined the J-YFAS 2.0's properties in a sample of healthy undergraduate students in Japan. The J-YFAS 2.0 had a one-factor structure and adequate convergent validity and reliability, like the YFAS 2.0 in other languages [15–20], whereas our results were not the same as hypothesized with regard to the associations of BMI and cognitive restraint in eating with the J-YFAS 2.0-diagnosed FA. The J-YFAS 2.0-diagnosed FA prevalence was 3.3% in our subjects. Similar findings were reported from Italian and Spanish young healthy samples [18,19].

A one-factor structure was confirmed for the J-YFAS 2.0, which is the same as for the English, German, French, Italian, and Spanish YFAS 2.0 [15–19]. Our findings did not strictly meet the Hu and Bentler criteria, i.e., RMSEA ≤ 0.06, CFI ≥ 0.95, TLI ≥ 0.95, and SRMR ≤ 0.08 [33]. However, one or more of the four indices do not often meet the criteria [34]. The French version of the YFAS 2.0 showed a CFI of 0.887 and RMSEA of 0.083 [17]. There is the criticism that the Hu and Bentler criteria may be too stringent [35]. Our fit indices did not deviate substantially from the Hu and Bentler criteria. We thus retained the one-factor structure of the J-YFAS 2.0. Regarding the reliability of the J-YFAS 2.0, KR-20 was 0.78. This suggests the acceptable internal consistency of the J-YFAS 2.0.

The J-YFAS 2.0-diagnosed FA prevalence was 3.3% in this study. A similar prevalence was observed in other developed countries: Italy (3.4%) and Spain (3.3%) [18,19]. The subjects' characteristics of these three studies bear some resemblance, which might account for the similar prevalence. They were mainly young (aged about 20) and normal-weight people. Like our study, the Italian study collected the subjects from a medical school. About 80% of the subjects were female in both the Spanish and our sample. On the other hand, a web-based survey found a much higher YFAS 2.0-diagnosed FA prevalence, 9.7%, among German-speaking university students with the similar age and BMI [16]. This could imply that not only biological characteristics but also cultural differences are associated with YFAS 2.0-diagnosed FA, although it is possible that the web-based survey received considerable attention from those with FA and obtained their participation. Similar to previous reports in the U.S. [15] and Italy [18], women exhibited a significantly greater YFAS 2.0-diagnosed FA prevalence than men in our study. This suggests a sex difference in YFAS 2.0-diagnosed FA, which should be further investigated in future studies.

The YFAS 2.0-diagnosed FA prevalence was high in overweight, obese, and underweight people in the U.S., Germany, France, Italy, Spain, and Egypt [15–21]. Contrary to these findings, both the current and highest attained BMI did not demonstrate an explicit association with the J-YFAS 2.0-diagnosed FA in the present study. We only found that the subjects with the highest attained BMI of 25.0–29.9 had a greater J-YFAS 2.0-diagnosed FA symptom count than those with the highest attained BMI of 17.0–24.9. However, its effect size was small—the η^2 was only 0.02. One possible reason for the finding might be the small numbers of our subjects with overweight, obesity, and extreme underweight. Current overweight and obesity were declared only by nearly 4% of the subjects. This reflected the fact that Japan has a much lower prevalence of overweight and obesity than other countries where the YFAS 2.0 has been validated [36]. Perhaps, some subjects in our study could have under-reported their BMI [37], although we arranged the categorical response options for BMI to minimize the shame that subjects may feel for self-reporting their actual BMI. Consequently, the low prevalence of overweight and obesity exerted a floor effect, diminishing the association between BMI, especially overweight and obesity, and the J-YFAS 2.0-diagnosed FA. Another possible reason is that the causes to affect BMI are multifactorial and different by region. We did not examine all causes that potentially affected BMI more strongly than FA. For instance, some researchers pointed out that social norms (pressure) might drive the young Japanese women's desire for slimming [38–40]. They could have more impact on BMI than FA among our subjects. Our subjects involved medical, nursing, and medical technology students. They must have a greater knowledge of health, nutrition, and exercise than the normal population, which may have skewed the association between BMI, especially the current BMI, and the J-YFAS 2.0-diagnosed FA. Development of the J-YFAS 2.0 improves the examination of FA in Japan where the prevalence of obesity is much lower than the western countries [36]. This may help elucidate our understanding of the impact of FA on body weight. Although FA was initially applied to understanding obesity, controversy remains over how much FA explains obesity [10,11].

Other variables hypothetically related to the convergent validity of the J-YFAS 2.0 showed significant associations with the J-YFAS 2.0-diagnosed FA as we expected. The TFEQ R-18 uncontrolled eating and emotional eating scores and desire to overeat were associated with the J-YFAS 2.0-diagnosed FA presence, severity, and symptom count in our study, as hypothesized based on the previous studies [15–19]. One study limitation is that we could not assess binge eating itself which was positively and moderately associated with the YFAS-diagnosed FA [41]. However, our findings regarding the desire to overeat and snacking would suggest the relationship between compulsive eating and FA. A desire to overeat forms a part of binge eating. Highly processed sweetened foods, which has been reported to be potentially related to FA [42–44], are often chosen for snacking in Japan [45]. We found that a high K6 score, which implied mood and anxiety disorders, is associated with the presence and higher symptom count of the J-YFAS 2.0-diagnosed FA. Some previous studies showed associations of psychopathological disorders [19] and depressive symptoms [18] with the YFAS 2.0-diagnosed FA.

A recent systematic review suggested a positive, moderate association of the YFAS-diagnosed FA with depression and anxiety [41]. Our finding was consistent with them.

Regarding the discriminant validity, we hypothesized that cognitive restraint in eating was a different entity from FA, referring to the idea that the YFAS 2.0 does not simply measure an intention and a failure to restrict food consumption [15,16]. Our findings exhibited an inconsistency in the association of cognitive restraint in eating with the FA presence, severity, and symptom count. In our sample, the association between cognitive restraint and the J-YFAS 2.0-diagnosed FA would not be so strong even if the association existed. Previous findings are also inconsistent regarding the association. It was reported in France and Italy that the YFAS 2.0-diagnosed FA was associated with a high level of cognitive restraint [17,18]. The Italian researchers mentioned the possibility that addictive-like eating and restricting food consumption could coexist in patients with anorexia nervosa [18]. We could not ascertain this possibility in our study since we did not examine whether the subjects suffered from anorexia. Further research would be necessary to examine the role of anorexia in the association between cognitive restraint and FA.

There are several limitations to the interpretation of our findings. First, the current study employed a convenience sample that was dominated by young, under- and normal-weight, female, healthy undergraduate students. For a generalization of the present findings, the J-YFAS 2.0 should be tested for different-age groups, obese individuals, and patients with eating disorders. Second, we were not able to include all kinds of validity and reliability. For instance, we did not address incremental validity and test-retest reliability. Third, we used our original questions to assess the desire to overeat and frequency of snacking. For instance, the Binge Eating Scale (BES) [46] and the Eating Behavior Patterns Questionnaire (EBPQ) [47] are the validated tools to evaluate binge eating and snacking, respectively. We did not use them since they were not translated into and validated in Japanese. This may limit the comparison of our results with the others. Finally, as FA has not yet been recognized in the DSM-5, we could not define the standard of psychiatrist-diagnosed FA.

As mentioned in the introduction, the conceptual construct of FA and the neurobiological changes underpinning it remain controversial [10,11]. Development of the J-YFAS 2.0 would facilitate research on FA in Japan where prevalence of overweight and obesity is much lower than the western countries [36]. This would contribute to specifying the conceptual construct of FA and the neurobiological changes related to FA.

5. Conclusions

The J-YFAS 2.0 had a one-factor structure and adequate convergent validity and reliability, like the YFAS 2.0 in other languages. Further studies are necessary to confirm the discriminant validity of the J-YFAS 2.0.

Supplementary Materials: The following are available online at http://www.mdpi.com/2072-6643/11/3/687/s1, Table S1: The Japanese version of the Yale Food Addiction Scale 2.0 (J-YFAS 2.0) (written in Japanese).

Author Contributions: M.T.K., A.O., A.N.G., and H.Y. designed the study plan. A.O., A.N.G., A.F., and H.Y. translated the English YFAS 2.0 into Japanese. M.T.K., A.O., A.F., M.M., A.M., Y.L., H.N., and H.Y. collected the data. M.T.K., A.O., and H.Y. analyzed the data and drafted the manuscript. A.N.G., A.F., M.M., A.M., Y.L., and H.N. critically reviewed and approved the manuscript. A.O. and H.Y. obtained the fund.

Funding: This research was funded by Fujita Health University. The funder had no role in the design of the study, the collection, analyses, or interpretation of data, writing the manuscript, or the decision to publish the results.

Acknowledgments: Data collection and processing were supported by Shunsuke Omura, Nozomi Furukawa, Yukino Onuma, Sho Nakahama, and Keigo Yamada.

Conflicts of Interest: The authors declare no conflict of interest. The content does not present the official views of the affiliations to which the authors belong.

References

1. Meule, A. Back by Popular Demand: A Narrative Review on the History of Food Addiction Research. *Yale J. Biol. Med.* **2015**, *88*, 295–302. [PubMed]
2. Schulte, E.M.; Smeal, J.K.; Gearhardt, A.N. Foods are differentially associated with subjective effect report questions of abuse liability. *PLoS ONE* **2017**, *12*, e0184220. [CrossRef] [PubMed]
3. Meule, A.; Hermann, T.; Kubler, A. Food addiction in overweight and obese adolescents seeking weight-loss treatment. *Eur. Eat. Disord. Rev. J. Eat. Disord. Assoc.* **2015**, *23*, 193–198. [CrossRef]
4. Meule, A.; Gearhardt, A.N. Five years of the Yale Food Addiction Scale: Taking stock and moving forward. *Curr. Addict. Rep.* **2014**, *1*, 193–205. [CrossRef]
5. Pursey, K.M.; Stanwell, P.; Gearhardt, A.N.; Collins, C.E.; Burrows, T.L. The prevalence of food addiction as assessed by the Yale Food Addiction Scale: A systematic review. *Nutrients* **2014**, *6*, 4552–4590. [CrossRef] [PubMed]
6. Brewerton, T.D. Food addiction as a proxy for eating disorder and obesity severity, trauma history, PTSD symptoms, and comorbidity. *Eat. Weight Disord.* **2017**, *22*, 241–247. [CrossRef]
7. Onaolapo, A.Y.; Onaolapo, O.J. Food additives, food and the concept of 'food addiction': Is stimulation of the brain reward circuit by food sufficient to trigger addiction? *Pathophysiology* **2018**, *25*, 263–276. [CrossRef]
8. Volkow, N.D.; Wise, R.A.; Baler, R. The dopamine motive system: Implications for drug and food addiction. *Nat. Rev. Neurosci.* **2017**, *18*, 741–752. [CrossRef]
9. Pelchat, M.L.; Johnson, A.; Chan, R.; Valdez, J.; Ragland, J.D. Images of desire: Food-craving activation during fMRI. *Neuroimage* **2004**, *23*, 1486–1493. [CrossRef]
10. Ziauddeen, H.; Farooqi, I.S.; Fletcher, P.C. Obesity and the brain: How convincing is the addiction model? *Nat. Rev. Neurosci.* **2012**, *13*, 279–286. [CrossRef]
11. Fletcher, P.C.; Kenny, P.J. Food addiction: A valid concept? *Neuropsychopharmacology* **2018**, *43*, 2506–2513. [CrossRef]
12. American Psychiatric Association (APA). Substance-related and addictive disorders. In *Diagnostic and Statistical Manual of Mental Disorders (DSM-5)*, 5th ed.; APA, Ed.; APA: Arlington, VA, USA, 2013; pp. 481–589.
13. Gearhardt, A.N.; Corbin, W.R.; Brownell, K.D. Preliminary validation of the Yale Food Addiction Scale. *Appetite* **2009**, *52*, 430–436. [CrossRef]
14. American Psychiatric Association (APA). Substance-related disorders. In *Diagnostic and Statistical Manual of Mental Disorders (DSM-IV)*, 4th ed.; APA, Ed.; APA: Arlington, VA, USA, 1994; pp. 175–272.
15. Gearhardt, A.N.; Corbin, W.R.; Brownell, K.D. Development of the Yale Food Addiction Scale Version 2.0. *Psychol. Addict. Behav.* **2016**, *30*, 113–121. [CrossRef] [PubMed]
16. Meule, A.; Müller, A.; Gearhardt, A.N.; Blechert, J. German version of the Yale Food Addiction Scale 2.0: Prevalence and correlates of 'food addiction' in students and obese individuals. *Appetite* **2017**, *115*, 54–61. [CrossRef] [PubMed]
17. Brunault, P.; Courtois, R.; Gearhardt, A.N.; Gaillard, P.; Journiac, K.; Cathelain, S.; Réveillère, C.; Ballon, N. Validation of the French Version of the DSM-5 Yale Food Addiction Scale in a Nonclinical Sample. *Can. J. Psychiatry* **2017**, *62*, 199–210. [CrossRef] [PubMed]
18. Aloi, M.; Rania, M.; Rodriguez Munoz, R.C.; Jimenez Murcia, S.; Fernandez-Aranda, F.; De Fazio, P.; Segura-Garcia, C. Validation of the Italian version of the Yale Food Addiction Scale 2.0 (I-YFAS 2.0) in a sample of undergraduate students. *Eat. Weight Disord.* **2017**, *22*, 527–533. [CrossRef] [PubMed]
19. Granero, R.; Jiménez-Murcia, S.; Gearhardt, A.N.; Agüera, Z.; Aymamí, N.; Gómez-Peña, M.; Lozano-Madrid, M.; Mallorquí-Bagué, N.; Mestre-Bach, G.; Neto-Antao, M.I.; et al. Validation of the Spanish Version of the Yale Food Addiction Scale 2.0 (YFAS 2.0) and Clinical Correlates in a Sample of Eating Disorder, Gambling Disorder, and Healthy Control Participants. *Front. Psychiatry* **2018**, *9*, 208. [CrossRef] [PubMed]
20. Fawzi, M.; Fawzi, M. Validation of an Arabic version of the Yale Food Addiction Scale 2.0. *East. Mediterr. Health J.* **2018**, *24*, 745–752. [CrossRef]
21. Hauck, C.; Weiß, A.; Schulte, E.M.; Meule, A.; Ellrott, T. Prevalence of 'Food Addiction' as Measured with the Yale Food Addiction Scale 2.0 in a Representative German Sample and Its Association with Sex, Age and Weight Categories. *Obes. Facts* **2017**, *10*, 12–24. [CrossRef] [PubMed]

22. Zhao, Z.; Ma, Y.; Han, Y.; Liu, Y.; Yang, K.; Zhen, S.; Wen, D. Psychosocial Correlates of Food Addiction and Its Association with Quality of Life in a Non-Clinical Adolescent Sample. *Nutrients* **2018**, *10*, 837. [CrossRef]
23. Chen, G.; Tang, Z.; Guo, G.; Liu, X.; Xiao, S. The Chinese version of the Yale Food Addiction Scale: An examination of its validation in a sample of female adolescents. *Eat. Behav.* **2015**, *18*, 97–102. [CrossRef] [PubMed]
24. Swarna Nantha, Y.; Abd Patah, N.A.; Ponnusamy Pillai, M. Preliminary validation of the Malay Yale Food Addiction Scale: Factor structure and item analysis in an obese population. *Clin. Nutr. ESPEN* **2016**, *16*, 42–47. [CrossRef]
25. Anai, A.; Ueda, K.; Harada, K.; Katoh, T.; Fukumoto, K.; Wei, C.-N. Determinant factors of the difference between self-reported weight and measured weight among Japanese. *Environ. Health Prev. Med.* **2015**, *20*, 447–454. [CrossRef] [PubMed]
26. Karlsson, J.; Persson, L.O.; Sjöström, L.; Sullivan, M. Psychometric properties and factor structure of the Three-Factor Eating Questionnaire (TFEQ) in obese men and women. Results from the Swedish Obese Subjects (SOS) study. *Int. J. Obes.* **2000**, *24*, 1715–1725. [CrossRef]
27. Anglé, S.; Engblom, J.; Eriksson, T.; Kautiainen, S.; Saha, M.-T.; Lindfors, P.; Lehtinen, M.; Rimpelä, A. Three factor eating questionnaire-R18 as a measure of cognitive restraint, uncontrolled eating and emotional eating in a sample of young Finnish females. *Int. J. Behav. Nutr. Phys. Act.* **2009**, *6*, 41. [CrossRef]
28. Adachi, Y.; Fujii, K.; Yamagami, T. Responses regarding restrained eating on the Three-Factor Eating Questionnaire and weight loss. *Jpn. J. Behav. Ther.* **1992**, *18*, 140–148.
29. Furukawa, T.A.; Kawakami, N.; Saitoh, M.; Ono, Y.; Nakane, Y.; Nakamura, Y.; Tachimori, H.; Iwata, N.; Uda, H.; Nakane, H.; et al. The performance of the Japanese version of the K6 and K10 in the World Mental Health Survey Japan. *Int. J. Methods Psychiatr. Res.* **2008**, *17*, 152–158. [CrossRef]
30. Cohen, J. A power primer. *Psychol. Bull.* **1992**, *112*, 155–159. [CrossRef]
31. Volker, M.A. Reporting Effect Size Estimates in School Psychology Research. *Psychol. Sch.* **2006**, *43*, 653–672. [CrossRef]
32. Lakens, D. Calculating and reporting effect sizes to facilitate cumulative science: A practical primer for t-tests and ANOVAs. *Front. Psychol.* **2013**, *4*, 863. [CrossRef]
33. Hu, L.-T.; Bentler, P.M. Cutoff criteria for fit indexes in covariance structure analysis: Conventional criteria versus new alternatives. *Struct. Equ. Model.* **1999**, *6*, 1–55. [CrossRef]
34. Bagozzi, R.P.; Yi, Y. Specification, evaluation, and interpretation of structural equation models. *J. Acad. Mark. Sci.* **2012**, *40*, 8–34. [CrossRef]
35. Barrett, P. Structural equation modelling: Adjudging model fit. *Pers. Individ. Dif.* **2007**, *42*, 815–824. [CrossRef]
36. Yatsuya, H.; Li, Y.; Hilawe, E.H.; Ota, A.; Wang, C.; Chiang, C.; Zhang, Y.; Uemura, M.; Osako, A.; Ozaki, Y.; et al. Global Trend in Overweight and Obesity and Its Association With Cardiovascular Disease Incidence. *Circ. J.* **2014**, *78*, 2807–2818. [CrossRef] [PubMed]
37. Elgar, F.J.; Roberts, C.; Tudor-Smith, C.; Moore, L. Validity of self-reported height and weight and predictors of bias in adolescents. *J. Adolesc. Health* **2005**, *37*, 371–375. [CrossRef] [PubMed]
38. Takimoto, H.; Yoshiike, N.; Kaneda, F.; Yoshita, K. Thinness Among Young Japanese Women. *Am. J. Public Health* **2004**, *94*, 1592–1595. [CrossRef] [PubMed]
39. Hayashi, F.; Takimoto, H.; Yoshita, K.; Yoshiike, N. Perceived body size and desire for thinness of young Japanese women: a population-based survey. *Br. J. Nutr.* **2007**, *96*, 1154–1162. [CrossRef]
40. Smith, A.R.; Joiner, T.E. Examining body image discrepancies and perceived weight status in adult Japanese women. *Eat. Behav.* **2008**, *9*, 513–515. [CrossRef] [PubMed]
41. Burrows, T.; Kay-Lambkin, F.; Pursey, K.; Skinner, J.; Dayas, C. Food addiction and associations with mental health symptoms: A systematic review with meta-analysis. *J. Hum. Nutr. Diet.* **2018**, *31*, 544–572. [CrossRef]
42. Schulte, E.M.; Avena, N.M.; Gearhardt, A.N. Which foods may be addictive? The roles of processing, fat content, and glycemic load. *PLoS ONE* **2015**, *10*, e0117959. [CrossRef]
43. Gearhardt, A.N.; Davis, C.; Kuschner, R.; Brownell, K.D. The Addiction Potential of Hyperpalatable Foods. *Curr. Drug Abuse Rev.* **2011**, *4*, 140–145. [CrossRef] [PubMed]
44. Lindgren, E.; Gray, K.; Miller, G.; Tyler, R.; Wiers, C.E.; Volkow, N.D.; Wang, G.J. Food addiction: A common neurobiological mechanism with drug abuse. *Front. Biosci. (Landmark Ed.)* **2018**, *23*, 811–836. [CrossRef] [PubMed]

45. Takeichi, H.; Taniguchi, H.; Fukinbara, M.; Tanaka, N.; Shikanai, S.; Sarukura, N.; Hsu, T.-F.; Wong, Y.; Yamamoto, S. Sugar intakes from snacks and beverages in Japanese children. *J. Nutr. Sci. Vitaminol. (Tokyo)* **2012**, *58*, 113–117. [CrossRef] [PubMed]
46. Gormally, J.; Black, S.; Daston, S.; Rardin, D. The assessment of binge eating severity among obese persons. *Addict. Behav.* **1982**, *7*, 47–55. [CrossRef]
47. Schlundt, D.G.; Hargreaves, M.K.; Buchowski, M.S. The Eating Behavior Patterns Questionnaire predicts dietary fat intake in African American women. *J. Am. Diet. Assoc.* **2003**, *103*, 338–345. [CrossRef] [PubMed]

© 2019 by the authors. Licensee MDPI, Basel, Switzerland. This article is an open access article distributed under the terms and conditions of the Creative Commons Attribution (CC BY) license (http://creativecommons.org/licenses/by/4.0/).

Article

Food Addiction Symptoms and Amygdala Response in Fasted and Fed States

Kirrilly M. Pursey [1,2], Oren Contreras-Rodriguez [3], Clare E. Collins [1,2], Peter Stanwell [1] and Tracy L. Burrows [1,*]

1. Faculty of Health and Medicine, The University of Newcastle, University Drive, Callaghan, NSW 2308, Australia; kirrilly.pursey@newcastle.edu.au (K.M.P.); Clare.Collins@newcastle.edu.au (C.E.O.); peter.stanwell@newcastle.edu.au (P.S.)
2. Priority Research Centre for Physical Activity and Nutrition, The University of Newcastle, University Drive, Callaghan, NSW 2308, Australia
3. Department of Psychiatry, Bellvitge Biomedical Research Institute (IDIBELL), and CIBERSAM, 08907 Barcelona, Spain; orencoro@gmail.com
* Correspondence: Tracy.Burrows@newcastle.edu.au; Tel.: +61-4921-5514

Received: 5 May 2019; Accepted: 4 June 2019; Published: 6 June 2019

Abstract: Few studies have investigated the underlying neural substrates of food addiction (FA) in humans using a recognised assessment tool. In addition, no studies have investigated subregions of the amygdala (basolateral (BLA) and central amygdala), which have been linked to reward-seeking behaviours, susceptibility to weight gain, and promoting appetitive behaviours, in the context of FA. This pilot study aimed to explore the association between FA symptoms and activation in the BLA and central amygdala via functional magnetic resonance imaging (fMRI), in response to visual food cues in fasted and fed states. Females ($n = 12$) aged 18–35 years completed two fMRI scans (fasted and fed) while viewing high-calorie food images and low-calorie food images. Food addiction symptoms were assessed using the Yale Food Addiction Scale. Associations between FA symptoms and activation of the BLA and central amygdala were tested using bilateral masks and small-volume correction procedures in multiple regression models, controlling for BMI. Participants were 24.1 ± 2.6 years, with mean BMI of 27.4 ± 5.0 kg/m^2 and FA symptom score of 4.1 ± 2.2. A significant positive association was identified between FA symptoms and higher activation of the left BLA to high-calorie versus low-calorie foods in the fasted session, but not the fed session. There were no significant associations with the central amygdala in either session. This exploratory study provides pilot data to inform future studies investigating the neural mechanisms underlying FA.

Keywords: Food addiction; Yale Food Addiction Scale; functional magnetic resonance imaging; basolateral amygdala

1. Introduction

There is increasing scientific interest in the possible role of "food addiction" (FA) underlying particular patterns of overeating, dietary relapse and weight gain in vulnerable individuals. Neuroimaging techniques, such as functional magnetic resonance imaging (fMRI), have provided insight into this phenomenon in humans. Visual food cues and consumption of palatable, energy-dense foods have been shown to activate reward-related brain circuits in humans in a similar way to substances of abuse [1]. Despite accumulating evidence supporting FA as a phenomenon in preclinical [2] and behavioural research [3], few studies have investigated the potential underlying neural substrates in humans. Many studies have used obesity as a proxy for addictive-like eating in lieu of a recognised assessment tool for FA, such as the Yale Food Addiction Scale (YFAS) [4]. Using obesity as a proxy for

FA could result in inconsistent neuroimaging findings as it is unclear as to the proportion of participants truly affected by addictive-like eating in these samples.

In one neuroimaging study using a recognised FA assessment tool, YFAS symptoms were associated with greater activation in response to a visual milkshake cue in brain areas encoding the reward value of foods and craving (amygdala, anterior cingulate cortex (ACC), medial orbitofrontal cortex (OFC), and the dorsolateral prefrontal cortex (DLPFC)) in young females [5]. In another study, while main effects were found activation in the amygdala, OFC, nucleus accumbens and inferior frontal cortex in response to the taste of sugar-sweetened beverages, no relationships were observed between YFAS symptoms and neural response in male and female adolescents [6]. The divergence in previous studies may be related to the limited range of food cues used, which have often been restricted to sweetened beverages [5,6]. These previous studies have also not studied participants in different motivational states (i.e., fasted and fed), which is important given the differences in responsivity in reward-related networks in these states [7,8].

The amygdala has been implicated in previous FA neuroimaging research [5] and plays a role in regulating the hedonic impact of salient stimuli and coordinating appetitive behaviours, with studies reporting selective sensitivity of the amygdala to food cues in the fasted state [9,10]. The amygdala has also been shown to integrate interoceptive states and sensory cues along the ventral visual stream [9] and has been implicated in drug cue reactivity and drug craving [11,12]. While the amygdala may play an important role in the context of FA, no studies have explored the potential role of distinct subregions of the amygdala. The basolateral amygdala (BLA) is of particular interest as it has been shown to drive external cues to the hypothalamic feeding centres in both humans and rats [13,14], consistent with the role in processing high-level sensory input and stimulus-value associations in humans [15]. In animal studies, the BLA has been implicated in reward-seeking behaviours in response to food-related stimuli [16] and relapse to food seeking [17]. In the satiated state, the BLA has also been associated with eating in the absence of homeostatic needs in rats [18] as well as predicting weight gain susceptibility in males and females [13]. In addition, the central amygdala has been reported to have a role in increasing reward saliency, modulating food consumption and promoting appetitive behaviours in mice [19]. While these previous studies of the BLA and central amygdala were not conducted in FA populations specifically, these findings suggest that there is a need to study the subregions of the amygdala in relation to FA in different motivational states.

This pilot study aimed to explore the association between YFAS assessed FA symptoms and activation in the central and BLA, assessed via functional MRI, in response to visual food cues in fasted and fed states. It was hypothesised that FA symptoms would be associated with greater activation in the BLA in response to high-calorie vs low-calorie foods in both the fasted and fed states, based on the findings of previous neuroimaging research [7,9,10].

2. Materials and Methods

2.1. Participants

Australian females aged 18–35 years were recruited to the current study from an existing pool of participants who had completed an online FA survey, which aimed to determine FA prevalence and associations with dietary intake in Australian adults [20]. At the end of the survey, participants could exit the survey with no further contact or could volunteer to be recontacted for future research. Full details regarding the FA survey are published elsewhere [20]. This study was approved by the University of Newcastle Human Research Ethics Committee (Approval number H-2012-0419) and was conducted in accordance with the ethical standards of the 1964 Helsinki Declaration.

Participants were eligible for the current study if they were female, aged 18–35 years, lived within a one-hour proximity to the imaging facility (Newcastle, NSW, Australia) and elected to be contacted for future research at the end of the online survey. Eligibility for this pilot study was restricted to females only in order to reduce potential inter-person variation in appetite and neural activation to visual food cues associated with sex-related differences [21]. Seventy-seven participants from the original survey volunteered for future research and were recontacted via email to participate in the fMRI component of the study. Of those recontacted, 35 responded that they were interested in participating in the fMRI study and were screened via telephone. Exclusion criteria for the current study included pregnancy, body mass >150 kg due to weight limitations for the MRI scanner, contraindications to MRI, left handedness, pre-existing medical or Axis 1 disorders, disordered eating behaviour, medications affecting appetite, history of substance abuse or head injury with loss of consciousness, allergy to any beverage ingredients, risk of adverse medical events as a result of fasting (e.g., diabetes), or inability to refrain from cigarette smoking. Seventeen participants did not meet the inclusion criteria for the current study, and five participants were not able to be contacted for screening, resulting in a final sample of thirteen participants. Written informed consent was obtained from all participants in the current study.

2.2. Procedures

The study procedure is outlined in Figure 1. Participants attended a single session where they underwent two fMRI scans of the brain, the first in the fasted state and the second in the fed state. Four hours prior to the first scan, participants consumed a standardised pre-fast meal (237 mL Ensure Plus; 1485 kJ/355 kcal). Participants then fasted for four hours, excluding water, to capture the hunger-state experienced when approaching the next meal. Participants were instructed to avoid caffeinated and alcoholic beverages, and smoking for twelve hours prior to the scan. Compliance with the pre-scan protocol was checked via verbal self-report upon arrival at the imaging facility prior to scanning. The session included anthropometric measurements, a demographic survey, hunger and image ratings, and the first of two fMRI scans. Following the first scan, participants drank a second Ensure Plus drink with compliance monitored by the research team. Participants again completed the hunger and image ratings before undertaking a second fMRI scan approximately 45 min following the completion of the second meal replacement beverage. Participants were not able to be scanned at the same time of day due to scheduling at the imaging facility.

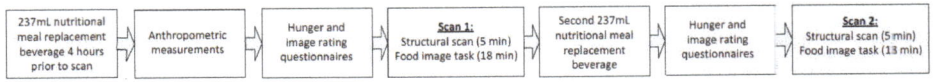

Figure 1. Flow diagram of the study procedure.

2.3. Measures

Demographics: Demographic data including sex, age, indigenous status, marital status and highest qualification were collected. Current and previous dieting history [22] was also assessed.

Yale Food Addiction Scale: The YFAS is a 25-item questionnaire which assesses addictive-like eating according to the diagnostic criteria for substance dependence. The YFAS has been shown to have adequate psychometric properties [4]. The tool includes two scoring options, a symptom score from 0–7 based on the DSM criteria for substance dependence, as well as a diagnosis of FA if ≥3 symptoms are reported as well as clinical impairment or distress. Cronbach α for the current sample was 0.93.

Anthropometrics: Height was taken using a BSM370 stadiometer, while weight and body composition (% body fat, fat free mass) was assessed using the InBody720 bioelectrical impedance analyser (Biospace, Seoul, Korea) using a standardised protocol. Body mass index (BMI) was subsequently calculated.

Hunger and Image Ratings: Participants completed a set of ratings related to hunger and appetite, as well as rating a subset of images shown to them in the MRI scanner ($n = 20$; $n = 10$ high-calorie, $n = 10$ low-calorie), using a 10 cm visual analogue scale. Participants completed the set of ratings in the fasted state at baseline (prior to Scan 1) and in the fed state (prior to Scan 2). Hunger and appetite ratings included self-reported (i) hunger, (ii) satiety, (iii) fullness and (iv) prospective food consumption [23]. For the image ratings, participants were asked to rate the (i) appeal of the food, (ii) desire to eat the food, (iii) whether the food increased their appetite and (iv) perceptions of prospective food consumption amount related to each of the food images.

2.4. Imaging Paradigm

Food images were chosen from a standardised database [24] and foods representative of the Australian food environment. Images were matched for brightness, complexity, resolution and size. Two groups of images were created according to nutrient composition, (i) appetising high-calorie foods (e.g., chocolate, chips, pizza, ice cream) and (ii) low-calorie foods (fruits and vegetables). The selected images were informed by the foods commonly reported in the original FA online survey. Food images were piloted on a sample of six university students independent of the study sample (mean age 27 years) regarding recognisability, familiarity and appeal. Those foods with low pilot ratings for recognisability and familiarity were excluded ($n = 27$). A final sample of 150 single food images were selected for the study including $n = 75$ high-calorie and $n = 75$ low-calorie foods. Nutritional composition of the food images can be found in Table S1.

Visual stimuli were presented in a block design format using Presentation software (Neurobehavioral Systems, Inc, Berkley, USA) and NordicNeuroLab sync box (NordicNeuroLab AS, Bergen, Norway), and consisted of two 18 min 20 s runs. Each run consisted of 15 epochs each of (i) low-calorie foods, (ii) high-calorie foods and (iii) fixation cross. Within each 20 s epoch of food images, five images were presented for four seconds each. A 20 s central fixation cross was presented as a control, and each epoch was separated by a four second blank screen. Each run consisted of 440 volumes over the scanning period. Images were presented onto an MRI compatible screen via a projector, and viewed via a mirror attached to the head coil. Standardised instructions to remain still focus on the visual stimuli were provided by the radiographer during the scan.

2.5. Imaging Data Collection

Structural and task-related functional scans were acquired using a Siemens 3-tesla MRI (Siemens AG, Germany). Structural images were acquired with a three-dimension (3D) magnetisation-prepared rapid gradient-echo (MP-RAGE) sequence with the following parameters: TE = 3.5 ms, TR = 2 s, 7° excitation flip angle, 160 slices with 1mm isotropic resolution. Task-related functional MR images were acquired using a T2*-weighted gradient-echo echo planar imaging (EPI) pulse sequence. The parameters were TE = 24 ms, TR = 2.5 s, 90° excitation flip angle, 47 axial slices 3 mm thick with a 0 mm interslice gap, in-plane resolution of 3×3 mm, covering from vertex to cerebellum. Axial slices were angled to the anterior cranial floor for to limit distortions that may impact on imaging of the orbito-frontal region.

2.6. Statistical Analysis

Participant characteristics, hunger and image ratings were analysed descriptively using Stata13. Differences in hunger and image ratings between the first and second scan were assessed using paired t-tests. Task-related functional MR images were processed and analysed using MATLAB version R2017a (The MathWorks Inc, Natick, Mass, USA) and Statistical Parametric Software (SPM12; The Welcome Department of Imaging Neuroscience, London, UK). Preprocessing steps involved motion correction, spatial normalization and smoothing using a Gaussian filter (FWHM 8 mm). The realigned functional sequences were coregistered to each participant's anatomical scan, which had been previously coregistered and normalized to the SPM-T1 template. Normalization parameters were then applied to the coregistered functional images, which were then resliced to a 2 mm isotropic resolution in Montreal Neurological Institute (MNI) space.

First-level (single-subject) SPM contrasts images were estimated for each of the fasted and fed scans for the following task main effects of interest: high-calorie > low-calorie foods, and low-calorie > high-calorie foods. Changes in brain activation between sessions were computed by the creation of the following contrasts: (Fasted scan (high-calorie > low-calorie foods) – Fed scan (high-calorie > low-calorie foods)) and (Fed scan (high-calorie > low-calorie foods) – Fasted scan (high-calorie > low-calorie foods)). The same contrasts were created to explore the between session effects on the activation of low-calorie > high calorie foods. Three regressors were used in these analyses to model conditions separately corresponding to high-calorie and low-calorie foods, and the fixation cross. Furthermore, to correct for subtle in-scanner movements from volume-to-volume scans, we identified the outlier scans present in the realigned task-related functional scans as determined using the CONN toolbox (Whitfield-Gabrieli and Nieto-Castanon, 2012). For each participant, the actual removal of outlier scans was completed by entering the subject-specific variables identifying the outlier scans (i.e., one regressor per outlier) in the first-level models as covariates of no interest, so these outlier scans were removed from these and subsequent analyses. We excluded one subject that had 19.8% of outlier scans. A hemodynamic delay of 4 s and a high-pass filter (1/120 Hz) were considered. The resulting first-level contrasts images were then carried forward to subsequent second-level random-effects (group) analyses. One-sample t-tests were used to assess the main task effects in the fasted and fed sessions. Multiple regression models were used to assess associations between FA symptoms and whole-brain activation under the main task effects in the fasted and fed sessions, and the change between sessions. These analyses were controlled for the subject-specific number of outlier scans, BMI, hunger and dieting when the association with FA symptoms was explored. Results were considered significant when surviving a $p > 0.001$ and a determined cluster-extent using Monte Carlo simulations implemented in the AlphaSim thresholding approach [25] (94 voxels and 135 voxels for the fasted and fed sessions, respectively incorporating a 2×2×2mm grey matter mask of 128,190 voxels). However, we specifically tested for associations between FA symptoms and the activation of the basolateral and central amygdala using bilateral masks and small-volume correction procedures in the multiple regression models. To that end, we defined bilateral basolateral and centromedial amygdala masks comprising 3.5 mm radial spheres cantered at left basolateral ($x = -26$, $y = -5$, $z = -23$) and right BLA ($x = 29$, $y = -3$, $z = -23$), and the left centromedial amygdala ($x = -19$, $y = -5$, $z = -15$) and right centromedial amygdala ($x = 23$, $y = -5$, $z = -13$) following previous research [26,27].

3. Results

Participant characteristics can be found in Table 1. The mean YFAS symptom score for the group was 4.1 ± 2.2 (range 1–7), with six participants classified as having a YFAS FA diagnosis. Hunger and image ratings are presented in Table 2. Hunger ratings were not significantly different between the fasted and fed scans ($p = 0.13$); however, ratings of fullness and satisfaction increased, and prospective food consumption decreased (all $p < 0.05$). Image ratings were significantly reduced from the fasted to fed scans ($p < 0.05$). FA symptoms were not associated with hunger or image ratings ($p > 0.05$)

Table 1. Participant characteristics at baseline ($n = 12$).

Characteristic	Total Sample
Age (years)	24.1 ± 2.6 (range 21–29)
Aboriginal or Torres Strait Islander background n (%)	0 (0)
Highest level of education n (%)	
Higher school certificate	2 (17)
Diploma	2 (17%)
University degree	8 (67%)
Body mass index (BMI) (kg/m^2)	27.4 ± 5.0 (range 21.7–39.0)
Body fat (%)	32.5 ± 9.8 (range 18.7–50.1)
Symptom score	4.1 ± 2.2 (range 1–7)
Current dieting n (%)	6 (50)
Previous dieting n (%)	12 (100)

Data is presented as mean ± SD unless otherwise specified.

Table 2. Hunger, appetite and image ratings in the fasted and fed states.

	Rating		p
	Fasted Scan	Fed Scan	
Hunger and appetite ratings			
Self-reported hunger	5.8 ± 1.9	4.2 ± 3.0	0.13
Self-reported satisfaction	3.1 ± 2.4	5.4 ± 2.6	0.03
Self-perceived fullness	2.1 ± 2.3	5.8 ± 3.1	0.003
Prospective food consumption amount	7.2 ± 1.8	4.8 ± 2.7	0.02
Image ratings			
Appeal of food	6.0 ± 1.6	4.5 ± 1.3	0.02
Desire to eat food	5.2 ± 1.7	3.8 ± 1.2	0.03
Effect of food in increasing appetite	4.5 ± 1.6	3.1 ± 1.1	0.02
Prospective consumption amount of food depicted in image	5.2 ± 1.3	3.2 ± 1.0	<0.001

Ratings completed using a 10 cm visual analogue scale. Results are presented as mean ± SD.

Brain Activation and Association with FA Symptoms

Visual cues of high-calorie vs low-calorie foods activated a cluster in the right amygdala, the left hippocampus, the fusiform gyrus, and in bilateral occipital cortex in the fasted session, and the left amygdala in the fed session (Table 3, Figure 2). No significant activations were recorded in the contrast low-calorie vs high-calorie foods in the fasted and fed sessions.

Table 3. Brain regions showing significant activation in response to the sight of high-calorie vs low-calorie foods in the fasted and fed sessions.

Brain Activation	Coordinates (x, y, z)	t-Value	CS
Fasted session			
Amygdala	14, −6, −18	6.1	111
Hippocampus	−36, −18, −16	6.7	243
Fusiform gyrus	32, −46, −16	5.2	127
Occipital cortex	42, −88, 6	8.3	434
	−34, −94, 2	7.9	165
Fed session			
Amygdala	−24, 14, −20	7.5	314

Coordinates are given in Montreal Neurological (MNI) Atlas space. The results for the fasted session surpassed a $p < 0.001$ and a cluster size (CS) of 94 voxels and 135 voxels, for the fasted and fed session, respectively. The t-values refer to the comparison of the activation of each of the listed brain regions to the high-calorie vs low-calorie food images.

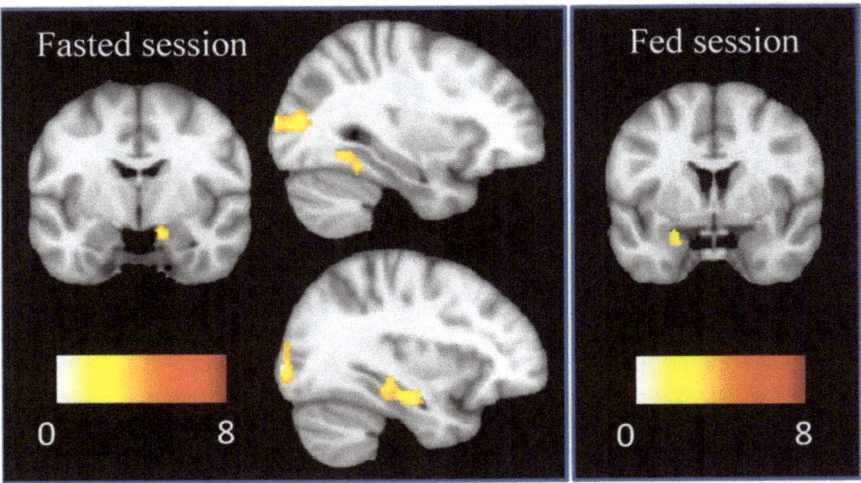

Figure 2. Increased brain activation to "high-calorie vs low-calorie" foods during the fasted and the fed sessions. Right side of the figure corresponds to the right hemisphere in the coronal views. In sagittal views, the upper figure corresponds to the right hemisphere, whereas the lower figure corresponds to the left hemisphere. Colour bar displays t-values for the comparison of activation in response to high-calorie vs low-calorie food cues.

No significant associations with FA symptoms were found using whole-brain corrections, but a significant association between the FA symptoms and higher activation of the left BLA to high-calorie vs low-calorie foods in the fasted session ($x = -26$, $y = -4$, $z = -26$, $t = 3.45$, 11 voxels, $p_{SVC-FWE<0.05} = 0.042$) were identified using small-volume correction procedures. Importantly, this association was found regardless of the body mass index of the participants, and it also remained statistically significant after controlling for hunger reported by the participants during the fasted session and current dieting status. No significant associations were identified between FA symptoms and the activation of the central amygdala in the fasted session, and the activation of the basolateral and central amygdala in the fed session (all $p_{SVC-FWE} > 0.05$). The change in the magnitude of activation from the fasted to the fed sessions of the left BLA showed a significant association with FA symptoms, controlling for BMI ($x = -26$, $y = -4$, $z = -26$, $t = 3.77$, 8 voxels, $p_{SVC-FWE<0.05} = 0.027$) (Figure 3).

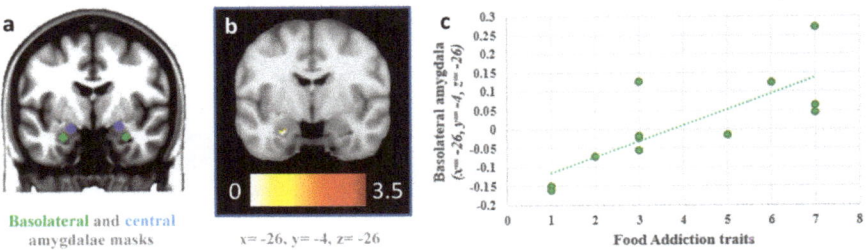

Figure 3. Association between the activation of the left basolateral amygdala and food addiction traits, (**a**) location of the basolateral (green) and central (blue) amygdala seeds, used as a mask for small-volume corrections; (**b**) left basolateral amygdala significantly associated with food addiction traits during the fasted session ($p_{SVC-FWE<0.05} = 0.042$); (**c**) scatter plot represents the correlation between the change in the activation of the left basolateral amygdala from the fasted to the fed sessions (y-axis) and food addiction traits (x-axis) ($x = -26$, $y = -4$, $z = -26$, $t = 3.77$, $p_{SVC-FWE<0.05} = 0.027$).

4. Discussion

This is the first study to use fMRI to investigate activation in distinct subregions of the amygdala in response to visual food cues in relation to FA symptoms in both fasting and fed states. In line with the hypothesis, activation in the BLA in response to high-calorie foods versus low-calorie foods was found to be associated with higher FA symptoms in the fasted state, which is consistent with previous research [7,9,10]. Furthermore, participants with greater FA symptoms showed higher activation in the BLA in the fasted than in the fed scans to the sight of high-calorie vs low-calorie foods. One possible mechanism consistent with this is that the amygdala boosts activation in the ventral stream [9,28], increasing food salience in the hunger state. Individual differences in BLA response in the hunger state to appetizing, high-calorie foods may therefore be associated with susceptibility to overeating.

The findings of the current study suggest that those with higher FA symptoms may be vulnerable to environmental food cues of energy-dense, nutrient-poor, high-calorie food when hungry compared to those with lower FA symptoms. This may lead those vulnerable individuals to greater desire to eat, hence leading to food seeking and consumption, which is consistent with previous research [16,17]. It is possible that FA may play a mediating role in non-homeostatic eating, dietary relapse and other longer-term issues such as weight gain. This may be an important consideration in future interventions targeting FA, suggesting a regular meal pattern that uses strategies to avoid long periods of fasting may be warranted. This may also suggest that further consideration to reducing environmental food cues (e.g., food advertisements) in treatment approaches may be warranted for those vulnerable to addictive-like eating.

The second hypothesis was not supported, with no significant associations found between FA symptoms and BLA activation in the fed state. This is divergent previous work investigating the role of the amygdala in eating in the absence of hunger and weight gain susceptibility [13]. However, the study by Sun and colleagues [13] did not recruit a sample to investigated FA, in contrast to the current study. These differences may also be due to the study samples (i.e., inclusion of males in the previous study) and use of different food cues (i.e., taste vs sight). Alternatively, this may be related to the standardised meal not significantly reducing self-reported hunger from baseline in the current study, although fullness was increased, and prospective food consumption was decreased. Of note, the standardised meal would have contributed a relatively small proportion of total daily energy intake (TEE) for participants (<30%). Future studies should therefore consider the use of a standardised meal with greater contribution to %TEE.

The strengths of the current study include the use of a recognised assessment tool for FA and investigation of participants in fasted and fed states. Food images were informed by previous research and selected from a standardised database [24], which was found to be a limitation of previous studies [21]. This current exploratory study is limited by the small sample size and inclusion of females only. Additionally, participants could not be scanned at the same time of day, however, a standardised meal replacement was provided to reduce inter-individual variation. Self-reported hunger was not significantly different between the fasted and fed states, however, other measures of appetite were significantly reduced. Future studies should consider standardised meals with greater calorie content, to ensure hunger is significantly different between the two conditions. As the scans were not conducted in a counterbalanced order, it is possible that the lack of associations may be attributed to a habituation effect. A further limitation is the lack of control condition in the cue reactivity task, and the addition of a control condition should be considered in future studies. Although participants were asked about disordered eating behaviours during the screening process, a clinical interview to identify eating disorder was not included in the baseline assessment. Future studies should use a clinical eating disorder assessment to better control for underlying disordered eating. This study used the original YFAS tool as the updated YFAS 2.0 tool had not been released at the time of the study. Future imaging studies should consider the use of the YFAS 2.0, which aligns with the DSM-5 criteria [29]. The current study also investigated associations between neural activation and FA symptoms, however, future studies may consider analysing according to the dichotomous YFAS diagnosis and in those with higher FA symptoms to better understand addictive-like eating.

5. Conclusions

This study demonstrates that activation of the BLA, which has been linked to reward-seeking behaviours and susceptibility to weight gain, was associated with FA symptoms in the fasted state. This study provides pilot data to inform future studies with larger sample sizes investigating the neural mechanisms associated with FA.

Supplementary Materials: The following are available online at http://www.mdpi.com/2072-6643/11/6/1285/s1, Table S1: Nutritional composition of food image groups per 100 g.

Author Contributions: Conceptualization, K.M.P., T.L.B., P.S. and C.E.C.; Analysis, O.C.-R., K.M.P.; Writing—Original Draft Preparation, K.M.P.; Writing—Review and Editing, T.L.B., O.C.-R., C.E.C., P.S.; Funding Acquisition, T.L.B. All authors have contributed to the development of the manuscript and have approved the final version.

Funding: This study was funded by a University of Newcastle, Faculty of Health and Medicine Pilot Grant obtained by T.L.B.

Acknowledgments: K.M.P. is supported by a Hunter Medical Research Institute Greaves Family Early Career Support Grant. O.C.-R. is funded by the postdoctoral contract "PERIS" (SLT006/17/00236) from the Catalan Government. T.L.B. is supported by University of Newcastle Brawn Research Fellowship.

Conflicts of Interest: The authors declare that they have no conflict of interest.

References

1. Kenny, P.J. Reward mechanisms in obesity: New insights and future directions. *Neuron* **2011**, *69*, 664–679. [CrossRef] [PubMed]
2. Avena, N.M.; Rada, P.; Hoebel, B.G. Evidence for sugar addiction: Behavioral and neurochemical effects of intermittent, excessive sugar intake. *Neurosci. Biobehav. Rev.* **2008**, *32*, 20–39. [CrossRef] [PubMed]
3. Pursey, K.M.; Stanwell, P.; Gearhardt, A.N.; Collins, C.E.; Burrows, T.L. The prevalence of food addiction as assessed by the Yale Food Addiction Scale: A systematic review. *Nutrients* **2014**, *6*, 4552–4590. [CrossRef] [PubMed]
4. Gearhardt, A.N.; Corbin, W.R.; Brownell, K.D. Preliminary validation of the Yale Food Addiction Scale. *Appetite* **2009**, *52*, 430–436. [CrossRef] [PubMed]
5. Gearhardt, A.N.; Yokum, S.; Orr, P.T.; Stice, E.; Corbin, W.R.; Brownell, K.D. Neural correlates of food addiction. *Arch. Gen. Psychiatry* **2011**, *68*, 808–816. [CrossRef]
6. Feldstein Ewing, S.W.; Claus, E.D.; Hudson, K.A.; Filbey, F.M.; Yakes Jimenez, E.; Lisdahl, K.M.; Kong, A.S. Overweight adolescents' brain response to sweetened beverages mirrors addiction pathways. *Brain Imaging Behav.* **2017**, *11*, 925–935. [CrossRef] [PubMed]
7. Siep, N.; Roefs, A.; Roebroeck, A.; Havermans, R.; Bonte, M.L.; Jansen, A. Hunger is the best spice: An fMRI study of the effects of attention, hunger and calorie content on food reward processing in the amygdala and orbitofrontal cortex. *Behav. Brain. Res.* **2009**, *198*, 149–158. [CrossRef]
8. Tomasi, D.; Volkow, N.D. Striatocortical pathway dysfunction in addiction and obesity: Differences and similarities. *Crit. Rev. Biochem. Mol. Biol.* **2013**, *48*, 1–19. [CrossRef]
9. LaBar, K.S.; Gitelman, D.R.; Parrish, T.B.; Kim, Y.-H.; Nobre, A.C.; Mesulam, M.M. Hunger selectively modulates corticolimbic activation to food stimuli in humans. *Behav. Neurosci.* **2001**, *115*, 493–500. [CrossRef]
10. Mohanty, A.; Gitelman, D.R.; Small, D.M.; Mesulam, M.M. The spatial attention network interacts with limbic and monoaminergic systems to modulate motivation-induced attention shifts. *Cereb. Cortex* **2008**, *18*, 2604–2613. [CrossRef]
11. Childress, A.R.; Mozley, P.D.; McElgin, W.; Fitzgerald, J.; Reivich, M.; O'Brien, C.P. Limbic activation during cue-induced cocaine craving. *Am. J. Psychiatry* **1999**, *156*, 11–18. [CrossRef] [PubMed]
12. Wilson, S.J.; Sayette, M.A.; Fiez, J.A. Prefrontal responses to drug cues: A neurocognitive analysis. *Nat. Neurosci.* **2004**, *7*, 211–214. [CrossRef] [PubMed]

13. Sun, X.; Kroemer, N.B.; Veldhuizen, M.G.; Babbs, A.E.; de Araujo, I.E.; Gitelman, D.R.; Sherwin, R.S.; Sinha, R.; Small, D.M. Basolateral amygdala response to food cues in the absence of hunger is associated with weight gain susceptibility. *J. Neurosci.* **2015**, *35*, 7964–7976. [CrossRef] [PubMed]
14. Petrovich, G.D.; Setlow, B.; Holland, P.C.; Gallagher, M. Amygdalo-hypothalamic circuit allows learned cues to override satiety and promote eating. *J. Neurosci.* **2002**, *22*, 8748–8753. [CrossRef] [PubMed]
15. Bzdok, D.; Laird, A.R.; Zilles, K.; Fox, P.T.; Eickhoff, S.B. An investigation of the structural, connectional, and functional subspecialization in the human amygdala. *Hum. Brain Mapp.* **2013**, *34*, 3247–3266. [CrossRef]
16. McLaughlin, R.J.; Floresco, S.B. The role of different subregions of the basolateral amygdala in cue-induced reinstatement and extinction of food-seeking behavior. *Neuroscience* **2007**, *146*, 1484–1494. [CrossRef] [PubMed]
17. Campbell, E.J.; Barker, D.J.; Nasser, H.M.; Kaganovsky, K.; Dayas, C.V.; Marchant, N.J. Cue-induced food seeking after punishment is associated with increased Fos expression in the lateral hypothalamus and basolateral and medial amygdala. *Behav. Neurosci.* **2017**, *131*, 155–167. [CrossRef]
18. Weingarten, H.P. Conditioned cues elicit feeding in sated rats: A role for learning in meal initiation. *Science* **1983**, *220*, 431–433. [CrossRef]
19. Douglass, A.M.; Kucukdereli, H.; Ponserre, M.; Markovic, M.; Gründemann, J.; Strobel, C.; Alcala Morales, P.L.; Conzelmann, K.-K.; Lüthi, A.; Klein, R. Central amygdala circuits modulate food consumption through a positive-valence mechanism. *Nat. Neurosci.* **2017**, *20*, 1384. [CrossRef]
20. Pursey, K.M.; Collins, C.E.; Stanwell, P.; Burrows, T.L. Foods and dietary profiles associated with 'food addiction' in young adults. *Addict. Behav. Rep.* **2015**, *2*, 41–48. [CrossRef]
21. Pursey, K.; Stanwell, P.; Callister, R.J.; Brain, K.; Collins, C.E.; Burrows, T.L. Neural responses to visual food cues according to weight status: A systematic review of functional magnetic resonance imaging studies. *Front. Nutr.* **2014**, *1*, 7. [CrossRef] [PubMed]
22. Meule, A.; Lutz, A.; Vogele, C.; Kubler, A. Self-reported dieting success is associated with cardiac autonomic regulation in current dieters. *Appetite* **2012**, *59*, 494–498. [CrossRef] [PubMed]
23. Flint, A.; Raben, A.; Blundell, J.E.; Astrup, A. Reproducibility, power and validity of visual analogue scales in assessment of appetite sensations in single test meal studies. *Int. J. Obes. Relat. Metab. Disord.* **2000**, *24*, 38–48. [CrossRef] [PubMed]
24. Blechert, J.; Meule, A.; Busch, N.A.; Ohla, K. Food-pics: An image database for experimental research on eating and appetite. *Front. Psychol.* **2014**, *5*, 617. [CrossRef] [PubMed]
25. Song, X.-W.; Dong, Z.-Y.; Long, X.-Y.; Li, S.-F.; Zuo, X.-N.; Zhu, C.-Z.; He, Y.; Yan, C.-G.; Zang, Y.-F. REST: A toolkit for resting-state functional magnetic resonance imaging data processing. *PLoS One* **2011**, *6*, e25031. [CrossRef] [PubMed]
26. Baur, V.; Hanggi, J.; Langer, N.; Jancke, L. Resting-state functional and structural connectivity within an insula-amygdala route specifically index state and trait anxiety. *Biol. Psychiatry* **2013**, *73*, 85–92. [CrossRef]
27. Pico-Perez, M.; Alonso, P.; Contreras-Rodriguez, O.; Martinez-Zalacain, I.; Lopez-Sola, C.; Jimenez-Murcia, S.; Verdejo-Garcia, A.; Menchon, J.M.; Soriano-Mas, C. Dispositional use of emotion regulation strategies and resting-state cortico-limbic functional connectivity. *Brain Imaging Behav.* **2018**, *12*, 1022–1031. [CrossRef]
28. Kravitz, D.J.; Saleem, K.S.; Baker, C.I.; Ungerleider, L.G.; Mishkin, M. The ventral visual pathway: An expanded neural framework for the processing of object quality. *Trends Cogn. Sci.* **2013**, *17*, 26–49. [CrossRef]
29. Gearhardt, A.N.; Corbin, W.R.; Brownell, K.D. Development of the Yale Food Addiction Scale Version 2.0. *Psychol. Addict. Behav.* **2016**, *30*, 113–121. [CrossRef]

© 2019 by the authors. Licensee MDPI, Basel, Switzerland. This article is an open access article distributed under the terms and conditions of the Creative Commons Attribution (CC BY) license (http://creativecommons.org/licenses/by/4.0/).

Article

Increasing Chocolate's Sugar Content Enhances Its Psychoactive Effects and Intake

Shanon L. Casperson [1,*], Lisa Lanza [2], Eram Albajri [2] and Jennifer A. Nasser [2,*]

[1] USDA, Agricultural Research Service, Grand Forks Human Nutrition Research Center, 2420 2nd Ave. North, Grand Forks, ND 58203-9034, USA
[2] College of Nursing and Health Professions, Drexel University, 1601 Cherry St MS31030 RM 389, Philadelphia, PA 19102-1320, USA; lisa.marcelalanza.lanza@drexel.edu (L.L.); eram.abdullah.albajri@drexel.edu (E.A.)
* Correspondence: shanon.casperson@ars.usda.gov (S.L.C.); jan57@drexel.edu (J.A.N.); Tel.: +1-701-795-8497 (S.L.C.); +1-267-359-5834 (J.A.N.)

Received: 29 January 2019; Accepted: 7 March 2019; Published: 12 March 2019

Abstract: Chocolate elicits unique brain activity compared to other foods, activating similar brain regions and neurobiological substrates with potentially similar psychoactive effects as substances of abuse. We sought to determine the relationship between chocolate with varying combinations of its main constituents (sugar, cocoa, and fat) and its psychoactive effects. Participants consumed 5 g of a commercially available chocolate with increasing amounts of sugar (90% cocoa, 85% cocoa, 70% cocoa, and milk chocolates). After each chocolate sample, participants completed the Psychoactive Effects Questionnaire (PEQ). The PEQ consists of questions taken from the Morphine-Benzedrine Group (MBG), Morphine (M,) and Excitement (E) subscales of the Addiction Research Center Inventory. After all testing procedures, participants completed the Binge Eating Scale (BES) while left alone and allowed to eat as much as they wanted of each of the different chocolates. We found a measurable psychoactive dose–effect relationship with each incremental increase in the chocolate's sugar content. The total number of positive responses and the number of positive responses on the E subscale began increasing after tasting the 90% cocoa chocolate, whereas the number of positive responses on the MBG and M subscales began increasing after tasting the 85% cocoa chocolate sample. We did not find a correlation between BES scores and the total amount of chocolate consumed or self-reported scores on the PEQ. These results suggest that each incremental increase in chocolate's sugar content enhances its psychoactive effects. These results extend our understanding of chocolate's appeal and unique ability to prompt an addictive-like eating response.

Keywords: chocolate; Addiction Research Center Inventory; sugar; craving; addictive-like eating; eating behavior

1. Introduction

Chocolate holds a special status in our society. Indeed, it is one of the most loved and craved, but problematic, foods [1]. Consuming chocolate evokes pleasant feelings, reduces tension, and improves mood [2,3]. Furthermore, chocolate elicits unique brain activity compared to other high-sugar and high-fat foods, recruiting brain structures that respond to craving-inducing stimuli, and is therefore more likely to provoke an addictive-like eating response [4]. The particular combination of cocoa, sugar, and fat in chocolate may play important, yet distinct, roles in chocolate's unique ability to elicit an addictive-like eating response. Smit et al. [5] demonstrated a role of the main psychopharmacological active constituents of cocoa in producing psychostimulant effects but determined that other attributes, such as sweetness and texture, may be more important. In a prior study, we observed effects of the percent cocoa and sugar contents on "desire to consume more chocolate", while fat content trended towards significance for this effect [6]. Defeliceantonio et al. [7] demonstrated a supra-additive effect of combining sugar and fat on food reward in humans, while others have demonstrated that the sugar component (in a sugar/fat combination) in particular is more effective at activating reward [8,9] and gustatory brain circuits [9]. Additionally, we have demonstrated that the highly reinforcing properties of sugar are difficult to overcome [10,11]. Taken together, chocolate's desirability appears to arise from the synergistic relationship among its components.

Fat and sugar are known to stimulate both the dopamine and the opioid neurotransmitter systems to regulate a food's rewarding potential [12–14]. The dopamine neurotransmitter system stimulates 'wanting', or the motivation to consume the food [12,14], whereas the opioid neurotransmitter system modulates the consumption of a desired food by amplifying 'liking', or the hedonic value, of the desired food [12,13]. Thus, the sight, smell, and taste of a highly palatable food such as chocolate work together to trigger motivational and hedonic reward mechanisms that result in the pursuit and consumption of the desired food. Subjective dopaminergic and opioidergic effects of food consumption can be differentiated using the Addiction Research Center Inventory (ACRI) [15]. We previously utilized the ARCI, specifically the Morphine-Benzedrine Group (MBG), Morphine (M), and Excitement (E) subscales, in a between-group design (groups identified by percent cocoa of the sample tasted), to provide indices of the psychoactive effects of chocolate that are associated with addictive-like eating [6]. We showed that simply tasting chocolate increases the number of positive responses on the MBG subscale of the ARCI, consistent with responses obtained on the MBG subscale after dopaminergic–opioidergic drug administration [16]. These results help explain chocolate's high reinforcing value, as foods that elevate feelings of euphoria have a greater reinforcing potential [17]. However, different individuals may specifically crave a particular combination of chocolate's main components (dark vs milk chocolate).

To continue our study of the psychoactive effects of chocolates varying in percent cocoa and sugar and fat content, we repeated our 2011 study using a within-subject design. Because some have postulated that "food addiction" is more of an "eating behavior addiction" similar to binge eating [18,19], we added the Binge Eating Scale (BES) [20] to our data collection. We hypothesized that self-reported scores on the MBG and E subscales of the ARCI would correlate with the particular combination of the main components (cocoa, sugar, and fat) of the chocolate and the amount of each of the different chocolates consumed after the tasting session. Consistent with our 2011 study, we expected no correlation between self-reported scores on the M subscale of the ARCI and the particular combination of chocolate's main components. In addition, we hypothesized that the amount of chocolate consumed after the tasting session would positively correlate with the increasing sugar content and decreasing cocoa and fat content of the chocolate. Furthermore, we hypothesized that BES scores would positively correlate with self-reported scores on the ARCI questionnaire and chocolate consumption.

2. Materials and Methods

2.1. Participants

Healthy adults (Table 1) were recruited from the greater Grand Forks, ND, and Philadelphia, PA, areas. Our participant population consisted of 57% non-overweight, 30% overweight, and 13% class 1 obese participants. Screening for study eligibility included height, weight, and a medical health history questionnaire. Exclusion criteria included: presence of food and non-food allergies; current status as a dieter; current or past metabolic illnesses (diabetes, renal failure, thyroid illness, hypertension); psychiatric, neurological, or eating disorders (schizophrenia, depression, Parkinson's Disease, Huntington's Disease, cerebral palsy, stroke, epilepsy, anorexia nervosa, or bulimia nervosa); taking prescription medications except for oral contraceptives or antihyperlipidemia agents. The study was approved by both the University of North Dakota and the Drexel University Institutional Review Boards and registered on clinicaltrials.gov as NCT03364413. Informed written consent was obtained from all participants prior to any study-related procedures.

Table 1. Participant characteristics. BMI, body mass index.

	Grand Forks	Philadelphia	Total
N (F/M)	20 (14/6)	10 (6/4)	30 (20/10)
Age, years	24.1 ± 6.8	26.1 ± 6.0	24.8 ± 6.5
Height, cm	168.7 ± 9.3	172.3 ± 11.1	169.9 ± 9.9
Weight, kg	72.1 ± 10.9	70.3 ± 11.1	70.7 ± 9.4
BMI, kg/m^2	25.3 ± 3.5	23.8 ± 4.3	24.8 ± 3.8

Values are means ± SD.

2.2. Experimental Procedures

Participants reported to the Center at which they were recruited 3–4 h postprandial to determine the psychoactive effect of consuming chocolate with varying amounts of fat, sugar, and cocoa (Table 2). Participants were instructed to have a light meal (e.g., sandwich with side salad or cup of soup) and to then refrain from eating or drinking (except water) before reporting for their study visit. Upon arrival, participants rated their appetite using 10 cm visual analog scales and completed a baseline Psychophysical Effects Questionnaire (PEQ; based on the ARCI) [6]. Participants were then presented with 5 g of commercially available chocolate varying in cocoa, sugar, and fat concentrations (Table 2). Chocolates were tested in order from least to most amount of sugar, and the PEQ was completed immediately after each chocolate tasting. Participants completed the Binge Eating Scale (BES) after the tasting session. Participants at the Grand Forks site were then left alone and allowed to eat as much as they wanted of each of the different chocolates.

Table 2. Characteristics of each 5 g chocolate sample.

Chocolate Type	kcal	Cocoa (%)	Sugar (g)	Fat (g)
Lindt® milk	29	38	2.4	1.9
Lindt® dark	30	70	1.5	2.4
Lindt® extra dark	29	85	0.6	2.3
Lindt® supreme dark	30	90	0.4	2.8

2.3. Questionnaires

Subjective homeostatic and hedonic hunger ratings were assessed using 100 mm visual analogue scales. Questions asked included: 1. How hungry do you feel; 2. How strong is your desire to eat; 3. How full do you feel; 4. How satisfied do you feel; 5. How much do you think you could eat right now; 6. Would you like to eat something sweet; and 7. Would you like to eat something fatty.

The PEQ is composed of 30 questions taken from the ARCI Morphine-Benzedrine Group (MBG), Morphine (M), and Excitement (E) subscales that assess subjective dopaminergic and opioidergic effects of psychoactive drugs [15]. The MBG subscale contains questions that center on feelings of well-being and euphoria, which correlate with the activation of both the dopaminergic and the opioidergic neurotransmitter systems. The M subscale focuses on attitude and physical sensations, and the E subscale relates to physical and psychological feelings of excitement, both of which correlate with activation of the dopaminergic neurotransmitter system [16]. Participants completed the PEQ before and after tasting each of the different chocolates.

The BES is composed of 16 questions used to assess the presence of binge eating behavior [20]. The BES contains questions that assess both behavioral manifestations of binge eating and of the feelings that either cue or follow a binge episode. Participants completed the BES after completing all testing procedures.

2.4. Anthropometric Measurements

Height, measured to the nearest 0.1 cm using a stadiometer (SECA Model 214, Hamburg, Germany), and body weight, measured using a calibrated digital scale to the nearest 0.1 kg (Fairbanks model 50735; Kansas City, MO, USA), were obtained at the end of the testing session.

2.5. Statistical Analysis

General linear models were used to compare general participant characteristics (i.e., sex, age, body mass index (BMI), hunger, testing site) and ARCI subscale scores of each chocolate type (defined by the percent cocoa in the chocolate sample tasted) and the amount of chocolate consumed. Tukey's post hoc analysis was used to determine differences. The threshold of significance was set at alpha = 0.05. JMP V14 (SAS Institute, Inc., Cary, NC, USA) was used for all analyses.

3. Results

3.1. Subjective Appetite Responses

On a scale from 0 (not at all) to 100 (very much), participants' rating of hunger was 58 ± 21 and of the desire to eat was 66 ± 28, while fullness was 25 ± 24 and feeling of being satisfied was 31 ± 21. There was no main effect of hunger or fullness on the total number of positive responses and on the number of positive responses on the MBG and M subscales; however, there was a main effect of hunger ($F(1,15) = 4.73$, $p = 0.046$) and fullness ($F(1,15) = 8.06$, $p = 0.012$) on self-reported scores on the E subscale of the PEQ.

Participants also rated their desire to eat foods with a specific taste profile on a scale from 0 (not at all) to 100 (very much). Participants' rating of the desire to eat something sweet was 79 ± 14 and that of the desire to eat something fatty was 48 ± 27. There was no main effect of the desire to eat something sweet or fatty on the total number of positive responses or on the number of positive responses on the MBG, M, and E subscales.

3.2. Psychophysical Effects Questionnaire

There was no main effect of testing site, sex, age, BMI, or BES on the total number of positive responses or on the number of positive responses on the MBG, M, and E subscales. Because of the direct correlation between the percent cocoa and sugar contents and the simultaneous changes of the fat and cocoa contents, we were not able to assess the direct effects of the individual chocolate components in this within-subject study design.

There was a significant main effect of chocolate type on the total number of positive responses ($F(4,106) = 30.10$, $p < 0.0001$) and on the number of positive responses on the MBG ($F(4,106) = 15.83$, $p < 0.0001$), M ($F(4,106) = 10.34$, $p < 0.0001$) and E ($F(4,106) = 16.57$, $p < 0.0001$) subscales. Tukey's post hoc analysis revealed slight differences in the effect of chocolate type on the total number of positive responses and on the number of positive responses on MBG, M, and E subscales.

The total number of positive responses (Figure 1) significantly increased after the consumption of all the different types of chocolate, with milk chocolate eliciting the greatest increase.

The number of positive responses on the MBG subscale (Figure 2) significantly increased after the consumption of the 85% cocoa chocolate and continued to increase in a dose-dependent manner in response to tasting each of the other chocolates. The number of positive responses on the MBG subscale after the consumption of the milk chocolate was significantly greater than after consumption of any of the other chocolates.

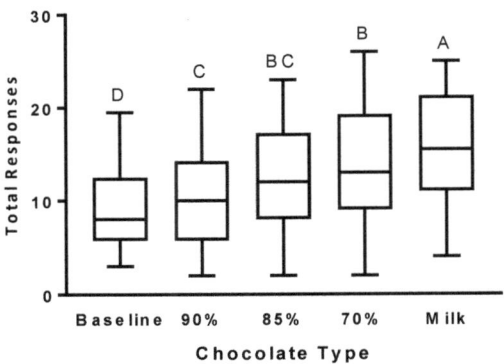

Figure 1. Total Psychophysical Effects Questionnaire (PEQ) scores after the consumption of chocolates differing in cocoa, sugar, and fat content. Self-reported scores on the PEQ questionnaire are presented as box and whiskers plots with the line representing the median, the box representing the 25th to 75th percentiles, and the whiskers representing the minimum to maximum values. Least-squares means are Baseline: 8.26 ± 1.16 (SE), 95% CI [5.93, 10.58]; 90% cocoa: 10.87 ± 1.10 (SE), 95% CI [8.65, 13.08]; 85% cocoa: 12.37 ± 1.10 (SE), 95% CI [10.15, 14.58]; 70% cocoa: 13.70 ± 1.10 (SE), 95% CI [11.49, 15.91]; Milk: 15.87 ± 1.10 (SE), 95% CI [13.65, 18.08]. Levels not connected by the same letter are significantly different.

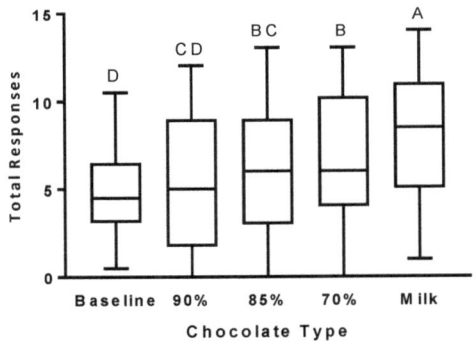

Figure 2. Morphine-Benzedrine Group (MBG) subscale scores after the consumption of chocolates differing in cocoa, sugar, and fat content. Self-reported scores on the MBG subscale are presented as box and whiskers plots with the line representing the median, the box representing the 25th to 75th percentiles, and the whiskers representing the minimum to maximum values. Least-squares means are Baseline: 4.28 ± 0.71 (SE), 95% CI [2.86, 5.69]; 90% cocoa: 5.37 ± 0.66 (SE), 95% CI [4.03, 6.70]; 85% cocoa: 6.10 ± 0.66 (SE), 95% CI [4.77, 7.43]; 70% cocoa: 6.63 ± 0.66 (SE), 95% CI [5.30, 7.97]; Milk: 7.97 ± 0.66 (SE), 95% CI [6.63, 9.30]. Levels not connected by the same letter are significantly different.

The number of positive responses on the M subscale (Figure 3) did not increase after the consumption of the 90% cocoa chocolate. A significant increase above baseline was not observed until participants consumed the 85% cocoa chocolate, with no further increases for each sequential chocolate consumption.

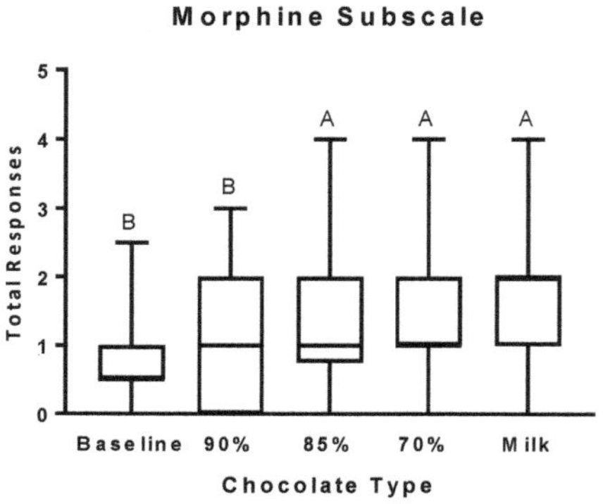

Morphine Subscale

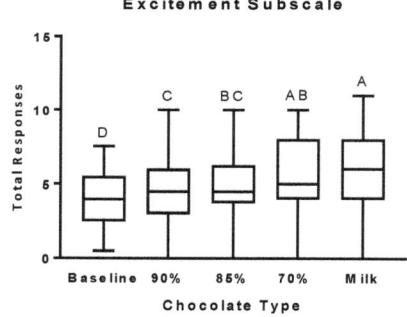

Figure 3. Morphine (M) subscale scores after the consumption of chocolates differing in cocoa, sugar, and fat content. Self-reported scores on the M subscale are presented as box and whiskers plots with the line representing the median, the box representing the 25th to 75th percentiles, and the whiskers representing the minimum to maximum values. Least-squares means are Baseline: 0.78 ± 0.20 (SE), 95% CI [0.37, 1.19]; 90% cocoa: 1.00 ± 0.19 (SE), 95% CI [0.62, 1.38]; 85% cocoa: 1.47 ± 0.19 (SE), 95% CI [1.09, 1.84]; 70% cocoa: 1.57 ± 0.19 (SE), 95% CI [1.19, 1.94]; Milk: 1.57 ± 0.19 (SE), 95% CI [1.19, 1.94]. Levels not connected by the same letter are significantly different.

The number of positive responses on the E subscale (Figure 4) significantly increased after the consumption of the 90% cocoa chocolate. As with the MBG subscale, the number of positive responses on the E subscale continued to increase in a dose-dependent manner in response to each incremental increase in the chocolate's sugar content and decrease in the percent cocoa and fat content.

Excitement Subscale

Figure 4. Excitement (E) subscale scores after the consumption of chocolates differing in cocoa, sugar, and fat content. Self-reported scores on the E subscale are presented as box and whiskers plots with the line representing the median, the box representing the 25th to 75th percentiles, and the whiskers representing the minimum to maximum values. Least-squares means are Baseline: 3.51 ± 0.47 (SE), 95% CI [2.56, 4.46]; 90% cocoa: 4.57 ± 0.44 (SE), 95% CI [3.68, 5.45]; 85% cocoa: 4.80 ± 0.44 (SE), 95% CI [3.92, 5.68]; 70% cocoa: 5.53 ± 0.44 (SE), 95% CI [4.65, 6.42]; Milk: 6.27 ± 0.44 (SE), 95% CI [5.38, 7.15]. Levels not connected by the same letter are significantly different.

3.3. Chocolate Consumption

There was a main effect of chocolate type on the amount of chocolate consumed ($p < 0.0001$). Overall, participants consumed significantly more milk chocolate (21 g ± 25 (SD)) than any other chocolate type (90% cocoa: 3 g ± 9; 85% cocoa: 2 g ± 4; 70% cocoa: 6 g ± 10 (SD)). There was no correlation between the amount of chocolate consumed and the total number of positive responses or the number of positive responses on the MBG, M, or E subscales. There was a main effect of hunger

(F(1,18) = 11.74, p = 0.003) on the total amount of chocolate consumed. Post hoc analysis revealed a main effect of hunger (F(1,17) = 8.64, p = 0.009) on the amount of milk chocolate consumed only. There was no main effect of BES score on the total amount of chocolate consumed.

4. Discussion

The current study aimed to extend our previous findings on the psychoactive effects of consuming chocolate varying in cocoa, sugar, and fat concentrations using a validated ARCI "drug effects" questionnaire in a within-subject design. The questions used simultaneously reflect alterations in motivation, mood, sensation, and perception, and, therefore, provide insight into the interrelation of these variables and chocolate consumption [16]. Our data indicate a measurable psychoactive dose–effect relationship with each incremental increase in the chocolate's sugar content and decrease in the percent cocoa and fat contents. Overall, there were an inverse dose–effect relationship with cocoa concentration and fat content and a positive dose–effect relationship with the sugar content of a chocolate. In addition, our data indicate that the dose–effect relationship of the different chocolates was slightly different for each ARCI subscale. Thus, the present study is the first to demonstrate a dose-dependent relationship between self-reported scores on the PEQ and chocolate consumption.

Increased feelings of well-being, euphoria, and physical and psychological feelings of excitement after chocolate consumption are consistent with chocolate's ability to modulate both the opioid and the dopamine neurotransmitter systems. Both human and animal research has demonstrated the reinforcing potential and comforting and mood-ameliorating effects of chocolate [2,3,21–23]. Contrary to our hypothesis, we did not find an association between self-reported scores on the MBG subscale and chocolate consumption. These results are also inconsistent with our previous study in which self-reported scores on the MBG subscale were associated with an increased desire to eat more chocolate [6]. A possible explanation for these findings is that subjective appetite measurements were obtained prior to the tasting session rather than after; however, participants were allowed to consume as much of the different chocolates as they wanted after the tasting session. On average, participants consumed 8 ± 15 g more chocolate than what was provided to them for the tasting session. Another possible explanation for this difference is that the smell and tasting of four different chocolates in the same session as opposed to only one type of chocolate could have satiated the desire to eat more chocolate. Massolt et al. [24] demonstrated that not only eating but simply smelling chocolate (85% cocoa) suppresses appetite. Additionally, Sørensen and Astrup [25] demonstrated that dark chocolate (70% cocoa) increases satiety and decreases the desire to eat something sweet more than milk chocolate. In the current study, participants consumed 15 g of dark chocolate before tasting the milk chocolate, and this could have decreased their appetite for more chocolate.

The sugar content, which plays a key role in chocolate's pleasurable taste and texture, is important in determining chocolate's reinforcing potential [26]. Research has shown that the added sugar component of a food is greatly associated with its reinforcing value [8,9,27]. We have also shown that the highly reinforcing properties of sugar are difficult to overcome [10,11]. The activation of sweet taste receptors, the speed at which the information about a food is delivered from the chemosensory and somatosensory neurons in the mouth to the brain, and the magnitude of the activation of the food reward system govern the reinforcing and rewarding effect of sugar [28,29]. Low et al. [30] recently reported that the average concentration at which sugar can be differentiated from water is 9 mass percent (m%); however, interindividual variability is large (reported range from 2 m% to 32 m%). The sugar content of the chocolates provided in this study were 8 m%, 13 m%, 30 m%, and 48 m% for the 90% cocoa, 85% cocoa, 70% cocoa, and milk chocolates, respectively. Therefore, our finding that consuming the 85% cocoa, 70% cocoa, and milk chocolates resulted in significant increases in self-reported scores on the PEQ, and each subscale, above baseline is supported by the fact that all of these chocolates were above the 9 m% detection threshold.

In agreement with our prior study [6], consuming milk chocolate elicited a greater increase, compared to all the other chocolates, in the total number of positive responses as well as a greater increase in positive responses on the MBG subscale. The sugar content of the milk chocolate, which was equivalent to the upper sucrose detection threshold reported by Low et al. [30], may explain these results, as well as the mass appeal of milk chocolate. Indeed, participants in the current study consumed an average of 21 ± 25 g of milk chocolate compared to 6 ± 10 g of the 70% cocoa and 2 ± 4 g of the 85% cocoa chocolates. Taken together, these results agree substantively with other indicators of reinforcement essential for motivational behavior [15,31] and support the "addictive-like" behavior response individuals can experience with chocolate, in particular milk chocolate, consumption.

The finding that consuming chocolate containing 90% cocoa increased the total number of positive responses and the number of positive responses on the E subscale is of interest, given its reported "bitter" taste. The bioactive compounds found in dark chocolate may explain these positive results. In two double-blind, placebo control studies, Smit et al. [5] demonstrated that the amount of methylxanthines (theobromine and caffeine) found in dark chocolate can produce psychostimulant effects. They found that the consumption of both encapsulated cocoa powder and a "typical portion" of dark chocolate increases energetic arousal ("energetic", "alert", etc. versus "tired", "sluggish"). This is consistent with our results demonstrating chocolate's ability to increase physical and psychological feelings of excitement. Interestingly, when examined separately, the amount of theobromine found in a "typical portion" of chocolate (200–300 mg) does not appear to play a psychopharmacological role [32], whereas the amount of caffeine (25–35 mg) is well above the previously reported stimulatory threshold (12.5 mg [33]). The amount of caffeine consumed from the 90% cocoa chocolate sample in the current study was 9 mg, well below the stimulatory threshold. It may be that the theobromine found in chocolate enhances the stimulatory effect of caffeine. Thus, it is plausible that the combination of theobromine and caffeine found in chocolate provides an additive psychostimulant effect. While the methylxanthines in chocolate provide psychostimulant effects, Smit et al. [5], as well as Michener and Rozin [26], reported no difference in chocolate cravings between the ingestion of cocoa powder versus placebo, whereas consuming chocolate, including white chocolate (albeit to a lesser extent), immediately reduced chocolate cravings. As mentioned above, another explanation for this finding is that the sugar content of the 90% cocoa chocolate (8 m%) is well above the lowest and close to the average sucrose detection threshold previously reported [30]. Taken together, these results confirm that the combination of the main chocolate constituents is necessary to produce its psychoactive effects.

As expected, more milk chocolate was consumed than any of the other chocolates, indicating its reinforcing potential. This is consistent with the greater number of positive responses on the PEQ. However, contrary to our hypothesis, we did not find an association between the total amount of chocolate consumed or the total number of positive responses on the PEQ and BES scores. A potential explanation for this finding is all of our study participants had a score less than or equal to 17 and thus would be classified as non-binge eaters [20]. The narrow range in BES scores (2–17) of our participants may not have provided the heterogeneity needed to reliably determine a correlation between BES scores and chocolate consumption or self-reported PEQ scores. Further studies are needed to determine the potential correlation between BES scores and chocolate consumption or self-reported PEQ scores. Additionally, the low BES scores may also explain why we did not find an association between self-reported scores on the MBG subscale and chocolate consumption.

This study is not without limitations. The direct correlation between the percent cocoa and sugar content and the simultaneous changes of the fat content in the chocolate samples did not allow us to assess the independent psychoactive effects of each component. However, using commercially available chocolate provides a "real-world" value to our observations. Furthermore, tasting the chocolate samples sequentially in a relatively short amount of time may have produced an additive effect on our outcomes, such as habituation or sensory-specific satiety [34]. For this study, each participant consumed a total of 20 g of chocolate (5 g/sample) compared to the 12.5 g provided in our previous study [6] with similar PEQ results. Therefore, it does not appear that our within-subject study design had a significant impact on our results. Lastly, the participants in this study tended to be healthy-weight individuals (57% of the participants), thus, the results may not reflect the response of overweight and obese individuals. Although we did not find a significant effect of BMI on self-reported scores on the PEQ or any of the subscales, research has shown that food reinforcement can vary significantly between lean, overweight, and obese individual [35–37] and individuals with an eating disorder [38,39]. Further research is needed to determine if there exist differential psychoactive effects of chocolate consumption between lean, overweight, and obese populations and individuals with an eating disorder.

Our prior work suggests that obese individuals who engage in binge eating may do so because of an inherent reward deficiency, resulting in overeating high-sugar foods. We previously found an increase in food reinforcement after the consumption of a liquid chocolate-based preload, indicating a "sensitization" response to foods that increase the dopaminergic response [39]. Davis et al. [38] found that obese individuals diagnosed with binge eating disorder have a "hyper-reactivity to the hedonic properties of food, coupled with the motivation to engage in appetitive behaviors." In women diagnosed with bulimia nervosa (BN), Bulik et al. [40] reported that food reinforcement decreases following deprivation. On the basis of these results, we posit that obese individuals with an eating disorder would demonstrate an overall increase in their self-reported scores on the PEQ due to reward sensitization. However, we would not anticipate a significant increase per se in obese individuals who do not exhibit a binge eating-related eating disorder.

Although questionnaires provide validated subjective measurements about food reward and eating behavior, the use of objective methodologies provide insight into the effect of highly palatable food on central dopamine activity and brain regions known to have a high density of dopamine neurons (positron emission tomography and functional magnetic resonance imaging, respectively). However, the cost of instrumentation, the expertise needed to operate the equipment and interpret the data, as well as the engineering and mechanical constraints of the scanners make objective measurements of the dopaminergic response to food challenging. A promising methodology, electroretinography (ERG), may provide a more efficient way to objectively assess central dopamine activity. ERG is a clinical ophthalmological procedure that overcomes the impediments of other neuroimaging methods. ERG records the electrical potential from the retina, which is dependent upon dopamine signaling [41,42], in response to light simulation. ERG has been used to show a negative correlation between self-reported cocaine craving and dopamine-mediated retinal signal [43,44] and a positive correlation between dopamine metabolite levels in the cerebrospinal fluid [45] and dopamine-mediated retinal signal. We have also used ERG to demonstrate a positive correlation between increased dopamine-mediated retinal signal with food stimulation and BES scores [46], consistent with positron emission tomography [47]. We are currently conducting research to extend those findings to chocolates varying in cocoa, sugar, and fat content. Results from this ongoing study should provide further evidence that ERG should be considered as a low-cost, non-invasive method for objective evaluation of the stimulating properties of food in conjunction with ARCI questionnaires.

5. Conclusions

Chocolate's ability to modulate both the opioidergic and the dopaminergic systems, evident by the significant increase in self-reported scores on the PEQ, is consistent with research demonstrating chocolate's reinforcing potential and comforting and mood-ameliorating effects [2,3,21–23]. These results help explain chocolate's high reinforcing value, as foods that elevate feelings of euphoria, as indicated by an increase in the number of positive responses on the MBG subscales, have a greater reinforcing potential [17]. Given the measurable psychoactive dose–effect relationship observed in this study, we posit that chocolate's ability to provoke "addictive-like" eating behavior is initiated by its bioactive compounds, and each incremental increase in added sugar further enhances these effects.

Author Contributions: The authors' responsibilities were as follows—S.L.C. and J.A.N.: conceived the project, developed the overall research plan, oversaw the study, wrote and edited the manuscript, and had primary responsibility for final content; S.L.C., L.L., and E.A.: collected the data, analyzed the data, and edited the manuscript; and all authors read and approved the final manuscript.

Funding: This work was supported by the Agricultural Research Services of the United States Department of Agriculture #5450-51530-051-00D (S.L.C.) and a competitive pilot grant from the Clinical Translational Research Institute (CTRI/DUCOM) to J.A.N. The role of the funding sponsors was to approve the study and the submission of this manuscript for publication (USDA) and approve funding for the grant proposing this research (CTRI/DUCOM).

Acknowledgments: The authors would like to thank Clint Hall and Joshua Severud for their assistance with the implementation of the protocol, data collection, and data entry, and Wei DU, MD, MS (Chair, Department of Psychiatry, Drexel University College of Medicine (DUCOM), for his insights on the physiological/pharmacological effects of addictive substances.

Conflicts of Interest: The authors have nothing to disclose. Mention of trade names or commercial products in this publication is solely for the purpose of providing specific information and does not imply recommendation or endorsement by the U.S. Department of Agriculture. The U.S. Department of Agriculture (USDA) prohibits discrimination in all its programs and activities on the basis of race, color, national origin, age, disability, and where applicable, sex, marital status, familial status, parental status, religion, sexual orientation, genetic information, political beliefs, reprisal, or because all or part of an individual's income is derived from any public assistance program. (Not all prohibited bases apply to all programs.) Persons with disabilities who require alternative means for communication of program information (Braille, large print, audiotape, etc.) should contact USDA's TARGET Center at (202) 720-2600 (voice and TDD). To file a complaint of discrimination, write to USDA, Director, Office of Civil Rights, 1400 Independence Avenue, S.W., Washington, D.C. 20250-9410, or call (800) 795-3272 (voice) or (202) 720-6382 (TDD). USDA is an equal opportunity provider and employer.

References

1. Schulte, E.M.; Avena, N.M.; Gearhardt, A.N. Which foods may be addictive? The roles of processing, fat content, and glycemic load. *PLoS ONE* **2015**, *10*, e0117959. [CrossRef]
2. Macht, M.; Dettmer, D. Everyday mood and emotions after eating a chocolate bar or an apple. *Appetite* **2006**, *46*, 332–336. [CrossRef] [PubMed]
3. Meier, B.P.; Noll, S.W.; Molokwu, O.J. The sweet life: The effect of mindful chocolate consumption on mood. *Appetite* **2017**, *108*, 21–27. [CrossRef]
4. Asmaro, D.; Liotti, M. High-caloric and chocolate stimuli processing in healthy humans: An integration of functional imaging and electrophysiological findings. *Nutrients* **2014**, *6*, 319–341. [CrossRef] [PubMed]
5. Smit, H.J.; Gaffan, E.A.; Rogers, P.J. Methylxanthines are the psycho-pharmacologically active constituents of chocolate. *Psychopharmacology* **2004**, *176*, 412–419. [CrossRef]
6. Nasser, J.A.; Bradley, L.E.; Leitzsch, J.B.; Chohan, O.; Fasulo, K.; Haller, J.; Jaeger, K.; Szulanczyk, B.; Del Parigi, A. Psychoactive effects of tasting chocolate and desire for more chocolate. *Physiol. Behav.* **2011**, *104*, 117–121. [CrossRef] [PubMed]
7. DiFeliceantonio, A.G.; Coppin, G.; Rigoux, L.; Edwin Thanarajah, S.; Dagher, A.; Tittgemeyer, M.; Small, D.M. Supra-additive effects of combining fat and carbohydrate on food reward. *Cell Metab.* **2018**, *28*, 33–44.e3. [CrossRef] [PubMed]
8. Epstein, L.H.; Carr, K.A.; Lin, H.; Fletcher, K.D. Food reinforcement, energy intake, and macronutrient choice. *Am. J. Clin. Nutr.* **2011**, *94*, 12–18. [CrossRef]
9. Stice, E.; Burger, K.S.; Yokum, S. Relative ability of fat and sugar tastes to activate reward, gustatory, and somatosensory regions. *Am. J. Clin. Nutr.* **2013**, *98*, 1377–1384. [CrossRef]
10. Casperson, S.L.; Johnson, L.; Roemmich, J.N. The relative reinforcing value of sweet versus savory snack foods after consumption of sugar- or non-nutritive sweetened beverages. *Appetite* **2017**, *112*, 143–149. [CrossRef]
11. Casperson, S.L.; Roemmich, J.N. Impact of dietary protein and gender on food reinforcement. *Nutrients* **2017**, *9*, 957. [CrossRef] [PubMed]
12. Berridge, K.C. 'Liking' and 'wanting' food rewards: Brain substrates and roles in eating disorders. *Physiol. Behav.* **2009**, *97*, 537–550. [CrossRef]
13. Pecina, S. Opioid reward 'liking' and 'wanting' in the nucleus accumbens. *Physiol. Behav.* **2008**, *94*, 675–680. [CrossRef] [PubMed]
14. Wise, R.A. Role of brain dopamine in food reward and reinforcement. *Philos. Trans. R. Soc. B Biol. Sci.* **2006**, *361*, 1149–1158. [CrossRef]
15. Balster, R.L.; Walsh, S.L. Addiction research center inventory. In *Encyclopedia of Psychopharmacology*; Stolerman, I.P., Ed.; Springer: Berlin/Heidelberg, Germany, 2010; p. 20.

16. Haertzen, C.A. Development of scales based on patterns of drug effects, using the addiction research center inventory (arci). *Psychol. Rep.* **1966**, *18*, 163–194. [CrossRef] [PubMed]
17. Schulte, E.M.; Smeal, J.K.; Gearhardt, A.N. Foods are differentially associated with subjective effect report questions of abuse liability. *PLoS ONE* **2017**, *12*, e0184220. [CrossRef]
18. Leigh, S.J.; Morris, M.J. The role of reward circuitry and food addiction in the obesity epidemic: An update. *Biol. Psychol.* **2018**, *131*, 31–42. [CrossRef]
19. Novelle, M.G.; Dieguez, C. Food addiction and binge eating: Lessons learned from animal models. *Nutrients* **2018**, *10*, 71. [CrossRef] [PubMed]
20. Gormally, J.; Black, S.; Daston, S.; Rardin, D. The assessment of binge eating severity among obese persons. *Addict. Behav.* **1982**, *7*, 47–55. [CrossRef]
21. Scholey, A.; Owen, L. Effects of chocolate on cognitive function and mood: A systematic review. *Nutr. Rev.* **2013**, *71*, 665–681. [CrossRef]
22. Bruinsma, K.; Taren, D.L. Chocolate: Food or drug? *J. Am. Diet. Assoc.* **1999**, *99*, 1249–1256. [CrossRef]
23. Parker, G.; Parker, I.; Brotchie, H. Mood state effects of chocolate. *J. Affect. Disord.* **2006**, *92*, 149–159. [CrossRef] [PubMed]
24. Massolt, E.T.; van Haard, P.M.; Rehfeld, J.F.; Posthuma, E.F.; van der Veer, E.; Schweitzer, D.H. Appetite suppression through smelling of dark chocolate correlates with changes in ghrelin in young women. *Regul. Pept.* **2010**, *161*, 81–86. [CrossRef]
25. Sorensen, L.B.; Astrup, A. Eating dark and milk chocolate: A randomized crossover study of effects on appetite and energy intake. *Nutr. Diabetes* **2011**, *1*, e21. [CrossRef]
26. Michener, W.; Rozin, P. Pharmacological versus sensory factors in the satiation of chocolate craving. *Physiol. Behav.* **1994**, *56*, 419–422. [CrossRef]
27. Avena, N.M.; Rada, P.; Hoebel, B.G. Evidence for sugar addiction: Behavioral and neurochemical effects of intermittent, excessive sugar intake. *Neurosci. Biobehav. Rev.* **2008**, *32*, 20–39. [CrossRef] [PubMed]
28. *Treatment for Stimulant Use Disorders*; Treatment Improvement Protocol (TIP) Series, No. 33; Center for Substance Abuse Treatment: Rockville, MD, USA, 1999.
29. Lee, A.A.; Owyang, C. Sugars, sweet taste receptors, and brain responses. *Nutrients* **2017**, *9*, 653. [CrossRef]
30. Low, J.Y.; McBride, R.L.; Lacy, K.E.; Keast, R.S. Psychophysical evaluation of sweetness functions across multiple sweeteners. *Chem. Senses* **2017**, *42*, 111–120. [CrossRef]
31. Noori, H.R.; Cosa Linan, A.; Spanagel, R. Largely overlapping neuronal substrates of reactivity to drug, gambling, food and sexual cues: A comprehensive meta-analysis. *Eur. Neuropsychopharmacol.* **2016**, *26*, 1419–1430. [CrossRef] [PubMed]
32. Baggott, M.J.; Childs, E.; Hart, A.B.; de Bruin, E.; Palmer, A.A.; Wilkinson, J.E.; de Wit, H. Psychopharmacology of theobromine in healthy volunteers. *Psychopharmacology* **2013**, *228*, 109–118. [CrossRef] [PubMed]
33. Smit, H.J.; Rogers, P.J. Effects of low doses of caffeine on cognitive performance, mood and thirst in low and higher caffeine consumers. *Psychopharmacology* **2000**, *152*, 167–173. [CrossRef] [PubMed]
34. Hetherington, M.; Rolls, B.J.; Burley, V.J. The time course of sensory-specific satiety. *Appetite* **1989**, *12*, 57–68. [CrossRef]
35. Temple, J.L.; Bulkley, A.M.; Badawy, R.L.; Krause, N.; McCann, S.; Epstein, L.H. Differential effects of daily snack food intake on the reinforcing value of food in obese and nonobese women. *Am. J. Clin. Nutr.* **2009**, *90*, 304–313. [CrossRef] [PubMed]
36. Volkow, N.D.; Wang, G.J.; Tomasi, D.; Baler, R.D. Obesity and addiction: Neurobiological overlaps. *Obes Rev.* **2013**, *14*, 2–18. [CrossRef]
37. Epstein, L.H.; Carr, K.A.; Lin, H.; Fletcher, K.D.; Roemmich, J.N. Usual energy intake mediates the relationship between food reinforcement and bmi. *Obesity* **2012**, *20*, 1815–1819. [CrossRef]
38. Davis, C.A.; Levitan, R.D.; Reid, C.; Carter, J.C.; Kaplan, A.S.; Patte, K.A.; King, N.; Curtis, C.; Kennedy, J.L. Dopamine for "wanting" and opioids for "liking": A comparison of obese adults with and without binge eating. *Obesity* **2009**, *17*, 1220–1225. [CrossRef]
39. Nasser, J.A.; Evans, S.M.; Geliebter, A.; Pi-Sunyer, F.X.; Foltin, R.W. Use of an operant task to estimate food reinforcement in adult humans with and without bed. *Obesity* **2008**, *16*, 1816–1820. [CrossRef]
40. Bulik, C.M.; Brinded, E.C. The effect of food deprivation on the reinforcing value of food and smoking in bulimic and control women. *Physiol. Behav.* **1994**, *55*, 665–672. [CrossRef]

41. Witkovsky, P.; Veisenberger, E.; Haycock, J.W.; Akopian, A.; Garcia-Espana, A.; Meller, E. Activity-dependent phosphorylation of tyrosine hydroxylase in dopaminergic neurons of the rat retina. *J. Neurosci.* **2004**, *24*, 4242–4249. [CrossRef]
42. Lavoie, J.; Illiano, P.; Sotnikova, T.D.; Gainetdinov, R.R.; Beaulieu, J.M.; Hebert, M. The electroretinogram as a biomarker of central dopamine and serotonin: Potential relevance to psychiatric disorders. *Biol. Psychiatry* **2014**, *75*, 479–486. [CrossRef]
43. Roy, M.; Smelson, D.A.; Roy, A. Abnormal electroretinogram in cocaine-dependent patients. Relationship to craving. *Br. J. Psychiatry* **1996**, *168*, 507–511. [CrossRef]
44. Smelson, D.A.; Roy, A.; Roy, M.; Tershakovec, D.; Engelhart, C.; Losonczy, M.F. Electroretinogram and cue-elicited craving in withdrawn cocaine-dependent patients: A replication. *Am. J. Drug Alcohol Abuse* **2001**, *27*, 391–397. [CrossRef]
45. Roy, A.; Roy, M.; Berman, J.; Gonzalez, B. Blue cone electroretinogram amplitudes are related to dopamine function in cocaine-dependent patients. *Psychiatry Res.* **2003**, *117*, 191–195. [CrossRef]
46. Nasser, J.A.; Del Parigi, A.; Merhige, K.; Wolper, C.; Geliebter, A.; Hashim, S.A. Electroretinographic detection of human brain dopamine response to oral food stimulation. *Obesity* **2013**, *21*, 976–980. [CrossRef]
47. Wang, G.J.; Geliebter, A.; Volkow, N.D.; Telang, F.W.; Logan, J.; Jayne, M.C.; Galanti, K.; Selig, P.A.; Han, H.; Zhu, W.; et al. Enhanced striatal dopamine release during food stimulation in binge eating disorder. *Obesity* **2011**, *19*, 1601–1608. [CrossRef]

© 2019 by the authors. Licensee MDPI, Basel, Switzerland. This article is an open access article distributed under the terms and conditions of the Creative Commons Attribution (CC BY) license (http://creativecommons.org/licenses/by/4.0/).

Discussion

Fat Addiction: Psychological and Physiological Trajectory

Siddharth Sarkar [1], Kanwal Preet Kochhar [2] and Naim Akhtar Khan [3,*]

1. Department of Psychiatry and National Drug Dependence Treatment Centre (NDDTC), All India Institute of Medical Sciences (AIIMS), New Delhi 110029, India; sidsarkar22@gmail.com
2. Department of Physiology, All India Institute of Medical Sciences (AIIMS), New Delhi 110029, India; kpkochhar6@gmail.com
3. Nutritional Physiology and Toxicology (NUTox), UMR INSERM U1231, University of Bourgogne and Franche-Comte (UBFC), 6 boulevard Gabriel, 21000 Dijon, France
* Correspondence: Naim.Khan@u-bourgogne.fr; Tel.: +33-3-80-39-63-12; Fax: + 33-3-80-39-63-30

Received: 9 October 2019; Accepted: 12 November 2019; Published: 15 November 2019

Abstract: Obesity has become a major public health concern worldwide due to its high social and economic burden, caused by its related comorbidities, impacting physical and mental health. Dietary fat is an important source of energy along with its rewarding and reinforcing properties. The nutritional recommendations for dietary fat vary from one country to another; however, the dietary reference intake (DRI) recommends not consuming more than 35% of total calories as fat. Food rich in fat is hyperpalatable, and is liable to be consumed in excess amounts. Food addiction as a concept has gained traction in recent years, as some aspects of addiction have been demonstrated for certain varieties of food. Fat addiction can be a diagnosable condition, which has similarities with the construct of addictive disorders, and is distinct from eating disorders or normal eating behaviors. Psychological vulnerabilities like attentional biases have been identified in individuals described to be having such addiction. Animal models have provided an opportunity to explore this concept in an experimental setting. This discussion sheds light on fat addiction, and explores its physiological and psychological implications. The discussion attempts to collate the emerging literature on addiction to fat rich diets as a prominent subset of food addiction. It aims at addressing the clinical relevance at the community level, the psychological correlates of such fat addiction, and the current physiological research directions.

Keywords: diet; fat; food addiction; obesity

1. Introduction

Over the last half a century, many developing countries have seen rapid socio-economic development, resulting in a move from a traditional to a modern way of life, including changes in local dietary and culinary profiles [1–3]. Abundance and easy availability of food, especially the one that is rich in fat and carbohydrate, have resulted in changes of dietary patterns and preferences. Right from early childhood in developing brains, the exposure and imprinting to high sugar, high salt and high fat food (rich in saturated and trans-fat), which is cheap and easily available, are impacting the health of younger population. Trans fat may lead to its greater consumption than polyunsaturated fat, as the latter is more quicker than the former to trigger satiety [4]. The changes in dietary intake profile with cultural and societal transitions have gained traction [5]. The dietary profile and constitution have a role in the etiopathogenesis of lifestyle-related diseases like obesity, metabolic syndrome, coronary artery disease, gut motility disorders, psychosomatic, autoimmune as well as degenerative disorders. Major transition, noticed during the last couple of years, has been an increasing use of sugar, processed food, beverages, animal-fat based food rich in trans-fats that have impacted human health [6–10].

In the recent years, there is a growing interest in the concept of food addiction from both clinical and applied nutritional research perspectives [9,11–14]. The increase in obesity, and associated metabolic syndrome and diabetes mellitus have called into the questions about factors leading to genesis of obesity. The imbalance of energy intake has been proposed as one of the reasons of increasing prevalence of obesity, though there are other several factors, i.e., epigenetics, psychological trauma, use of medications and dieting, that may increase body weight gain. why do some individuals consume excess of certain types of food (including fat-rich food)? Hence, the phenomenon of addiction to food has been suggested to be one of the mechanisms. Food addiction can refer to a variety of substrates, but fat and sugars have been considered the typical prototype food items to which individuals develop addiction. The occurrence of distinct features of salience and inability to control intake of specific types of foods have been considered similar to addiction to other psychoactive substances. The adverse consequences of uncontrolled fat intake on the body metabolism have been documented [15,16]. This has implications in intervention modules for addressing this problem and promoting healthy lifestyles [17,18]. Yet, the understanding of fat addiction as a concept is still under evolution, and progress is being made to characterise and discern the psychology and physiology behind this condition.

Recent studies suggest that fat has its own metabolic, physiological and nutritional profiles, which are distinct from other macronutrients [19–21]. Fat accords palatability and organoleptic properties to food, and is consumed across all ages from infancy through adulthood to elderly. Evolutionarily, this confers survival benefit due to high energy density, more so in cultures with thrifty genotypes [22–24]. In recent years, the benefits of mono-unsaturated fatty acid and controlled amounts of saturated fat intake have been revisited, especially in the context of benefits in cognitive functioning, synaptic connectivity, and membrane stability for both brain and heart health [25,26].

In this context, taste for fat has been proposed as the sixth taste modality in recent years [27]. The interactions between fatty acids and specific receptors in taste bud cells elicit physiological changes that are implicated in dietary fat preference via the activation of tongue-brain-gut axis. This phenomenon has an implication in the genesis of obesity as oro-sensory detection of nutrients determines the 'liking' and 'wanting' of food products. It has been proposed that there are two components of eating behavior, represented in neuronal circuits, i.e., emotional (hedonic and affective) and metabolic (homeostatic). Obesity may arise due to the imbalance of these two eating motives.

Research on food addiction or eating addiction has not paid distinct attention to specific nutrients like dietary fat [9,10,12,13]. The concept of fat addiction would have important psychological determinants like motivation, depression, anxiety and reasoning that merit cautious evaluation. There is cognitive appraisal that makes an individual "like" and "want" a specific food product, and the reward obtained from the food is cognitively processed as well. Hence, the intertwined psychological and physiological aspects of addiction towards fat rich food must be considered and understood further. There is a lack of comprehensive synthesis of literature to provide an account of fatty food addiction. In this paper, we have aimed at providing an overview of the construct, the clinical relevance, the psychological correlates, and the current physiological research trajectory in the emerging area of fat addiction as a subset of food addiction. Wherever specific literature with regard to fat is not available, evidence related to food addiction will be alluded to. However, our main emphasis is to discuss about fat addiction as this phenomenon might lead to high dietary fat intake and, consequently, to obesity. The term "high fat" in this article would mean the diet where the calories brought by fat are more than 40% of total dietary calories as most of industrialized countries recommend to respect this limit. As we have mentioned in the title, our main emphasis is to shed light on fat addiction and we have excluded other addictive behaviors like sweet addiction.

2. The Construct of Fat Rich Food Addiction

Addressing the issue of obesity would require improved knowledge of pathophysiological and neurobehavioral mechanisms. This would help better target behaviors which predispose individuals to obesity [28]. Schmidt and Campbell argue that disordered eating cannot remain "brainless" [29], and the "psychological constructs", that define aberrant consumptive patterns of food, are relevant. In this regard, addiction to food explains hedonistic excess and uncontrolled consumption of food items which are associated with adverse consequences.

Addiction towards fat rich diet relates to the overall definition of addiction. Addiction has been conceptualized as a maladaptive pattern of substance intake or behavior that signifies neurobiological changes and is associated with adverse consequences. The nosological systems providing nomenclature to diagnosis has moved on from abuse and dependence to substance use disorders in DSM5. The criteria based evaluation of the cluster of symptomatology helps provide with coherent account of the disorder, and categorize individuals who meet a threshold for diagnosis and consequent potential treatment.

2.1. Defining Fatty Food Addiction in the Context of Nutrient Intake

Food addiction shares some of the commonalities with drug addiction like craving, bingeing and tolerance [30]. The DSM5 criteria for substance use disorders have been adapted and explored in the context of food addiction [31,32]. The 11 criteria for substance use disorders can be applicable to individuals with addiction to lipid dense foods (especially trans and saturated fat). The empirically supported criteria describe a substance (food) often taken in larger amounts or over a longer period of time than that was intended, persistent desire or unsuccessful efforts to cut down or control substance use (food), and continued use despite knowledge of having a persistent or recurrent physical or psychological problem. The plausible features include great deal of time being spent in activities necessary to obtain or use the substance (food) or recover from its effects, recurrent substance (food) use resulting in a failure to fulfil major role obligations, continued use despite having persistent or recurrent social or interpersonal problems, important social, occupational, or recreational activities are given up or reduced, and tolerance. What might be difficult to clearly clinically elicit are withdrawal (while differentiating from energy deficit), and recurrent use in physically hazardous situations. As with different substances, each of the criteria is endorsed to different extent by a sample of participants.

The diagnostic constructs related to food addiction include binge eating disorder and an eating disorder not otherwise specified. Binge eating disorder is characterised by repeated ingestion of eating in large amounts of food in a short amount of time, followed by intense guilt and attempts to either remove the food (by vomiting or using laxatives) or compensatory behaviors to increase the energy expenditure [33,34]. On the other hand, eating disorder not otherwise specified is a diagnostic rubric that resembles anorexia nervosa or binge eating disorder, but does not fulfil the diagnostic thresholds for these disorders. These disorders may have some overlap with food addiction from a phenomenological and behavioral perspective, but the constructs themselves are distinct. It has been seen that individuals with binge eating disorders have greater rates of food addiction, than expected by chance [35,36], though at the same time, not all individuals with binge eating disorders would have food addiction [36,37].The main point of divergence lies in the focus of the constructs: food addiction lays emphasis on the salience and loss of control of hedonic eating behaviors, while eating disorders are accompanied by intense immediate guilt after excessive food consumption and efforts are made to get rid of (effects of) the ingested food quickly.

2.2. Clinical and Epidemiological Implications of Addiction towards Fat

While limited literature has looked at addiction to fat rich foods per se, there is enough evidence that has ascertained the occurrence rate and determinants of food addiction in the community and clinical samples [38,39]. The questionnaires used to assess food addiction generally incorporate fat as a component of food that the respondents are asked to think about, when they answer the questions. The Yale Food Addiction Scale is perhaps the most commonly used instrument for the assessment of food addiction. The weighted mean prevalence of food addiction according to this instrument was 19.9% [38]. The prevalence of food addiction was high in women with obesity [38]. Also, food addiction was higher in clinical samples, as compared to community samples [37,38]. Food addiction was high in subjects that were either obese, or suffered from eating disorders. High scores of food addiction were associated with high depressive symptoms, food craving and impulsivity. Food addiction has not only been related to negative mood states, but also with poorer quality of life [40]. It has been seen that individuals with food addiction had higher dietary fat intake as compared to those without food addiction [41]. Similarly, Pursey et al reported that the subjects with high food addiction scores had high percentage of consumption of saturated fat [42]. Thus, food addiction provides a paradigm for identification of individuals with skewed dietary profiles with other psychological vulnerabilities, which might require concomitant attention.

Food addiction has also been studied in those individuals who have undergone bariatric surgery which is generally indicated for people with severe obesity [43]. The rates of food addiction in bariatric surgery population go down after the surgery. In one study, the proportion of individuals with food addiction reduced to 2% post-surgery from 32% pre-surgery [44]. Another long term follow-up suggested that the rates of food addiction reduced from 57.8% to 7.2% at 6 months and to 13.7% at 12 months after surgery [45]. In pre-operative cases of bariatric surgery, the dietary intervention is less effective in individuals with food addiction [46]. It has been seen that food addiction in bariatric surgery patients was associated with greater levels of depression, anxiety and binge eating episodes, though it did not predict the degree of weight loss. Thus, it seems that food addiction has some clinical prognostic influence with surgical intervention outcomes.

2.3. Measurement Approaches

Currently the standard of practice for determination of food addiction has been the diagnostic cut-off from the Yale Food Addiction Scale (YFAS) [47]. The YEAS is a 25 item self-reported questionnaire based scale that assesses various features of food addiction. There are two items that assess for clinically significant impairment or distress. The instrument looks at the past year pattern of food intake and includes fatty foods like steak, bacon, hamburgers, cheeseburgers, pizza, and French fries as one of the representative group of foods that are mentioned in the questionnaire. The instrument has become standard of use in the field of food addiction. The instrument has adequate internal reliability, good convergent validity and good discriminant validity. The instrument has been adapted for use in children [48]. The instrument has also been translated into several other languages like Chinese, French and Malay [49–51]. A newer version of the scale (YFAS 2.0) has been developed considering the changes in conjunction with the DSM5 [52]. The instrument has been used in studies of epidemiology, etiology, nosology and interventions of food addiction. While the YFAS addressed food as a whole, assessing fat addiction separately may have implications for interventions. This could be in terms of the type of food products that are focused upon in the intervention modules that are developed. This would have also a corollary for the investigation procedures that can include assessment of salience and behavioral neuroplasticity (eye tracking and neuroimaging) for different types of food products (fat rich versus carbohydrate rich, sweet versus savory fatty food) that are implicated in food addiction.

Other self-reported scales and questionnaires for assessment of aspects of food addiction are also available and have been validated, though they rely on features like craving and eating patterns. These include Eating Behaviors Questionnaire [53], Food Cravings Questionnaire [54], Eating Behaviors Patterns Questionnaire [55], and Power of Food Scale [56]. Many of these questionnaires are self-reported, i.e., the individual reads through the questions and responds through them. The responses are thereafter graded and interpreted based upon the cut-offs from the population scores.

3. Psychological Correlates of Addiction to Fat Rich Diets

3.1. Attentional Biases and Cognitive Functioning

Research has been carried out towards attentional biases and psychological processing in individuals with food addiction. Obese as compared to lean teens showed less activation of prefrontal regions (dorsolateral prefrontal cortex, ventral lateral prefrontal cortex) when trying to inhibit responses to high-calorie food images which suggest behavioral evidence of reduced inhibitory control [57]. Adults who had greater dorsolateral prefrontal cortex activation when instructed to "resist craving" after viewing food images had better weight loss success following gastric bypass surgery [58]. This suggests that visual cue induction paradigms have relevance to assessment of how food images are processed centrally.

Rodrigue et al. [59] compared those with higher and lower food addiction scores on cognitive processes of planning, inhibition, cognitive flexibility and error processing. The investigators found that high food addiction group differed from the low food addiction group only in terms of inhibition/cognitive flexibility scaled scores, but not in individual scores. The authors infer that though basic level processes are intact, individuals with higher food addiction scores experience greater difficulties in more challenging context where they had to simultaneously keep in mind to inhibit a behavior and switch their mind-set when the task required it. This might make it difficult for them to anticipate the long-term consequences of behavior. Also, individuals with symptoms of food addiction made more errors as the interference task became challenging, suggesting that those with food addiction might have greater difficulty in detecting and monitoring errors. Another study compared error monitoring among individuals with food addiction and healthy controls using the Eriksen flanker task [60]. The results suggested that food addiction group had higher number of errors on the flanker task, implying impaired performance monitoring and cognitive control, as seen with other addictions. In a study that included women with obesity, food addiction severity levels were negatively correlated with overall scores on the Iowa Gambling Task, which measures decision making capacity [61]. Also, those with food addiction had attentional deficits as reflected by more omissions and perseveration errors on the Continuous Performance Task. On the other hand, Blume et al [62] compared response inhibition, attention, decision-making, and impulsivity among four groups of individuals, i.e., obesity and food addiction; obesity and binge eating disorder; obesity/food addiction and binge eating disorder; and obesity only. The authors did not find food addiction to be related to altered executive functioning.

Ruddock et al. [63] evaluated the attentional bias using eye tracking while showing pictures of chocolate among individuals with self-perceived food addiction in design that evaluated state factors like hunger or expectancy of reward or having food addiction. The authors found that the expectancy of receiving chocolate as reward was associated with attentional bias, while hunger state or having self-perceived food addiction was not associated with attentional bias toward food related cues. In another eye tracking paradigm, sad mood induction through showing of a video of child passing away with cancer was associated with attentional bias towards unhealthy food among those with food addiction, but such a change did not occur in those without food addiction [64]. This suggests that emotional cues may impel or prime those with food addiction towards specific food types. In another study, Gearhardt et al. [65] studied food-related visual attention and dwell time of food stuff among obese and overweight women. The authors reported that hunger was associated with attentional bias toward sweets, and trend level attentional bias towards fried (fatty) foods. On the other hand, hunger was associated with shorter dwell time on fried food. Taken together, literature suggests that hunger may be an important component that may influence attentional biases in individuals with food addiction. We acknowledge that though addiction has gained traction, fat addiction is an emerging concept, and nevertheless needs debate and discussion to inform lifestyle practices and research directions.

3.2. Craving and Liking

Craving and liking are related, but represent distinct terms that are linked to food addiction. While craving refers to desire or urge to eat a food item, liking refers to qualitative and affective evaluation of food [66–68]. Liking for fat has been evaluated in a large web-based study to examine the determinants of dietary patterns and nutritional status [69]. The investigators reported that individuals with a strong liking for fat had high total energy and fat intake, and high consumption of saturated fats, meat, butter, sweetened cream desserts and croissant-like pastries. Such individuals also consumed low quantities of fiber, fruits, vegetables and yogurt. It was highlighted that increased liking for fat, especially fat-and-salt liking, was associated with a lower intake of fruit and vegetables.

Gearhardt et al. [70] assessed craving for 180 food items among a sample of 105 obese or overweight women. The authors found that those with greater symptomatology of food addiction had higher craving ratings for fatty foods. However, as BMI increased, the craving decreased. In contrast to craving in this study, high fat content was not associated with high liking for food product, suggesting a dichotomy between craving and liking.

4. Understanding the Physiological and Neurobiological Processes of Fat Food Addiction

There have been considerable advances in understanding the mechanisms of addiction for food rich in lipids. Some of them were conducted on animals, particularly rodent models. Other directions of research, for example, genetics and neuroimaging have explored the origin of addiction towards fat and other palatable foods in human participants [71,72]. The reward pathway (schematically shown in Figure 1) is intricately linked to understanding the addiction to fatty food, though some differences have been reported in food addiction and substance use disorders [73].

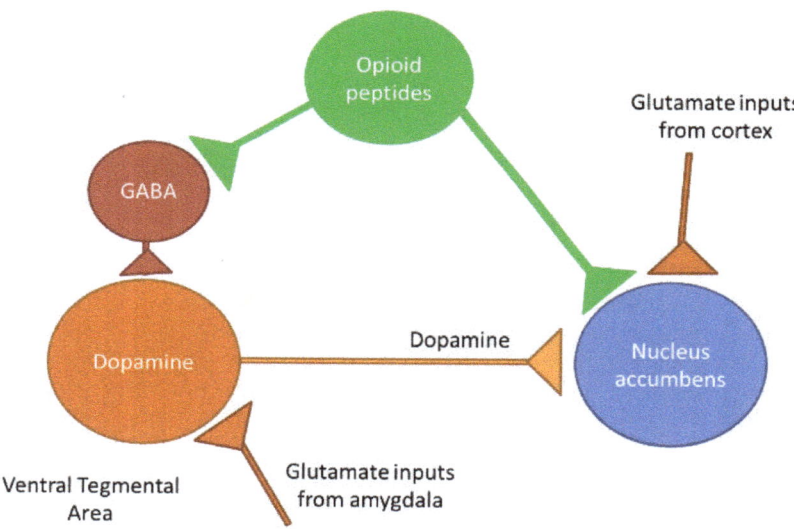

Figure 1. Schematic representation of the reward pathway. The figure shows the interplay between different neurons where the nucleus accumbens seems to be the central player, receiving the projection of dopaminergic, glutamatergic and opioidergic neurons. The model for food addiction might be quite different and is under examination [74].

4.1. Animal Models for Understanding the Addiction to Fat Rich Foods

The advantage of animal models is that they are able to develop addiction to fat as the diets given in animal models are homogenous [75–77]. This is not possible in human studies. The high-fat diet that generally comprises of 45% of energy from lipids is used to trigger obesity in rodents [78]. However, none of the experimental high-fat diets resembles closely to human diet fatty acid composition, though they are efficient to induce obesity.

Initially, Avena et al. [79] developed a model for sugar addiction which showed patterns of binge eating, withdrawal symptoms, and neurochemical changes similar to those observed with opiate addiction. The phenotype of animal was created by restricting the frequency, duration or access to sugar. Subsequently, fat models of bingeing have been developed in conjunction with carbohydrates, wherein corn oil is used as a reinforcing food item. However, the features of opiate-like withdrawal were not noted by Avena et al. [80] when animals were deprived of food after fat bingeing. This observation suggests that fat addiction may have different phenomenological aspects than the addiction to sugars. An alternate explanation could be that fat addiction might be more closely aligned to behavioral addiction like gambling disorder, while addiction to sugar rich food might be more closely aligned to substance use disorders. The development of animal models has the potential to advance the field substantially, by enabling to better understand the neurobiological alterations, and to assess the changes in bingeing behaviors with medications or other interventions [81]. Yet, one needs to be cognizant of the fact that translation of human behavior of food consumption is much more complex than animals, and is influenced by socio-economic and political environment, and the determinants like cost, availability and marketing. Furthermore, it is possible that modeling addiction in animals (especially rodents) might differ substantially from clinical situation [8]. It has been argued that simplistic experiments would need critical reflection about translational validity of patterns of eating behavior and food choice from animals to humans.

4.2. Neurotransmitters Including Dopamine

Animal studies have suggested implication of dopamine in the nucleus accumbens in the rat model of addictive behaviors towards fat [82]. Hence, microdialysis samples were taken in this model before, during and after sham feeding with corn oil. The study found an increase in dopamine in the sham licking group leading to the inference that corn oil increases dopamine concentrations in the nucleus accumbens in a manner similar to those induced by sucrose. In another study, low concentration of non-esterified fatty acid (linoleic acid) increased the dopamine levels in the nucleus accumbens and amygdala in a manner equivalent to those resulting from corn oil in the brain's reward system [83].

Dela Cruz and colleagues studied the expression of c-Fos in reward circuit areas in rats which were exposed to sugars and fats [84,85]. The authors reported c-Fos like immunoreactivity after consumption of corn oil solutions, isocaloric glucose and fructose, in the dopaminergic mesotelencephalic nuclei (ventral tegmental area) and projections (infralimbic and prelimbic medial prefrontal cortex, basolateral and central-cortico-medial amygdala, core of nucleus accumbens as well as the dorsal striatum), but not in the nucleus accumbens shell. This signified transcriptional activation of the dopaminergic pathway with exposure to certain nutrients including fat.

Dela Cruz et al aimed at investigating whether dopamine antagonists (D1 receptor antagonist SCH23390 and D2 receptor antagonist raclopride) attenuated the development of fat conditioned flavour preference among rats [86]. These investigators reported that, as compared to sucrose, the D1 and D2 receptor antagonists were not able to attenuate the fat conditioned flavor preference. They further suggested that fat addiction in rats could possibly have distinct mechanisms than sugars, which involved the post ingestive phase.

The role of opioid receptors and fatty food addiction has also been explored [76]. It has been seen that after injection of morphine, a mu-opioid receptor agonist, rats preferred fats over carbohydrates when both were available. Intra-accumbens administration of opioid agonists increased the consumption of fats, and the effect was blocked by the administration of naltrexone, an opioid antagonist [87]. The opioid receptors have been implicated in not only the 'liking' process, but also the 'wanting' process of excessive food consumption, and the effects are blocked by opioid antagonists.

Endocannabinoid system is another neurotransmitter system studied in relation with animal model of excessive fat consumptive behavior. Ward et al. [88] studied male, cannabinoid (CB1) knockout mice which were trained to respond to the sweet reinforcer (Ensure) or corn oil. The authors suggest that CB1 receptor antagonism selectively attenuated reinstatement of responding for Ensure. Interestingly, the genetic deletion of the CB1 receptor did not attenuate reinstatement of corn-oil seeking. The authors suggest that either CB1 receptor system does not play an equivalent role in modulating conditioned seeking or corn oil may serve as a robust reinforcer. Additionally, Brissard et al. [89] found that invalidation of CB1R gene was related to lower levels of fat preference among mice, and similar results were obtained after using rimonabant, a cannabinoid receptor antagonist. The authors reported that fat taste perception was mediated through calcium signaling and GLP-1 secretion in lingual taste bud cells. Peterschmitt et al. [90] looked at the link between the gustatory and the reward pathway with regard to fat intake. The authors observed that lipid taste perception was based upon the systematic activation of the major cerebral structures of the canonical gustatory pathway and was intricately linked to the reward pathway through the ventral tegmental area.

4.3. Neuroimaging Correlates

Though literature exists on the neuroimaging correlates of obesity [91,92], studies on the neuroimaging of food addiction have gradually started to come up. Gearhardt et al. [93] assessed the blood oxygen level-dependent functional magnetic resonance imaging (fMRI) activation in response to receipt and anticipated receipt of palatable food (chocolate milkshake) among adolescent female participants. The investigators demonstrated that food addiction scores correlated with greater activation in the anterior cingulate cortex, medial orbitofrontal cortex and amygdala, consequent to anticipated receipt of food. The participants with high food addiction scores had enhanced activation of dorsolateral prefrontal cortex and caudate, but less activation in lateral orbitofrontal cortex in response to anticipated receipt of food. These findings underscore the similarity of food addiction to other types of addictions, especially in relation to involvement of the reward pathway.

Hsu et al. [94] assessed response inhibition and error processing among subjects with obesity and sweet food addiction by fMRI. Women with obesity and food addiction had a higher score for impulsivity and lower brain activation (processing response inhibition over the right rolandic operculum and thalamus) than controls. The activation during error processing over the left insula, precuneus, and bilateral putamen were higher in the subjects with obesity and sweet food addiction than controls. These findings suggest that women with obesity and sweet food addiction have impaired rolandic operculum activation.

A further study looked at the relationship of food addiction and functional connectivity in the brain during fasting and fed state [95]. The authors found that high number of symptoms of food addiction were associated with ventral caudate-hippocampus hyperconnectivity in the fasted scan only. However, a significant reduction of this connectivity was observed in the fed scans, suggesting that heightened connectivity in the ventral striatum during a fasted state corroborated reward prediction signals, further lending credence to the involvement of the reward pathway.

A schematic representation of the neurobiological relationship of fatty food intake, mediated through gustatory signaling and reward pathway, is presented in Figure 2.

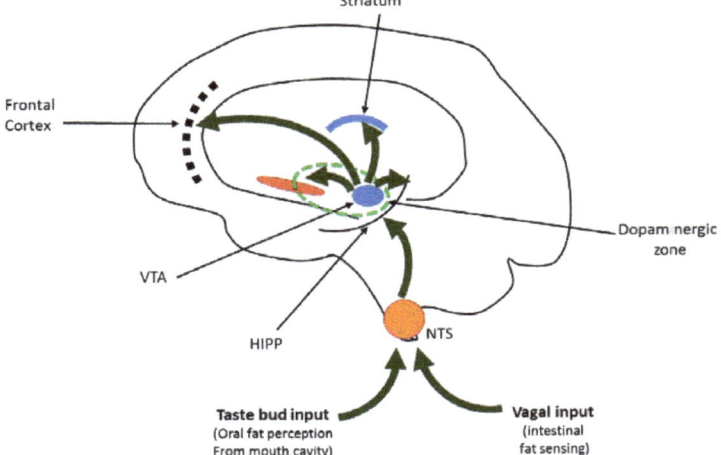

Figure 2. Relationship of food intake and reward pathway. The figure shows that the gustatory memory for fat and its implication would depend on the cues coming from taste bud cells, localized in the lingual papillae, and vagal nerve information from intestinal lipid sensing. Both kinds of information will ascend to different parts of the brain via NTS. Hippocampus will be involved in the learning of palatability of fat, and communicate to VTA which is sending its afferences to frontal cortex, striatum and other parts of the brain. Indeed, the dopaminergic zone covers VTA and NA. NTS: nucleus tractus solitaris; HIPP: hippocampus; VTA: ventral tegmental area.

4.4. Genetics Underpinnings

Several studies have also looked at the genetic associations of food addiction. A study evaluated whether a composite index of elevated dopamine signaling, a multilocus genetic profile score (MLGP) could segregate between those with food addiction and normal eating behavior [96]. The authors observed that MLGP score was high in subjects with food addiction, and it correlated positively with binge eating, food cravings, and emotional overeating. This finding supported the view that dopamine signaling genetic profile was different in subjects with food addiction.

Pedram et al. [97] studied food addiction in the Newfoundland population and observed the major allele A of rs2511521 located in DRD2 and the minor allele T of rs625413 located in TIR domain containing adaptor protein (TIRAP) to be significantly associated with food addiction. A study on the Asian American college students assessed the relationship of food addiction and a dopamine-resistant receptor (DRD2) polymorphism [98]. The authors reported that DRD2 A1 allele among Asian Americans (versus A2 allele) was associated with greater carbohydrate craving, but not fat craving. Cornelis et al. [99] presented genome wide analysis of food addiction in more than 9000 women with European ancestry. This study showed two loci significant at genome-wide level (17q21.31 and 11q23.4), but they did not have any obvious roles in eating behavior. The study did not find any candidate single nucleotide polymorphism or gene for drug addiction to be significantly associated with food addiction after correction for multiple testing.

There is accruing literature that suggests that reduced fat taste perception may contribute to increased fat consumption and, consequently, to obesity [100], and this might be influenced by the genetic polymorphisms. Studies from USA, Algeria and Tunisia seem to suggest that rs1761667-AA genotype of CD36 receptor is associated with obesity, and high thresholds for oro-sensory detection of dietary lipids [101–103]. Interestingly, Plesnik et al. [104] reported that another variant of CD36, i.e., rs1527483 SNP, was associated with greater body weight in young Czech participants. Thus, the taste threshold and preference for fat, mediated through specific genetic polymorphisms, may determine fat-eating behaviors that may lead to fat addiction.

5. Conclusions, Limitations and Future Directions

Addiction to food products, especially those rich in fat has received attention in recent decades. Figure 3 depicts the overall associations and implications of addiction to fat replete diets. The construct of food addiction has undergone sufficient scrutiny, and means and measures have been developed to reliably assess this condition. Fat as a component of food addiction itself has yet to find its niche, but has possible implications for the control and prevention of obesity. Research has elaborated on the attentional biases and cognitive functioning in individuals with food addiction, and has pitched varied findings. Animal models of food addiction, especially those which have used fat as a substrate, have expanded the scope of the field and have given an armamentarium of options for understanding the condition and interventional choices. Neuroimaging and genetic studies have also progressed, enriching the field.

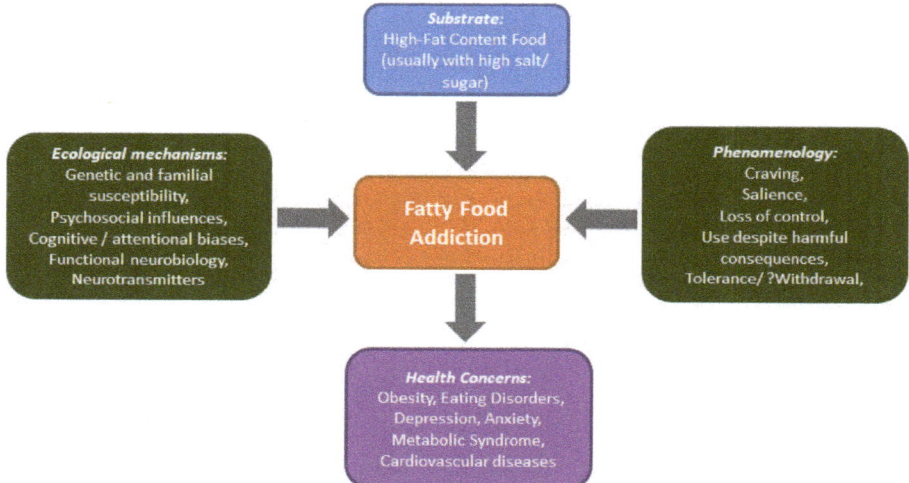

Figure 3. Schematic representation of addiction to fat.

Some of the limitations of the present paper should be born in mind while considering different observations. The present findings have not been synthesized as a systematic review, but are rather in the form of a narrative review. The advantage of narrative review is that broad range of findings can be presented to provide the reader with various dimension of the topic, but it may not be able to present all relevant literature in the field. The concept of food addiction, similar or different from other (substance/drug) addiction, as a continuum of behavioral addiction has been debated. Additionally, segregation of food addiction into a specific macronutrient based fat addiction may be difficult to operationalize clinically, as food products generally contain multiple elements together (fats, sugars and salt). Furthermore, food addiction as a concept has been criticized as pathologizing a normal behavior, and some researchers have questioned the validity and nomenclature of the construct [8].

Future research investigations are required to look at the stability of addiction to fat rich food over longitudinal course. The etiological understanding would be strengthened by foray into multi-modal assessment incorporating neuroimaging, genetic and psychological domains. Another aspect would be determining a threshold for fat composition in the food to qualify for fat addiction. Neurobiological studies would be strengthened if they incorporate the neuroimaging responses to palatable food taste, and cues (including visual cues). Neurobiological correlates of fat addiction and its persistence can be elicited by developing studies for individual nutrient components as well as combinations in sweet and savory food and linking it to markers of obesity. It would be pertinent to see how other neuropsychological functions like motivation, sensory processing in various domains and working memory interact with reward and homeostatic systems in controlling the various phenomenological aspects of fat addiction. Also, relationship of addictive behavior with intervention outcomes (for example, for obesity) needs to be looked into. The social impact of food addiction from a macro policy level, and the lived experience of the individual with 'food addiction' would help better understand the condition. Also, attempts to enhance the awareness of this condition and the harmful impact of trans-fat and saturated fat, coupled with greater funding to understand and address this issue would help both for primary and secondary prevention. Of course, the overarching aim would be to provide with relevant prevention for at-risk population, and suitable interventions for affected individuals.

Author Contributions: Conceptualization, S.S., K.P.K., and N.A.K.; review of literature, S.S., K.P.K., and N.A.K.; data curation, S.S.; writing—original draft preparation, S.S.; writing—review and editing, K.P.K. and N.A.K.; project administration, N.A.K.

Funding: This research received no external funding

Conflicts of Interest: The authors declare no conflict of interest.

References

1. Popkin, B.M.; Adair, L.S.; Ng, S.W. Global nutrition transition and the pandemic of obesity in developing countries. *Nutr. Rev.* **2012**, *70*, 3–21. [CrossRef] [PubMed]
2. Kochhar, K.P. Dietary spices in health and diseases: I. *Ind. J. Physiol. Pharmacol.* **2008**, *52*, 106–122.
3. Kochhar, K.P. Dietary spices in health and diseases (II). *Ind. J. Physiol. Pharmacol.* **2008**, *52*, 327–354.
4. Erlanson-Albertsson, C. Fat-rich food palatability and appetite regulation. In *Fat Detection: Taste, Texture, and Post Ingestive Effects*; CRC Press/Taylor & Francis: Boca Raton, FL, USA, 2010.
5. Drewnowski, A. Obesity, diets, and social inequalities. *Nutr. Rev.* **2009**, *67* (Suppl. 1), S36–S39. [CrossRef]
6. Popkin, B.M. Global nutrition dynamics: The world is shifting rapidly toward a diet linked with noncommunicable diseases. *Am. J. Clin. Nutr.* **2006**, *84*, 289–298. [CrossRef]
7. Popkin, B.M.; Gordon-Larsen, P. The nutrition transition: Worldwide obesity dynamics and their determinants. *Int. J. Obes.* **2004**, *28*, S2–S9. [CrossRef]
8. Hebebrand, J.; Albayrak, Ö.; Adan, R.; Antel, J.; Dieguez, C.; de Jong, J.; Leng, G.; Menzies, J.; Mercer, J.G.; Murphy, M.; et al. "Eating addiction", rather than "food addiction", better captures addictive-like eating behavior. *Neurosci. Biobehav. Rev.* **2014**, *47*, 295–306. [CrossRef]
9. Gordon, E.L.; Ariel-Donges, A.H.; Bauman, V.; Merlo, L.J. What Is the Evidence for "Food Addiction?" A Systematic Review. *Nutrients* **2018**, *10*, 477. [CrossRef]
10. Lerma-Cabrera, J.M.; Carvajal, F.; Lopez-Legarrea, P. Food addiction as a new piece of the obesity framework. *Nutr. J.* **2015**, *15*, 5. [CrossRef]
11. Bassareo, V.; Gambarana, C. Food and Its Effect on the Brain: From Physiological to Compulsive Consumption. *Front. Psychiatry* **2019**, *10*, 209. [CrossRef]
12. Fernandez-Aranda, F.; Karwautz, A.; Treasure, J. Food addiction: A transdiagnostic construct of increasing interest. *Eur. Eat. Disord. Rev. J. Eat. Disord. Assoc.* **2018**, *26*, 536–540. [CrossRef] [PubMed]
13. Pelchat, M.L. Food addiction in humans. *J. Nutr.* **2009**, *139*, 620–622. [CrossRef] [PubMed]
14. Shriner, R.; Gold, M. Food addiction: An evolving nonlinear science. *Nutrients* **2014**, *6*, 5370–5391. [CrossRef] [PubMed]
15. Golay, A.; Bobbioni, E. The role of dietary fat in obesity. *Int. J. Obes. Relat. Metab. Disord. J. Int. Assoc. Study Obes.* **1997**, *21* (Suppl. 3), S2–S11.
16. Smilowitz, J.T.; German, J.B.; Zivkovic, A.M. Food Intake and Obesity: The Case of Fat. In *Fat Detection: Taste, Texture, and Post Ingestive Effects*; Frontiers in Neuroscience; Montmayeur, J.-P., le Coutre, J., Eds.; CRC Press/Taylor & Francis: Boca Raton, FL, USA, 2010; ISBN 978-1-4200-6775-0.
17. Meule, A. A Critical Examination of the Practical Implications Derived from the Food Addiction Concept. *Curr. Obes. Rep.* **2019**, *8*, 11–17. [CrossRef]
18. Cassin, S.E.; Buchman, D.Z.; Leung, S.E.; Kantarovich, K.; Hawa, A.; Carter, A.; Sockalingam, S. Ethical, Stigma, and Policy Implications of Food Addiction: A Scoping Review. *Nutrients* **2019**, *11*, 710. [CrossRef]

19. Khan, N.A.; Besnard, P. Oro-sensory perception of dietary lipids: New insights into the fat taste transduction. *Biochim. Biophys. Acta* **2009**, *1791*, 149–155. [CrossRef]
20. Mattes, R.D. Fat taste and lipid metabolism in humans. *Physiol. Behav.* **2005**, *86*, 691–697. [CrossRef]
21. Drewnowski, A.; Mennella, J.A.; Johnson, S.L.; Bellisle, F. Sweetness and food preference. *J. Nutr.* **2012**, *142*, 1142S–1148S. [CrossRef]
22. Sellayah, D.; Cagampang, F.R.; Cox, R.D. On the evolutionary origins of obesity: A new hypothesis. *Endocrinology* **2014**, *155*, 1573–1588. [CrossRef]
23. Reddon, H.; Patel, Y.; Turcotte, M.; Pigeyre, M.; Meyre, D. Revisiting the evolutionary origins of obesity: Lazy versus peppy-thrifty genotype hypothesis. *Obes. Rev. Off. J. Int. Assoc. Study Obes.* **2018**, *19*, 1525–1543. [CrossRef] [PubMed]
24. Genné-Bacon, E.A. Thinking evolutionarily about obesity. *Yale J. Biol. Med.* **2014**, *87*, 99–112. [PubMed]
25. Clifton, P.M.; Keogh, J.B. A systematic review of the effect of dietary saturated and polyunsaturated fat on heart disease. *Nutr. Metab. Cardiovasc. Dis. NMCD* **2017**, *27*, 1060–1080. [CrossRef] [PubMed]
26. Power, R.; Prado-Cabrero, A.; Mulcahy, R.; Howard, A.; Nolan, J.M. The Role of Nutrition for the Aging Population: Implications for Cognition and Alzheimer's Disease. *Annu. Rev. Food Sci. Technol.* **2019**, *10*, 619–639. [CrossRef] [PubMed]
27. Besnard, P.; Passilly-Degrace, P.; Khan, N.A. Taste of Fat: A Sixth Taste Modality? *Physiol. Rev.* **2016**, *96*, 151–176. [CrossRef]
28. Val-Laillet, D.; Aarts, E.; Weber, B.; Ferrari, M.; Quaresima, V.; Stoeckel, L.E.; Alonso-Alonso, M.; Audette, M.; Malbert, C.H.; Stice, E. Neuroimaging and neuromodulation approaches to study eating behavior and prevent and treat eating disorders and obesity. *NeuroImage Clin.* **2015**, *8*, 1–31. [CrossRef]
29. Schmidt, U.; Campbell, I.C. Treatment of eating disorders can not remain "brainless": The case for brain-directed treatments. *Eur. Eat. Disord. Rev. J. Eat. Disord. Assoc.* **2013**, *21*, 425–427. [CrossRef]
30. Rogers, P.J. Food and drug addictions: Similarities and differences. *Pharmacol. Biochem. Behav.* **2017**, *153*, 182–190. [CrossRef]
31. Hone-Blanchet, A.; Fecteau, S. Overlap of food addiction and substance use disorders definitions: Analysis of animal and human studies. *Neuropharmacology* **2014**, *85*, 81–90. [CrossRef]
32. Meule, A.; Gearhardt, A.N. Food addiction in the light of DSM-5. *Nutrients* **2014**, *6*, 3653–3671. [CrossRef]
33. Schreiber, L.R.N.; Odlaug, B.L.; Grant, J.E. The overlap between binge eating disorder and substance use disorders: Diagnosis and neurobiology. *J. Behav. Addict.* **2013**, *2*, 191–198. [CrossRef] [PubMed]
34. Citrome, L. A primer on binge eating disorder diagnosis and management. *CNS Spectr.* **2015**, *20* (Suppl. 1), 44–50. [CrossRef] [PubMed]
35. Carter, J.C.; Van Wijk, M.; Rowsell, M. Symptoms of "food addiction" in binge eating disorder using the Yale Food Addiction Scale version 2.0. *Appetite* **2019**, *133*, 362–369. [CrossRef] [PubMed]
36. Linardon, J.; Messer, M. Assessment of food addiction using the Yale Food Addiction Scale 2.0 in individuals with binge-eating disorder symptomatology: Factor structure, psychometric properties, and clinical significance. *Psychiatry Res.* **2019**, *279*, 216–221. [CrossRef] [PubMed]
37. Burrows, T.; Skinner, J.; McKenna, R.; Rollo, M. Food Addiction, Binge Eating Disorder, and Obesity: Is There a Relationship? *Behav. Sci. Basel Switz.* **2017**, *7*, 54. [CrossRef] [PubMed]
38. Pursey, K.M.; Stanwell, P.; Gearhardt, A.N.; Collins, C.E.; Burrows, T.L. The prevalence of food addiction as assessed by the Yale Food Addiction Scale: A systematic review. *Nutrients* **2014**, *6*, 4552–4590. [CrossRef] [PubMed]
39. Penzenstadler, L.; Soares, C.; Karila, L.; Khazaal, Y. Systematic Review of Food Addiction as Measured With the Yale Food Addiction Scale: Implications for the Food Addiction Construct. *Curr. Neuropharmacol.* **2018**, *16*, 520. [CrossRef]
40. Zhao, Z.; Ma, Y.; Han, Y.; Liu, Y.; Yang, K.; Zhen, S.; Wen, D. Psychosocial Correlates of Food Addiction and Its Association with Quality of Life in a Non-Clinical Adolescent Sample. *Nutrients* **2018**, *10*, 837. [CrossRef]
41. Ayaz, A.; Nergiz-Unal, R.; Dedebayraktar, D.; Akyol, A.; Pekcan, A.G.; Besler, H.T.; Buyuktuncer, Z. How does food addiction influence dietary intake profile? *PLoS ONE* **2018**, *13*, e0195541. [CrossRef]
42. Pursey, K.M.; Collins, C.E.; Stanwell, P.; Burrows, T.L. Foods and dietary profiles associated with "food addiction" in young adults. *Addict. Behav. Rep.* **2015**, *2*, 41–48. [CrossRef]
43. Ivezaj, V.; Wiedemann, A.A.; Grilo, C.M. Food addiction and bariatric surgery: A systematic review of the literature. *Obes. Rev. Off. J. Int. Assoc. Study Obes.* **2017**, *18*, 1386–1397. [CrossRef] [PubMed]

44. Pepino, M.Y.; Stein, R.I.; Eagon, J.C.; Klein, S. Bariatric surgery-induced weight loss causes remission of food addiction in extreme obesity. *Obes. Silver Spring Md.* **2014**, *22*, 1792–1798. [CrossRef] [PubMed]
45. Sevinçer, G.M.; Konuk, N.; Bozkurt, S.; Coşkun, H. Food addiction and the outcome of bariatric surgery at 1-year: Prospective observational study. *Psychiatry Res.* **2016**, *244*, 159–164. [CrossRef]
46. Guerrero Pérez, F.; Sánchez-González, J.; Sánchez, I.; Jiménez-Murcia, S.; Granero, R.; Simó-Servat, A.; Ruiz, A.; Virgili, N.; López-Urdiales, R.; Montserrat-Gil de Bernabe, M.; et al. Food addiction and preoperative weight loss achievement in patients seeking bariatric surgery. *Eur. Eat. Disord. Rev. J. Eat. Disord. Assoc.* **2018**, *26*, 645–656. [CrossRef]
47. Gearhardt, A.N.; Corbin, W.R.; Brownell, K.D. Preliminary validation of the Yale food addiction scale. *Appetite* **2009**, *52*, 430–436. [CrossRef]
48. Gearhardt, A.N.; Roberto, C.A.; Seamans, M.J.; Corbin, W.R.; Brownell, K.D. Preliminary validation of the Yale Food Addiction Scale for children. *Eat. Behav.* **2013**, *14*, 508–512. [CrossRef] [PubMed]
49. Chen, G.; Tang, Z.; Guo, G.; Liu, X.; Xiao, S. The Chinese version of the Yale Food Addiction Scale: An examination of its validation in a sample of female adolescents. *Eat. Behav.* **2015**, *18*, 97–102. [CrossRef]
50. Nantha, Y.S.; Patah, N.A.A.; Pillai, M.P. Preliminary validation of the Malay Yale Food Addiction Scale: Factor structure and item analysis in an obese population. *Clin. Nutr. ESPEN* **2016**, *16*, 42–47. [CrossRef]
51. Brunault, P.; Ballon, N.; Gaillard, P.; Réveillère, C.; Courtois, R. Validation of the French version of the Yale Food Addiction Scale: An examination of its factor structure, reliability, and construct validity in a nonclinical sample. *Can. J. Psychiatry* **2014**, *59*, 276–284. [CrossRef]
52. Gearhardt, A.N.; Corbin, W.R.; Brownell, K.D. Development of the Yale Food Addiction Scale Version 2.0. *Psychol. Addict. Behav. J. Soc. Psychol. Addict. Behav.* **2016**, *30*, 113–121. [CrossRef]
53. Wardle, J.; Guthrie, C.A.; Sanderson, S.; Rapoport, L. Development of the children's eating behaviour questionnaire. *J. Child Psychol. Psychiatry* **2001**, *42*, 963–970. [CrossRef] [PubMed]
54. Cepeda-Benito, A.; Gleaves, D.H.; Williams, T.L.; Erath, S.A. The development and validation of the state and trait food-cravings questionnaires. *Behav. Ther.* **2000**, *31*, 151–173. [CrossRef]
55. Schlundt, D.G.; Hargreaves, M.K.; Buchowski, M.S. The eating behavior patterns questionnaire predicts dietary fat intake in African American women. *J. Am. Diet. Assoc.* **2003**, *103*, 338–345. [PubMed]
56. Cappelleri, J.C.; Bushmakin, A.G.; Gerber, R.A.; Leidy, N.K.; Sexton, C.C.; Karlsson, J.; Lowe, M.R. Evaluating the Power of Food Scale in obese subjects and a general sample of individuals: Development and measurement properties. *Int. J. Obes.* **2009**, *33*, 913–922. [CrossRef] [PubMed]
57. Batterink, L.; Yokum, S.; Stice, E. Body mass correlates inversely with inhibitory control in response to food among adolescent girls: An fMRI study. *NeuroImage* **2010**, *52*, 1696–1703. [CrossRef] [PubMed]
58. Goldman, R.L.; Canterberry, M.; Borckardt, J.J.; Madan, A.; Byrne, T.K.; George, M.S.; O'Neil, P.M.; Hanlon, C.A. Executive control circuitry differentiates degree of success in weight loss following gastric-bypass surgery. *Obes. Silver Spring Md.* **2013**, *21*, 2189–2196. [CrossRef]
59. Rodrigue, C.; Ouellette, A.-S.; Lemieux, S.; Tchernof, A.; Biertho, L.; Bégin, C. Executive functioning and psychological symptoms in food addiction: A study among individuals with severe obesity. *Eat. Weight Disord. EWD* **2018**, *23*, 469–478. [CrossRef]
60. Franken, I.H.A.; Nijs, I.M.T.; Toes, A.; van der Veen, F.M. Food addiction is associated with impaired performance monitoring. *Biol. Psychol.* **2018**, *131*, 49–53. [CrossRef]
61. Steward, T.; Mestre-Bach, G.; Vintró-Alcaraz, C.; Lozano-Madrid, M.; Agüera, Z.; Fernández-Formoso, J.A.; Granero, R.; Jiménez-Murcia, S.; Vilarrasa, N.; García-Ruiz-de-Gordejuela, A.; et al. Food addiction and impaired executive functions in women with obesity. *Eur. Eat. Disord. Rev. J. Eat. Disord. Assoc.* **2018**, *26*, 574–584. [CrossRef]
62. Blume, M.; Schmidt, R.; Hilbert, A. Executive Functioning in Obesity, Food Addiction, and Binge-Eating Disorder. *Nutrients* **2018**, *11*, 54. [CrossRef]
63. Ruddock, H.K.; Field, M.; Jones, A.; Hardman, C.A. State and trait influences on attentional bias to food-cues: The role of hunger, expectancy, and self-perceived food addiction. *Appetite* **2018**, *131*, 139–147. [CrossRef] [PubMed]
64. Frayn, M.; Sears, C.R.; von Ranson, K.M. A sad mood increases attention to unhealthy food images in women with food addiction. *Appetite* **2016**, *100*, 55–63. [CrossRef] [PubMed]

65. Gearhardt, A.N.; Treat, T.A.; Hollingworth, A.; Corbin, W.R. The relationship between eating-related individual differences and visual attention to foods high in added fat and sugar. *Eat. Behav.* **2012**, *13*, 371–374. [CrossRef] [PubMed]
66. Havermans, R.C. "You Say it's Liking, I Say it's Wanting … ". On the difficulty of disentangling food reward in man. *Appetite* **2011**, *57*, 286–294. [CrossRef] [PubMed]
67. Mela, D.J. Why do we like what we like? *J. Sci. Food Agric.* **2001**, *81*, 10–16. [CrossRef]
68. Pelchat, M.L. Of human bondage: Food craving, obsession, compulsion, and addiction. *Physiol. Behav.* **2002**, *76*, 347–352. [CrossRef]
69. Méjean, C.; Deglaire, A.; Kesse-Guyot, E.; Hercberg, S.; Schlich, P.; Castetbon, K. Association between intake of nutrients and food groups and liking for fat (The Nutrinet-Santé Study). *Appetite* **2014**, *78*, 147–155. [CrossRef]
70. Gearhardt, A.N.; Rizk, M.T.; Treat, T.A. The association of food characteristics and individual differences with ratings of craving and liking. *Appetite* **2014**, *79*, 166–173. [CrossRef]
71. Volkow, N.; Wang, G.J.; Fowler, J.S.; Tomasi, D.; Baler, R. Food and drug reward: Overlapping circuits in human obesity and addiction. In *Brain Imaging in Behavioral Neuroscience*; Springer: Berlin, Germany, 2011; pp. 1–24.
72. Fortuna, J.L. Sweet preference, sugar addiction and the familial history of alcohol dependence: Shared neural pathways and genes. *J. Psychoact. Drugs* **2010**, *42*, 147–151. [CrossRef]
73. Ahmed, S.H.; Lenoir, M.; Guillem, K. Neurobiology of addiction versus drug use driven by lack of choice. *Curr. Opin. Neurobiol.* **2013**, *23*, 581–587. [CrossRef]
74. Ziauddeen, H.; Farooqi, I.S.; Fletcher, P.C. Obesity and the brain: How convincing is the addiction model? *Nat. Rev. Neurosci.* **2012**, *13*, 279–286. [CrossRef] [PubMed]
75. Morgan, D.; Sizemore, G.M. Animal models of addiction: Fat and sugar. *Curr. Pharm. Des.* **2011**, *17*, 1168–1172. [CrossRef] [PubMed]
76. Novelle, M.G.; Diéguez, C. Food Addiction and Binge Eating: Lessons Learned from Animal Models. *Nutrients* **2018**, *10*, 71. [CrossRef] [PubMed]
77. De Jong, J.W.; Vanderschuren, L.J.M.J.; Adan, R.A.H. Towards an animal model of food addiction. *Obes. Facts* **2012**, *5*, 180–195. [CrossRef]
78. Marques, C.; Meireles, M.; Norberto, S.; Leite, J.; Freitas, J.; Pestana, D.; Faria, A.; Calhau, C. High-fat diet-induced obesity Rat model: A comparison between Wistar and Sprague-Dawley Rat. *Adipocyte* **2016**, *5*, 11–21. [CrossRef]
79. Avena, N.M. Examining the addictive-like properties of binge eating using an animal model of sugar dependence. *Exp. Clin. Psychopharmacol.* **2007**, *15*, 481–491. [CrossRef]
80. Avena, N.M.; Rada, P.; Hoebel, B.G. Sugar and fat bingeing have notable differences in addictive-like behavior. *J. Nutr.* **2009**, *139*, 623–628. [CrossRef]
81. Wong, K.J.; Wojnicki, F.H.W.; Corwin, R.L.W. Baclofen, raclopride, and naltrexone differentially affect intake of fat/sucrose mixtures under limited access conditions. *Pharmacol. Biochem. Behav.* **2009**, *92*, 528–536. [CrossRef]
82. Liang, N.-C.; Hajnal, A.; Norgren, R. Sham feeding corn oil increases accumbens dopamine in the rat. *Am. J. Physiol. Regul. Integr. Comp. Physiol.* **2006**, *291*, R1236–R1239. [CrossRef]
83. Adachi, S.; Endo, Y.; Mizushige, T.; Tsuzuki, S.; Matsumura, S.; Inoue, K.; Fushiki, T. Increased levels of extracellular dopamine in the nucleus accumbens and amygdala of rats by ingesting a low concentration of a long-chain Fatty Acid. *Biosci. Biotechnol. Biochem.* **2013**, *77*, 2175–2180. [CrossRef]
84. Dela Cruz, J.A.D.; Coke, T.; Bodnar, R.J. Simultaneous Detection of c-Fos Activation from Mesolimbic and Mesocortical Dopamine Reward Sites Following Naive Sugar and Fat Ingestion in Rats. *J. Vis. Exp. JoVE* **2016**. [CrossRef]
85. Dela Cruz, J.A.D.; Coke, T.; Karagiorgis, T.; Sampson, C.; Icaza-Cukali, D.; Kest, K.; Ranaldi, R.; Bodnar, R.J. c-Fos induction in mesotelencephalic dopamine pathway projection targets and dorsal striatum following oral intake of sugars and fats in rats. *Brain Res. Bull.* **2015**, *111*, 9–19. [CrossRef]
86. Dela Cruz, J.A.D.; Icaza-Cukali, D.; Tayabali, H.; Sampson, C.; Galanopoulos, V.; Bamshad, D.; Touzani, K.; Sclafani, A.; Bodnar, R.J. Roles of dopamine D1 and D2 receptors in the acquisition and expression of fat-conditioned flavor preferences in rats. *Neurobiol. Learn. Mem.* **2012**, *97*, 332–337. [CrossRef] [PubMed]

87. Zhang, M.; Gosnell, B.A.; Kelley, A.E. Intake of high-fat food is selectively enhanced by mu opioid receptor stimulation within the nucleus accumbens. *J. Pharmacol. Exp. Ther.* **1998**, *285*, 908–914. [PubMed]
88. Ward, S.J.; Walker, E.A.; Dykstra, L.A. Effect of cannabinoid CB1 receptor antagonist SR141716A and CB1 receptor knockout on cue-induced reinstatement of Ensure and corn-oil seeking in mice. *Neuropsychopharmacol. Off. Publ. Am. Coll. Neuropsychopharmacol.* **2007**, *32*, 2592–2600. [CrossRef] [PubMed]
89. Brissard, L.; Leemput, J.; Hichami, A.; Passilly-Degrace, P.; Maquart, G.; Demizieux, L.; Degrace, P.; Khan, N.A. Orosensory Detection of Dietary Fatty Acids Is Altered in $CB_1R^{-/-}$ Mice. *Nutrients* **2018**, *10*, 1347. [CrossRef] [PubMed]
90. Peterschmitt, Y.; Abdoul-Azize, S.; Murtaza, B.; Barbier, M.; Khan, A.S.; Millot, J.-L.; Khan, N.A. Fatty Acid Lingual Application Activates Gustatory and Reward Brain Circuits in the Mouse. *Nutrients* **2018**, *10*, 1246. [CrossRef] [PubMed]
91. Patriarca, L.; Magerowski, G.; Alonso-Alonso, M. Functional neuroimaging in obesity. *Curr. Opin. Endocrinol. Diabetes Obes.* **2017**, *24*, 260–265. [CrossRef] [PubMed]
92. Brooks, S.J.; Cedernaes, J.; Schiöth, H.B. Increased prefrontal and parahippocampal activation with reduced dorsolateral prefrontal and insular cortex activation to food images in obesity: A meta-analysis of fMRI studies. *PLoS ONE* **2013**, *8*, e60393. [CrossRef]
93. Gearhardt, A.N.; Yokum, S.; Orr, P.T.; Stice, E.; Corbin, W.R.; Brownell, K.D. The Neural Correlates of "Food Addiction". *Arch. Gen. Psychiatry* **2011**, *68*, 808–816. [CrossRef]
94. Hsu, J.-S.; Wang, P.-W.; Ko, C.-H.; Hsieh, T.-J.; Chen, C.-Y.; Yen, J.-Y. Altered brain correlates of response inhibition and error processing in females with obesity and sweet food addiction: A functional magnetic imaging study. *Obes. Res. Clin. Pract.* **2017**, *11*, 677–686. [CrossRef] [PubMed]
95. Contreras-Rodriguez, O.; Burrows, T.; Pursey, K.M.; Stanwell, P.; Parkes, L.; Soriano-Mas, C.; Verdejo-Garcia, A. Food addiction linked to changes in ventral striatum functional connectivity between fasting and satiety. *Appetite* **2019**, *133*, 18–23. [CrossRef] [PubMed]
96. Davis, C.; Loxton, N.J.; Levitan, R.D.; Kaplan, A.S.; Carter, J.C.; Kennedy, J.L. "Food addiction" and its association with a dopaminergic multilocus genetic profile. *Physiol. Behav.* **2013**, *118*, 63–69. [CrossRef] [PubMed]
97. Pedram, P.; Zhai, G.; Gulliver, W.; Zhang, H.; Sun, G. Two novel candidate genes identified in adults from the Newfoundland population with addictive tendencies towards food. *Appetite* **2017**, *115*, 71–79. [CrossRef]
98. Yeh, J.; Trang, A.; Henning, S.M.; Wilhalme, H.; Carpenter, C.; Heber, D.; Li, Z. Food Cravings, Food Addiction, and a Dopamine-Resistant (DRD2 A1) Receptor Polymorphism in Asian American College Students. *Asia Pac. J. Clin. Nutr.* **2016**, *25*, 424–429.
99. Cornelis, M.C.; Flint, A.; Field, A.E.; Kraft, P.; Han, J.; Rimm, E.B.; van Dam, R.M. A genome-wide investigation of food addiction. *Obes. Silver Spring Md.* **2016**, *24*, 1336–1341. [CrossRef]
100. Khan, A.S.; Murtaza, B.; Hichami, A.; Khan, N.A. A cross-talk between fat and bitter taste modalities. *Biochimie* **2019**, *159*, 3–8. [CrossRef]
101. Love-Gregory, L.; Abumrad, N. CD36 genetics and the metabolic complications of obesity. *Curr. Opin. Clin. Nutr. Metab. Care* **2011**, *14*, 527–534. [CrossRef]
102. Mrizak, I.; Šerý, O.; Plesnik, J.; Arfa, A.; Fekih, M.; Bouslema, A.; Zaouali, M.; Tabka, Z.; Khan, N.A. The a allele of cluster of differentiation 36 (CD36) SNP 1761667 associates with decreased lipid taste perception in obese Tunisian women. *Br. J. Nutr.* **2015**, *113*, 1330–1337. [CrossRef]
103. Melis, M.; Carta, G.; Pintus, S.; Pintus, P.; Piras, C.A.; Murru, E.; Manca, C.; Di Marzo, V.; Banni, S.; Tomassini Barbarossa, I. Polymorphism rs1761667 in the CD36 Gene Is Associated to Changes in Fatty Acid Metabolism and Circulating Endocannabinoid Levels Distinctively in Normal Weight and Obese Subjects. *Front. Physiol.* **2017**, *8*, 8. [CrossRef]
104. Plesník, J.; Serý, O.; Khan, A.; Bielik, P.; Khan, N.A. The rs1527483, but not rs3212018, CD36 polymorphism associates with linoleic acid detection and obesity in Czech young adults. *Br. J. Nutr.* **2018**, *119*, 1–7. [CrossRef]

© 2019 by the authors. Licensee MDPI, Basel, Switzerland. This article is an open access article distributed under the terms and conditions of the Creative Commons Attribution (CC BY) license (http://creativecommons.org/licenses/by/4.0/).

Review

Food Addiction: Implications for the Diagnosis and Treatment of Overeating

Rachel C. Adams [1,*], Jemma Sedgmond [1], Leah Maizey [1], Christopher D. Chambers [1] and Natalia S. Lawrence [2]

1. CUBRIC, School of Psychology, Cardiff University, Maindy Road, Cardiff CF24 4HQ, UK
2. School of Psychology, College of Life and Environmental Sciences, University of Exeter, Exeter EX4 4QG, UK
* Correspondence: adamsrc1@cardiff.ac.uk; Tel.: +44-(0)29-2087-0365

Received: 15 July 2019; Accepted: 21 August 2019; Published: 4 September 2019

Abstract: With the obesity epidemic being largely attributed to overeating, much research has been aimed at understanding the psychological causes of overeating and using this knowledge to develop targeted interventions. Here, we review this literature under a model of food addiction and present evidence according to the fifth edition of the Diagnostic and Statistical Manual (DSM-5) criteria for substance use disorders. We review several innovative treatments related to a food addiction model ranging from cognitive intervention tasks to neuromodulation techniques. We conclude that there is evidence to suggest that, for some individuals, food can induce addictive-type behaviours similar to those seen with other addictive substances. However, with several DSM-5 criteria having limited application to overeating, the term 'food addiction' is likely to apply only in a minority of cases. Nevertheless, research investigating the underlying psychological causes of overeating within the context of food addiction has led to some novel and potentially effective interventions. Understanding the similarities and differences between the addictive characteristics of food and illicit substances should prove fruitful in further developing these interventions.

Keywords: food addiction; overeating; obesity; impulsivity; reward sensitivity; cognitive training; neuromodulation

1. Introduction

In 2003, obesity was declared a global epidemic by the World Health Organisation [1], and the prevalence of overweight and obesity in both developed and developing countries continues to increase [2,3]. In 2016, 39% of adults were estimated to be overweight and 13% to be obese [4]. Overweight and obesity present a substantial economic burden; in the UK, the total direct and indirect costs are expected to reach £37.2 billion by 2025 [5]. One of the common explanations for the increase in obesity over recent decades is the environment and, in particular, the availability of highly varied, palatable and fattening foods—which have been considered to be addictive [6–9]. While many individuals manage to resist these temptations and maintain a healthy weight, obese individuals have been shown to have a preference for such energy-dense foods compared to healthy-weight individuals [10–12]. The critical question is why some individuals are able to resist overeating while others cannot; what is the evidence for 'food addiction' and how can this be used to inform interventions for overeating.

The concept of 'food addiction' has been evident in the media and general public for some time and is gaining increasing interest in the scientific literature [13]. There are now numerous reviews discussing the diagnostic, neurobiological and practical aspects of food addiction, with arguments both for and against its utility and validity [14–20]. This surge of interest comes with the perspective that addiction can be conceptualised as a loss of control over intake for a particular substance or behaviour without the need to focus purely on psychoactive substances [21,22]. The fifth edition of the Diagnostic and Statistical Manual [23] acknowledged this shift in perspective, with the addition of gambling disorder as the first behavioural addiction. Acceptance of this disorder was based on evidence that gambling can produce behavioural symptoms that parallel those of substance addiction and can activate the same neural reward circuits as drugs of abuse [24,25]. There is now a large body of research documenting similar observations for overeating and obesity. Moreover, treatments developed for addictive disorders have also shown some efficacy for the treatment of obesity and overeating. These findings highlight how a model of food addiction may help us to understand elements of overweight/obesity beyond a simple lack of willpower and can also be used to inform effective interventions and policy [26–30].

Food addiction has not yet been recognised in the DSM; however, the similarities between some feeding and eating disorders and substance-use disorders (SUDs) have been acknowledged. These similarities include the experience of cravings, reduced control over intake, increased impulsivity and altered reward-sensitivity. Binge eating disorder (BED) and bulimia nervosa (BN) have been proposed as phenotypes that may reflect these similarities to the greatest extent [31–34]. Both BED and BN are characterised by recurrent episodes of binge eating in which large quantities of food are consumed in a short time accompanied by feelings of a lack of control, despite physical and emotional distress. Reports of food addiction have been shown to be particularly high amongst these individuals [32,35,36]. Food addiction has also been acknowledged with a standardised 'diagnostic' tool—the Yale Food Addiction Scale (YFAS) [37,38]. The YFAS is a questionnaire that parallels the diagnostic criteria for SUDs. The scale has so far been shown to exhibit good internal reliability as well as convergent, discriminant and incremental validity [37–40].

In this review, we first discuss the DSM-5 diagnostic criteria for SUDs to summarise evidence for food addiction. These criteria are defined as 'a cluster of cognitive, behavioural and physiological symptoms' [23]. More specifically, the following categories are considered: impaired control, social impairment, repeated use despite negative consequences and physiological criteria. However, it should be noted that the physiological criteria of tolerance and withdrawal—for which there is less evidence in relation to food—are not necessary for a diagnosis of SUD. The DSM-5 also states that although changes in neural functioning are a key characteristic of SUDs, the diagnosis is based on a pathological pattern of behaviours. Hence, we discuss the diagnostic criteria initially, followed by a review of neurobiological evidence. We then explore the question of how this information can be, and has been, applied to interventions for overeating.

1.1. Impaired Control

Taking larger amounts of the substance for longer periods than intended has been cited as one of the most commonly reported symptoms in overweight/obese and BED individuals [41,42]. Excessive and uncontrolled eating also forms the definition of binge eating in BED [23]. Although bingeing can be a planned behaviour, it has been shown that planned binges still result in a greater intake than initially intended [41]. Binge eating has also been documented in non-clinical samples [43,44]; however, in these individuals, occasions of impaired control are more likely to reflect unintentional snacking and excessive portion sizes [8,41,45].

Unsuccessful efforts to restrict food intake are also well documented, with many dieters failing to maintain their diet or even gaining weight in the long term [46–51]. In their paper reviewing evidence for refined food addiction (i.e., processed foods with high levels of sugars or sweeteners, refined carbohydrates, fat, salt and caffeine), Ifland et al. [52] report that 'Every refined food addict reports a series of attempts to cut back on eating. They have used a variety of techniques' (pg. 521). Curtis and Davis [41] also report similar anecdotes in women with BED who describe avoiding certain trigger foods to control their binges.

The third criterion of time spent obtaining, using and recovering from substance use also translates to BED and BN. These individuals may spend a lot of their time thinking about, engaging in and recovering from binge episodes. As mentioned earlier, bingeing is often a planned behaviour which may require a great deal of effort to purchase and store foods ready for a binge episode [41]. In addition, the criteria for BED emphasise the time spent bingeing, with the number of binge episodes per week determining the severity of the disorder [23]. Moreover, these individuals often experience physical and emotional distress following a binge eating episode. Recovery from food consumption has also been reported in self-identified food addicts with references to feeling sleepy or 'hung-over' [52,53].

Although evidence for food addiction directly related to the DSM-5 diagnostic criteria for impaired control is largely anecdotal, there is a considerable amount of empirical evidence for an association between overeating/obesity and impaired control generally. Two aspects of self-regulatory failure that are particularly pertinent in the case of substance use and overeating are impulsivity and reward sensitivity [54–56].

1.1.1. Impulsivity

Although impulsivity is a multi-faceted construct, it can be defined broadly as the tendency to think and act without sufficient forethought, which often results in behaviour that is discordant with one's long-term goals. The role of impulsivity in SUDs is well documented [35,57–60]. Many studies have reported higher impulsivity levels with increasing substance use across a wide range of questionnaires and behavioural tasks, and for a variety of different substances [61–66]. For example, Noël et al. [67] performed a series of behavioural tasks assessing the ability to suppress irrelevant responses (response inhibition) and irrelevant information (proactive interference) in a group of detoxified alcohol-dependent individuals and matched healthy controls. They found a statistically significant group difference for all three tests assessing response inhibition but no differences for proactive interference.

Impulsivity has also been implicated in overeating and obesity [54,68–71]. Overweight/obese individuals score higher on self-reported [72–74] and behavioural measures of impulsivity [75–77], whereas those high in self-control have been shown to be less likely to give in to temptation [78–80] and are more likely to maintain a healthy diet and engage in physical exercise [81–83] Impulsivity scores have also been shown to predict poor food choices [84] and correlate positively with food consumption [85–87]. For example, Guerrieri et al. [87] found that, in a sample of healthy-weight women, those with higher impulsivity scores ate more candy during a 'bogus' taste test than those with lower impulsivity scores. Churchill and Jessop [88] also showed a predictive relationship between impulsivity and snacking on high-fat foods over a two-week period. Scores on the YFAS have also been associated with various measures of impulsivity, such as motor and attentional impulsivity, mood-related impulsivity and delay discounting [89,90].

1.1.2. Reward Sensitivity

A heightened general sensitivity to reward has also been linked to both substance use and overeating [69,77,91–93]. In the food literature, self-report measures of reward sensitivity have revealed associations with BMI, food craving and preferences for foods high in fat and sugar [93–95]. Using two behavioural tasks, Guerrieri et al. [69] measured reward sensitivity and response inhibition in children aged 8–10. They subsequently measured food intake in a bogus taste test when the foods were either varied or monotonous. Their results revealed that reward-sensitive children consumed significantly more calories than non-reward sensitive children only when the food was varied. There was no effect of response inhibition on food intake, nor any interaction with variety; however, unlike reward sensitivity, deficient response inhibition was associated with being overweight. The authors suggested that reward sensitivity may play a causal role in overeating, whereas deficient inhibitory control may be more of a maintaining factor. This fits well with findings from a study demonstrating a role of reward sensitivity in the early onset of heroin use and a role of impulsivity in escalating use [92,96].

There is also evidence to suggest that reward sensitivity may decrease with more prolonged or established overeating, with studies showing anhedonia, or hypo-sensitivity to reward, in obese participants [97–100]. For example, Davis et al. [97] demonstrated that although overweight women were more sensitive to reward than healthy-weight women, those who were obese were significantly *less* reward sensitive than overweight women. Importantly, the earlier mentioned association between reward sensitivity and increased BMI was found in a sample of mainly healthy-weight women, with only 1% classified as obese [93]. Although there is a great deal of evidence to suggest that sensitivity to reward plays a role in substance abuse and overeating, the causal direction of this relationship remains unclear. On the one hand, increasing reward sensitivity may lead to overeating by increasing motivation towards pleasurable activities, such as consuming energy-dense foods that elicit dopamine and opioid activation. On the other hand, decreased reward sensitivity may cause individuals to seek out rewarding activities as a form of 'self-medication' in order to boost dopamine functioning (i.e., addictive behaviour is the result of a 'reward deficiency syndrome') [101,102]. These two arguments, and the relevant neuroimaging literature, are discussed further below (see the Neurobiological Similarities section below) and in more detail by Burger and Stice [103].

Burger and Stice [103] offer several theories for how these two causal directions combine to explain obesity. They propose that high sensitivity to reward may initially cause individuals to over-consume palatable foods, but this sensitivity is then modified over time as the brain's reward system adapts and shows divergent changes in food motivation ('wanting') versus hedonic pleasure ('liking'). According to Robinson and Berridge's [104–106] incentive-sensitisation theory, repeated intake results in an increased incentive value for these foods and their associated cues, which may be subjectively experienced as excessive wanting or craving. Moreover, this theory argues that with repeated presentations of palatable foods, the hedonic pleasure derived from consuming the food will decrease due to neural habituation, while the anticipation of reward increases. Hence, a vicious cycle emerges in which the individual will experience less pleasure from the food ('liking'), but will simultaneously experience an increased desire ('wanting') for the food, driving further food seeking and consumption [107–109] (see Figure 1). The experience of intense cravings is the third criterion of impaired control and is another symptom of substance addiction that can be readily applied to overeating and obesity.

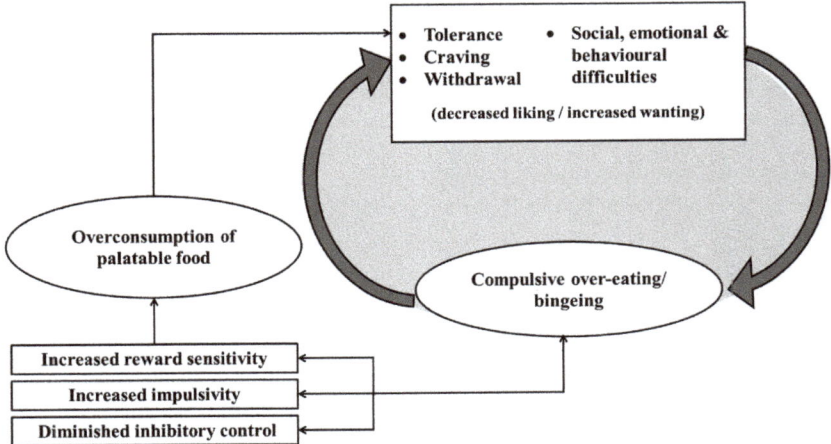

Figure 1. The proposed cycle of 'food addiction'. Initial vulnerability for the over-consumption of palatable food is marked by increased impulsivity and reward sensitivity, as well as a diminished capacity for inhibitory control. As a consequence of overconsumption, individuals experience tolerance, craving and withdrawal, along with a range of social, emotional and behavioural difficulties such as weight stigmatisation and feelings of guilt and shame. With repeated consumption of these foods, the individual is likely to habituate to the hedonic properties of the food, resulting in reduced enjoyment or liking. These changes are also accompanied by an increased desire or 'wanting' for the food [104–108]. In an attempt to relieve these symptoms, the individual 'self-medicates' by increasing food consumption, which can result in compulsive or binge eating behaviour, thus creating a cycle of addiction. It should be noted that the extent to which each of these mechanisms is experienced varies considerably across individuals. In particular, initial vulnerability to addiction may be related to individual differences in reward sensitivity, impulsivity and inhibitory control [110–113].

1.2. Craving

The term 'food craving' typically refers to an intense desire to consume a specific food [114,115]. Food cravings appear to be very common with reports of 100% of young women and 70% of young men experiencing a craving for at least one food in the past year [116,117]. The most commonly reported craved food is chocolate, although cravings for carbohydrates and salty snacks are also common [118–122]. The prevalence of food cravings has prompted the development of several standardised questionnaires that measure food cravings with a good degree of internal consistency and construct validity [123–127], including a specific questionnaire just for chocolate (Attitudes to Chocolate Questionnaire) [128]. Recurrent food cravings are of interest in relation to food addiction as they have been associated with binge eating, increased food intake and increased BMI [124,127,129–132]. Increased reports of food craving have also been demonstrated in individuals who score highly on measures of self-reported food addiction [133–135] and those with BED and BN [136–138]. Furthermore, just as drug craving is associated with an increased likelihood of relapse [139–141], food craving has been linked to poor dieting success [142–144].

Further support for the similarity between drug and food craving is evident in the findings of cue-reactivity research. The aphorism that cravings are most likely to occur in the presence of substance-related stimuli has been well documented, with cue-exposure paradigms showing significant effects of drug-related cues on self-reported and physiological measures of craving [145–148]. Similarly, exposure to food cues has also been shown to increase food cravings [149,150] and a recent systematic review of 45 studies (involving 3292 participants) concluded that 'food cue-reactivity' (physiological, neural and subjective reward-related responses to food cues) reliably and prospectively predicts both energy intake and weight gain, particularly over the longer-term, accounting for ~11% (7%–26%) of variance in these outcomes [129]. Food cue-induced craving is especially prevalent among binge eaters and those with BED [151,152] in whom it has been correlated with binge eating frequency and BMI [153]. It is possible, therefore, that certain individuals are more susceptible to cue-induced cravings, and also that this susceptibility may transfer across different substances. Both Mahler and de Wit [147] and Styn et al. [148] found a significant correlation between cue-induced cigarette craving and cue-induced food craving in smokers, suggesting a common mechanism. Cue-induced craving is also believed to strengthen with repeated consumption, fueling the vicious circle shown in Figure 1.

1.3. Social Impairment

Overeating and obesity have been associated with poor social functioning, especially among children and adolescents. When assessing quality of life with child and parent-proxy reports, social functioning is significantly lower for obese compared to healthy-weight children and is inversely correlated with BMI [154–156]. Poor social functioning in overweight children may be partly due to the overt victimisation and teasing experienced as a direct result of their weight status [157,158]. Hayden-Wade et al. [159] found that the degree of teasing experienced by overweight children was positively correlated with loneliness, an increased preference for isolative activities and a lower preference for social activities. This preference for being alone, along with the emotional difficulty of being victimised, fuels a vicious cycle as these circumstances are likely to promote further overeating and binge-eating—which, in turn, leads to increased weight gain and further teasing [42,160] (see Figure 1).

Weight stigmatisation may also affect interpersonal friendships and romantic relationships in adulthood with reports of discriminatory attitudes and behaviours in occupational [161,162] and romantic settings [158,162,163]. For example, Chen and Brown [164] reported that when making sexual choices about a partner, both male and female college students ranked an obese individual as the least liked. In a study focusing on the psychosocial correlates of food addiction, Chao et al. [165] found that, compared to control participants, those who met the YFAS criteria scored lower on physical, mental and social aspects of health-related quality of life. Social impairments were related to self-esteem, sexual life, public distress and work. Interpersonal problems have also been associated with binge eating—a relationship which is likely to be bidirectional [166,167].

1.4. Repeated Use Despite Negative Consequences

It has been noted that due to its increase in prevalence and associated comorbidities, obesity now appears to be a greater threat to the burden of disease than smoking [168]. The physical and psychological effects of overweight and obesity are well documented and include, but are not limited to, depression, an increased risk of diabetes, hypertension, cardiovascular disease and some cancers [169–177]. With pervasive warnings regarding the consequences of overeating, from the media, government, and the medical profession, it seems fair to assume that most overweight and obese individuals are aware of the negative outcomes associated with their dietary behaviour [41,52]. Critically, even those who have undergone weight loss treatment often fail to lose weight or gain weight following intervention [46,48,50,51]. Continued overeating also occurs in those who have received bariatric surgery with patients showing continued snacking and poor food choices [178,179]. There is, therefore, considerable evidence to support continued overeating despite negative consequences.

1.5. Physiological Criteria

Tolerance to a substance occurs when the same amount of the substance has an increasingly diminished effect with repeated use. This effect usually results in escalated use as the individual increases their dosage in order to recreate the original experience. There is some evidence of food tolerance in animal models of sugar addiction. Rats given intermittent and excessive access to sugar solution increase their intake significantly over time, and this is accompanied by neurochemical changes that are similar to those seen in drug abuse [180,181]. In humans, there is some indication that tolerance to sugar may occur in the first few years of life. The effectiveness of sucrose as an analgesic in young infants is reported to diminish after 18 months of age as sugar consumption increases [182–185]. The possibility of such early tolerance to palatable foods and the methodological difficulties of diet restriction in humans makes finding empirical evidence of tolerance in adults difficult and unlikely. However, statistics indicating increased consumption and portion sizes for these foods provide indirect evidence of tolerance to high-fat/high-sugar foods at a population level [52,186], and also at an individual level based on anecdotal reports. For example, Pretlow [42] found that 77% of overweight poll respondents reported eating more now than when they originally became overweight. Furthermore, in response to a follow-up question asking why they believed that they ate more, 15% indicated that they were less satisfied by food. Hetherington et al. [109] also found that when participants were provided with chocolate for three weeks, they increased their intake over time while simultaneously reporting a reduction in food liking.

Withdrawal is the second physiological criterion for substance abuse and is defined by the presence of physical or psychological symptoms in response to substance deprivation, or the use of the substance in order to relieve these symptoms. Evidence of withdrawal has also been found in the aforementioned animal models of sugar addiction. Under conditions of sugar deprivation, these animals show withdrawal symptoms similar to those seen with morphine and nicotine withdrawal, including physical symptoms of teeth chattering, forepaw tremor, head shaking and reduced body temperature [187,188] as well as increased aggression [189] and anxiety [190]. There are also anecdotal reports of withdrawal-like symptoms in humans, including persistent cravings and negative affects when attempting to reduce food intake [42,191], as well as the tendency to eat to avoid the emotional symptoms associated with withdrawal such as fatigue, anxiety and depression [52]. Using the YFAS, withdrawal symptoms (such as agitation, anxiety, or other physical symptoms) have been reported in up to 50% of individuals with obesity and BED [35].

2. Neurobiological Similarities between Palatable Foods and Drugs of Abuse

Just as altered brain functioning has been reported in SUDs, overeating and obesity have also been associated with changes in the neural processing of the motivational properties of food. This includes changes in systems coding the hedonic and rewarding aspects of the substance, as well as the systems involved in controlling these motivations [103,192–194]. Volkow and colleagues [195–199] have proposed a common model for addiction and obesity that involves two neural circuits that are both modulated by dopamine—increased reward sensitivity and diminished inhibitory control [70].

2.1. Neurobiology of Reward Sensitivity

Addictive drugs directly affect the mesolimbic dopamine system (MDS), which is thought to mediate the processing of motivational salience, pleasure and reward [200]. Animal studies have shown that, similar to drugs of abuse, palatable foods are capable of triggering dopamine release in the nucleus accumbens (NAc) and ventral tegmental area (VTA) [181,201–203]. Furthermore, activity in the MDS has been linked to the amount of food ingested and its rewarding properties [204,205]. However, distinct patterns of neuronal firing in the NAc to food and illicit substances have also been reported [206,207]. Increased activation of this reward system has also been shown in human participants during the presentation of food cues and meal consumption [96,208–211]. For example, Stoeckel et al. [212] demonstrated that when viewing images of high-calorie foods, obese women showed significantly greater activation in a number of regions associated with reward, compared to healthy-weight women. Obese participants have also demonstrated increased responsivity to food in gustatory and somatosensory regions [213,214], suggesting a heightened sensitivity to palatable food that may contribute to overeating and obesity.

Although an increased sensitivity to reward may initially drive individuals to consume calorific foods, it has been speculated that compulsive eating may develop as the pleasure derived from these foods diminishes with increased tolerance (see Figure 1). It has been argued that, just as with drugs of abuse, the chronic consumption of such rewarding foods may cause the downregulation of dopamine receptors in order to compensate for their overstimulation [215–217]. Decreased striatal dopamine receptor availability has frequently been observed in individuals with substance addictions [218–222], whereas increased receptor availability has been shown to have a protective role against alcoholism [223,224]. It has also been shown that striatal D2 receptor availability is significantly lower in severely obese individuals compared to controls and is significantly and negatively correlated with BMI [99,100].

It has been argued, therefore, that a reduction in dopamine receptor availability may subsequently cause or exacerbate overeating as a form of 'self-medication' in which the individual attempts to compensate for a diminished experience of reward [100,225–227] (see Figure 2). For example, Geiger et al. [228] found that rats fed on a cafeteria-style diet showed reduced baseline levels of mesolimbic dopamine activity. This activity was stimulated by cafeteria foods but not by their regular chow, thus suggesting that a preference for palatable food may develop as a consequence of its ability to increase dopamine release compared to other, less palatable, foods. Animal studies have also demonstrated causal effects of D2 receptor agonists and antagonists on overeating. The administration of D2 antagonists has been shown to increase meal size, meal duration and body weight, whereas treatment with D2 agonists can reduce hyperphagia and prevent weight gain [229–231]. The effects of such pharmaceutical interventions in humans, however, have been fairly mixed. The use of antipsychotic medication which blocks D2 receptors is typically associated with weight gain [232] and some D2 agonists have been found to reduce body weight [233]. A recent trial, however, found no effect of the dopamine agonist cabergoline on preventing weight regain [234,235] and there is some evidence that D2 agonists can promote weight gain in patients with anorexia nervosa [236]. More encouragingly, studies with gastric bypass patients have demonstrated increased D2 receptor availability following weight loss, indicating that the effects of overeating on dopamine receptor downregulation may be reversible [237–239].

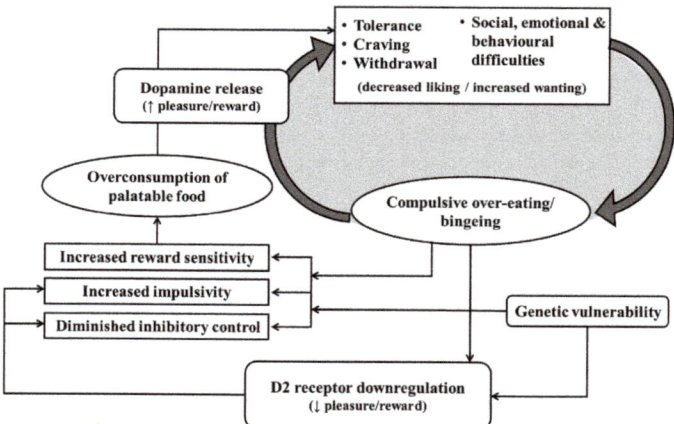

Figure 2. The proposed cycle of 'food addiction' including the role of dopamine. When palatable food is consumed, the brain releases the hormone dopamine (alongside other neurotransmitters such as opioids). Over time, this increase in dopamine leads to the downregulation of dopamine receptors, causing individuals to experience a reduction in pleasure during palatable food consumption. This decrease in pleasure, combined with symptoms of tolerance, craving, withdrawal and other social, emotional and behavioural difficulties, results in the individual engaging in compensatory behaviour by increasing food consumption. As a consequence, food consumption may become compulsive, thus creating a cycle of food addiction.

2.2. Neurobiology of Inhibitory Control

Dopamine receptor availability in obese individuals has also been shown to correlate positively with metabolism in prefrontal regions involved in inhibitory control (specifically the dorsolateral prefrontal cortex [DLPFC], medial orbitofrontal cortex [mOFC] and anterior cingulate gyrus, as well as the somatosensory cortices) [99]. Similar findings have been observed in healthy-weight participants, who demonstrated a positive correlation between dopamine receptor availability and inhibitory control performance on the stop-signal task [240]. Volkow et al. [99] hypothesised that altered dopamine functioning may play a role in overeating not only through altering the rewarding properties of food but also by reducing inhibitory control. A significant negative correlation between BMI and prefrontal activity has also been reported [75,241,242] along with reduced prefrontal activation following a meal in obese men and women [243–245]. Conversely, successful dieting has been positively associated with frontal activation [246–249].

In a study of healthy women, Lawrence et al. [96] reported an association between food cue reactivity in the NAc and later snack consumption [117]. They also found that this reactivity was associated with increased BMI for individuals who reported low self-control. The authors proposed a 'dual hit' of increased reward motivation and poor self-control in predicting increased food intake [250]. Similarly, reductions in frontal grey matter volume have also been linked to increased BMI, poor food choices and related deficits in executive functioning [251–258]. These findings are reflective of a growing literature on the cognitive dysfunction associated with drug abuse and obesity, although research indicates that the causal relationship is bidirectional [76,259–263].

Although it has been hypothesised that overeating is initially caused by a hyper-responsive reward circuitry and maintained by the subsequent degradation of this system [103], there is also evidence to suggest that some individuals may be genetically vulnerable to an impaired capacity for reward and inhibitory control. Genetics studies have revealed that both drug users and obese individuals have a significantly greater prevalence of the *TaqI* A1 allele polymorphism which can cause a 30%–40% reduction in striatal D2 receptors [213,264–269]. In addition, this polymorphism has been associated with behavioural measures of impulsivity and low reward sensitivity [270–272]. It has also been linked to low grey matter volume in the anterior cingulate cortex (ACC) [273], an area which is believed to be involved in executive control and reward expectancy [240,274,275], and has been shown to be active during resistance of cigarette craving [276]. Together these findings demonstrate that overeating and SUDs may share a common neurobiological mechanism involving altered dopamine functioning that subsequently disrupts mechanisms involved in reward sensitivity and inhibitory control.

Our review, considering each of the DSM-5 criteria for SUDs in isolation, suggests that there is considerable evidence for food addiction. Whether an individual meets clinical diagnostic criteria under an SUD model, and the severity of the disorder, however, is dependent on an individual presenting a number of symptoms (mild: two to three symptoms; moderate: four to five symptoms; severe: six or more symptoms). Studies utilising the YFAS (which uses diagnostic criteria for SUDs) have certainly suggested that a substantial proportion of the general population meet the diagnostic cut-off for food addiction (15%–20%), with approximately 11% of the population being classified as 'severe' [38,276]. The prevalence of food addiction in those with BED and BN has been reported as much higher, with estimates of 92% for BED and 96%–100% for BN [32,277,278]. Acknowledging the potential prevalence of food addiction, we next discuss a range of treatments for overeating that have been informed by the similarities between SUDs and overeating.

3. Treatment Implications

One of the greatest potential advantages of identifying the similarities between substance addictions and overeating is the development of effective interventions. The standard approach to weight loss, involving maintaining a healthy diet and physical exercise, is often associated with poor adherence rates and overall weight gain [46–51,279]. One possible reason for the ineffectiveness of dieting is that it is treating the outcome of overeating and not the underlying cause. Approaches that target increased impulsivity and reduced self-control may have more success. For example, Hall, Fong, Epp and Elias [280] showed that executive function on the go/no-go task (a measure of response inhibition) predicted unique variance for dietary behaviour and physical exercise, and also moderated the association between intentions and behaviour [117,281]. This suggests that individuals who are more capable of controlling their impulsive actions are more likely to successfully meet their goals. This also implies that techniques to improve such abilities may prove to be effective tools for aiding weight loss.

3.1. Cognitive Interventions

Increased motivation for illicit substances has been associated with several cognitive biases including attentional biases [282–287], approach biases [283,284,288–290] and affective biases [291–294]. One method for reducing this motivation, therefore, has been to use training tasks that are designed to reduce these cognitive biases, and recently, these training tasks have been explored as potential interventions for overeating.

Heightened attentional biases towards food have been demonstrated across various populations, including those with disordered eating patterns [295–300] and those who are overweight or obese [301–304]. Just as the addiction literature has explored whether attentional biases can be manipulated to reduce substance intake, this approach has also been explored with food consumption, although with mixed results [305–309]. Hardman et al. [305] trained undergraduate students on the visual probe task to either attend or avoid images of cake and stationery. They found a modest increase in attentional bias for the attend-cake group but no effects of bias training, for either group, on hunger or food consumption, suggesting that any attentional biases with food may be particularly difficult to modify. Using a female-only sample, Kemps et al. [306] demonstrated significant effects of a similar dot-probe training on attentional bias and food consumption. Effects on attention were found to generalise to novel pictures, however, effects on food intake were specific to the trained food and were undermined by participants consuming more of an equally unhealthy novel food. More recent results hold some promise for attentional bias modification, indicating that training can be used to decrease immediate calorie consumption in overweight and obese women [310] and can increase the consumption of healthy foods [311]. Multiple training sessions have also demonstrated that effects can persist beyond the period of training for up to a week [312].

There is also a small body of evidence demonstrating an approach tendency towards food for individuals with disordered eating [295,313,314], high trait food craving [315] and those who are overweight or obese [316–318]. For example, Veenstra and de Jong [313] showed that those who scored highly on a measure of dietary restraint (a measure for the chronic, cognitive limitation of food intake) were significantly faster to move a manikin towards than away from images of food. Using a different measure of approach bias, Kemps, Tiggemann, Martin and Elliott [319] also found that participants who liked chocolate were significantly faster at pairing images of chocolate with approach words compared to avoid words. Furthermore, they also demonstrated that participants who were trained to pair images of chocolate with either approach or avoid words increased and decreased their approach bias, respectively. The approach group also demonstrated a significant increase in chocolate cravings. Although the avoid group showed a decrease in reported craving, this finding was not statistically significant from baseline [320]. Similar training protocols have also been shown to be effective at reducing action tendencies towards high-calorie foods and reducing trait- and cue-induced craving in participants with subclinical eating disorders [321].

The use of addictive substances can also be motivated by the positive affect associated with them; therefore, reducing such an affective bias should discourage their use. Studies using evaluative (or affective) conditioning have shown initial promise. Here, the evaluation of a conditioned stimulus (CS) can be modified by consistently pairing it with a valenced unconditioned stimulus (US) [322,323]. In the food literature, negatively valenced stimuli are typically used to reduce the implicit liking of unhealthy foods [324–328]. Further, evaluative conditioning has been found to lead to more favourable food choices in some studies. For example, Walsh and Kiviniemi [329] found that participants were three times more likely to select fruit over a granola bar after receiving evaluative conditioning training where positive (relative to negative or neutral) words and images were paired with images of fruit. Similarly, Hollands et al. [325] showed that participants were more likely to select fruit over an unhealthy snack when snack images were repeatedly paired with negative body images compared to a blank screen. Interestingly, this effect was moderated by implicit attitudes towards the snack foods; participants with more favourable attitudes at baseline showed the greatest change in subsequent behaviour. Although these studies have involved healthy participants, this latter finding, in particular, suggests that evaluative conditioning may be an appropriate intervention for those with disordered eating who show strong preferences for unhealthy foods. However, the effects of evaluative conditioning training on reducing unhealthy food consumption are currently unclear. To date, only a limited number of studies have included post-intervention follow-up, and while some have found reduced consumption in the week following training [330], others have failed to show immediate effects [327,328,330]. It is likely that training effects are dependent on baseline strength of liking for specific foods [325,330], the specificity of the US used [331], and awareness of the CS-US contingencies [332–334]. To establish the therapeutic benefits of evaluative conditioning training, studies in overweight and disordered eating groups are required.

Another approach to cognitive training is to reduce such biases indirectly through tasks such as response inhibition training. Response inhibition refers to our ability to interrupt or override impulsive reactions in accordance with new information, and plays a key role in goal-directed behaviour [335–340]. Deficient response inhibition has been linked not only to the use of different addictive substances [70] but also to the severity of use [61,64,341], poor treatment outcomes [342] and likelihood of relapse [343]. Houben and Wiers [344] have also shown that positive implicit attitudes towards alcohol are only related to alcohol consumption when inhibitory control is low. These results suggest that an increased ability to inhibit responses may enable an individual to exert self-control over their behaviour, even when they possess strong implicit preferences [117].

Similar findings have also been replicated with overeating and obesity. Obese individuals have been shown to demonstrate less efficient response inhibition than their healthy-weight counterparts [69,251,345,346] and poor inhibitory control has been associated with increased unhealthy food consumption [86,347,348], high BMI [75,85,349,350], food cravings [351], unhealthy food choices [84,352] and binge-eating [353]. Moreover, as in the addiction literature, inhibitory control has also been shown to interact with implicit attitudes towards food, thus indicating that effective response inhibition may play a protective role against strong implicit preferences for unhealthy foods [80,117,250].

Simple tasks designed to train response inhibition to relevant cues or contexts have been shown to reduce gambling behaviour and alcohol consumption [354–359]—although available evidence suggests that the longevity of such effects may be limited [360]. These training tasks have also been adapted to train response inhibition to food stimuli and are showing encouraging effects across a range of eating-related behaviours including food consumption [361–365], food choices [365–371] and even weight loss [372–375]. For example, Lawrence et al. [376] trained participants to inhibit their responses towards either images of unhealthy snack foods (active group) or non-food items (control group). After four training sessions, they found that, compared to the control group, individuals in the active group showed reduced energy intake (220 kcal less per 24 h food diary) and reduced liking for unhealthy foods. Furthermore, participants in the active group showed significant weight loss; showing objectively measured weight loss of 0.7 kg after 2 weeks and self-reported weight loss of more than two kilograms after a six month follow up (2.66% decrease).

The cognitive training paradigms discussed above show promise but are currently in the early phases of testing. Before such training methods can be taken forward to clinical trials, researchers should further explore the effects of different experimental protocols with the aim of developing the most effective training techniques. One aspect of training that is likely to be important in determining successful behavior change is training performance. For example, the proportion of successful inhibitions on an inhibition training task and accuracy during attentional bias training have both been shown to moderate efficacy [308,375]. To establish whether effects can be long-lasting, we need to consider repeated testing sessions, personalised training stimuli and combining training techniques to simultaneously reduce cognitive biases and increase executive control [312,369,377,378]. Understanding the mechanisms that underlie such effects could also prove crucial. For example, the effects of inhibition training on food consumption may be due to the devaluation of inhibited stimuli. Several studies have shown that repeatedly pairing a stimulus with the inhibition of a response can reduce how much the image is liked or how attractive it was perceived to be [371,376]. Such devaluation could be the result of action conflict or inherent links between avoidance and aversion [371,379–382]. Designing interventions that promote automatic associations between stimuli and action tendencies may, therefore, prove fruitful, especially if training is performed accurately, personalised and delivered across multiple sessions. Combining cognitive training tasks with prefrontal brain stimulation is another avenue worthy of investigation. Brain stimulation methods have the potential to augment learning effects [383] and can also be used to reduce food consumption and craving in isolation; these methods are discussed below.

3.2. Neuromodulation Interventions

Non-surgical brain stimulation techniques have also been explored for their potential benefits in reducing craving and addictive behaviours by altering neural activity and increasing dopamine [384–388]. The most commonly applied stimulation methods are transcranial magnetic stimulation (TMS) and transcranial direct current stimulation (tDCS). These methods are used in awake participants and are generally considered to be safe when administered within recommended guidelines [389–394].

The use of TMS involves the delivery of electromagnetic pulses that penetrate the skull to induce electric current in the underlying cortex and cause short-term changes in cortical excitability. The modulation of cortical excitability can last beyond the period of stimulation by delivering trains of pulses, a technique known as repetitive TMS (rTMS) [395]. When applied to the DLPFC, rTMS has been shown to effectively reduce cravings for cigarettes, alcohol and drugs of abuse, especially when applied for multiple sessions [396–400]. The DLPFC is an area involved extensively in inhibitory control [401–404] and stimulation of this region may act to boost self-control, potentially by increasing dopamine release in the caudate nucleus [405,406].

Reductions in substance craving have also been demonstrated with stimulation of the DLPFC using tDCS [114,407–412]. The use of tDCS involves the application of a weak (typically 1–2mA) direct electrical current to the scalp via a pair of electrodes. The effect of tDCS on brain activity is dependent on the stimulation polarity; anodal stimulation is thought to increase cortical excitability by neuronal depolarisation whereas cathodal stimulation is believed to decrease excitability by hyperpolarising neurons [413–418]. Long-lasting effects on resting membrane potential have been shown with longer stimulation durations, for example, 13 min of anodal tDCS has been shown to increase motor cortical excitability for up to 90 min [419]. Compared to TMS, tDCS is a weaker form of stimulation with fewer incidental artefacts and is considered to be safer and more appropriate for reliable double-blinding [387,420,421]. It is also thought that tDCS can be used to potentiate learning [422], and may effectively enhance the effects of the aforementioned cognitive interventions [378,423].

These stimulation methods are currently being investigated for their potential to reduce food craving and consumption [424–429]. Using rTMS to the left DLPFC, Uher et al. [430] found an increase in cue-induced craving for palatable foods in the group who experienced sham stimulation but not the active group. However, no effect was found on ad-libitum food consumption, although this result may have been due to the limited time period (5 min) causing participants in both groups to consume a large amount of calories. Using a similar methodology, Van den Eynde et al. [431] demonstrated an increase in craving scores in the sham group, but a decrease in craving scores for the active group in a sample of participants with bulimic-type eating disorders. In addition, active rTMS was associated with a reduction in binge-eating episodes in the following 24 h period. However, blinding was only partially successful in this study with most participants correctly guessing whether they were receiving active or sham rTMS. In a later study, Barth et al. [432] used a within-subjects design with an improved sham condition in which they matched the perceived pain of active rTMS with scalp electrodes. They found an equal reduction in cravings for both conditions and attributed this effect to the experience of pain rather than prefrontal stimulation.

As mentioned, tDCS is believed to involve a more appropriately matched sham condition, especially when participants receive active stimulation for a short initial period [420,421]. When stimulating the DLPFC bilaterally using tDCS, Fregni et al. [433] found a significant increase in cue-induced craving, measured before and after stimulation, in the sham condition and a significant reduction when participants received anodal right/cathodal left stimulation. Compared to the sham condition, active stimulation was also associated with a reduction in food intake during an ad-libitum eating phase. Although the authors did not assess blinding in this study, they did report equal occurrences of mild adverse effects across conditions. Using the same montage, Goldman et al. [424] and Lapenta et al. [426] also found the same reduction in food craving, and an early meta-analysis revealed a medium effect-size favouring active over sham stimulation in the reduction of cravings [434]. However, as the number of studies utilizing tDCS in the exploration of its effects on food craving has increased, evidence of efficacy has weakened. A more recent meta-analysis, including eight experiments, found no effect of tDCS on food craving [386], and subsequent research, including a large pre-registered experiment, has also failed to replicate findings for both food craving and consumption [435,436].

Another neuromodulation intervention, which is worthy of a brief mention and gaining in popularity for the treatment of SUDs, is real-time fMRI (rt-fMRI) neurofeedback training. Neurofeedback training involves providing participants with feedback of their neural response to certain cues and instructing them to increase or decrease their response, so that they may gain volitional control over specific brain regions. In the treatment of SUDs, neurofeedback training typically involves increasing activity in control regions, such as the prefrontal cortex, or decreasing activity in regions associated with craving, such as the ACC. For example, it has been shown that decreasing activity in the ACC with rt-fMRI neurofeedback is significantly correlated with decreased nicotine craving in smokers [437–439]. Using a similar technique with electroencephalography (EEG) has also shown improvements in cravings, drug use and treatment outcomes for a range of different substances [440–443]. Although in its early days, the application of neurofeedback training to food consumption and obesity has already been proposed [444–446]; recent findings have also suggested that neurofeedback may be another method of decreasing activity in motivation- and reward-related regions [447] and increasing activity within critical prefrontal regions such as the DLPFC [448,449].

3.3. Therapeutic Interventions

Therapeutic interventions such as Overeaters Anonymous and cognitive behavioural therapy have taken a more holistic approach to the treatment of obesity. Overeaters Anonymous (OA) is based directly on the 12-step programme developed by Alcoholics Anonymous. The OA organisation promotes the central belief that obesity is a symptom of 'compulsive overeating', which is an addictive-like illness with physical, emotional and spiritual components [450]. Individuals are required to acknowledge that compulsive overeating is beyond their willpower to overcome and, therefore, they must attempt to control their intake by avoiding certain foods and surrendering to a 'higher power'. Just like Alcoholics Anonymous, OA involves group meetings for individuals to share their feelings and experiences. Although the way in which this programme influences outcomes is unclear [53], the group meetings may act to alleviate feelings of isolation and instead foster a sense of community. As discussed earlier, due to the feelings of shame and guilt and the weight teasing experienced, overweight and obesity are associated with a preference for isolative activities [159]. This social isolation can subsequently exacerbate overeating, creating a vicious cycle [42,160]. It is possible, therefore, that OA acts to break this cycle by providing a supportive and encouraging social environment. However, due to the anonymous nature of OA, there has been little research conducted on its efficacy and it is not understood exactly how OA affects overeating and the extent to which it may do so.

Cognitive behavioural therapy (CBT), on the other hand, is a therapeutic approach which is extensively informed by research. CBT requires patients to critically evaluate the thoughts, feelings and behaviours that result in maladaptive behaviour and then modify them through therapy. This therapy allows patients to recognise potential triggers and develop appropriate coping strategies. CBT interventions have been effective in the treatment of substance addictions [451] and have also demonstrated their potential in the treatment of obesity [452,453] and BED [453–456]. However, it has been argued that the success of treating overeating and BED with CBT refutes the food addiction model [30]. The food addiction model applied in OA requires complete avoidance of so-called trigger foods, thereby acting to increase dietary restraint, whereas a reduction in dietary restraint has been shown to moderate the increased effectiveness of CBT on binge eating in a sample of patients with BN [457]. The focus of CBT is to replace dysfunctional eating with more normalised eating behaviour, therefore, favouring moderation and flexibility rather than absolute restraint.

4. Conclusions

As the prevalence of obesity continues to increase and traditional weight loss methods appear to be largely unsuccessful, researchers and clinicians have begun to consider the addictive potential of food. There is a substantial body of evidence demonstrating the similarities between addictive drugs and food on reward and control pathways in the brain and subsequent behaviour such as craving and impulsivity. There is also limited evidence to indicate that in some circumstances, overeating meets the physiological criteria of substance dependence, although more research is necessary to determine the validity of these symptoms in human participants. More research is also required for other behavioural criteria such as social impairment and repeated use despite negative consequences, as the evidence to date is largely anecdotal. However, meeting the physiological criteria for addiction is not necessary for a DSM diagnosis, and as food is a legal substance, just like caffeine, tobacco and alcohol, not all criteria associated with SUDs [23] readily translate to food addiction. Nevertheless, the criterion of withdrawal in SUDs has been associated with clinical severity and the number of symptoms that an individual endorses is used to determine the disorder's overall severity [23].

With a number of these criteria having a limited application to food addiction, a clinical diagnosis appears unlikely in most cases of overeating; however, using the YFAS, it has been estimated that approximately 11% of the general population meet the criteria for a 'severe' food addiction [38]. It should also be made clear that the concept of food addiction does not equate with obesity. Obesity is a multifactorial condition determined by genetic, environmental, biological and behavioural components. For the majority of cases, obesity is caused by a steady increase in excess energy intake and it is not characterised by a compulsive drive for food consumption. Instead, it is thought that the concept of food addiction applies most appropriately to individuals with BED and BN [31,32,277,278].

Despite there being considerable parallels between substance use and compulsive overeating, there is still some concern regarding the use and validity of the term 'food addiction', which is unlikely to apply to the majority of cases [17]. There is also concern over the use of such terminology in the wider social context and whether the term may do more harm than good. While most people would believe that an addiction model reduces individual responsibility, it has also been argued that attributing the problem to a minority of individuals also reduces corporate responsibility [28,458]. As the majority of the population would not be considered 'food addicts', there would be less pressure for the food industry to reduce marketing or to promote healthier alternatives. Likewise, any environmental interventions to reduce access and availability may also seem less critical with a food addiction model.

There are also implications of such terminology for the diagnosed individual. Obesity is already associated with significant social stigmatisation [157–159,161–164] and an additional 'addict' label, which may invoke stereotypes of a person who is untrustworthy and inferior [459], may only serve to heighten the problem [460–462]. DePierre et al. [460] found that when an individual was labelled as an 'obese food addict' they were more stigmatised than when they received either label in isolation ('food addict' or 'obese'). However, a study investigating the effect of an addiction model on public perceptions found that it actually reduced stigma, blame and perceived psychopathology [463,464], suggesting that it may be beneficial in reducing weight-related prejudice. The 'addicted' individual described in the study was viewed as being less at fault for their weight. Although, it is unclear whether the fault then lies with the individual's biology (i.e., certain individuals are prone to becoming 'food addicts') or the industry that continues to promote potentially addictive foods. Although it is almost certain to be a combination of both entities, demonstrating that certain foods can be addictive should increase corporate responsibility and pressure on the food industry to regulate the availability, advertising and nutritional content of such palatable foods [9,458].

Despite these issues and concerns, it has also been acknowledged that for some individuals, 'food addiction' may be the most appropriate diagnosis for their symptoms and it may help to inform their treatment [34]. The available evidence suggests, therefore, that some individuals *are* capable of experiencing an addictive-type relationship with food, although the majority of individuals who compulsively overeat are unlikely to receive such a diagnosis. Considering the underlying causes of impulsive overeating has also led to the development of some exciting and potentially effective interventions. While there are differences between the addictive characteristics of food and illicit substances, there are many parallels that should not be ignored. These parallels have contributed greatly to our current knowledge of compulsive overeating and potential treatments. Both the similarities and differences should encourage more research, which is necessary to determine the extent and potential impact of such a disorder. Until then, the idea of 'food addiction' is expected to remain hotly debated [14,19,20].

Author Contributions: R.C.A. conceived, drafted and finalised the manuscript. J.S., L.M., C.D.C. and N.S.L. made substantial contributions to the manuscript drafts and final approval of the manuscript. All the authors agree to be accountable for all aspects of the work in ensuring that questions related to the accuracy or integrity of any part of the work are appropriately investigated and resolved.

Funding: This research was supported by grants held by C.D.C. from the Biotechnology and Biological Sciences Research Council [BB/K008277/1] and the European Research Council [Consolidator grant 647893 CCT].

Conflicts of Interest: The authors declare no competing interests.

References

1. WHO. *Obesity and Overweight*; World Health Organization: Geneva, Switzerland, 2003.
2. Global Burden of Disease Collaborative Network. *Global Burden of Disease Study. Obesity Prevalence 1990–2013*; Institute for Health Metrics and Evaluation (IHME): Seattle, WA, USA, 2014.
3. Ng, M.; Fleming, T.; Robinson, M.; Thomson, B.; Graetz, N.; Margono, C.; Mullany, E.C.; Biryukov, S.; Abbafati, C.; Abera, S.F.; et al. Global, regional, and national prevalence of overweight and obesity in children and adults during 1980–2013: A systematic analysis for the Global burden of disease study 2013. *Lancet* **2014**, *384*, 766–781. [CrossRef]
4. WHO. *Obesity and Overweight*; World Health Organization: Geneva, Switzerland, 2018.
5. Butland, B.; Jebb, S.; Kopelman, P.; McPherson, K.; Thomas, S.; Mardell, J.; Parry, V. *Tackling Obesities: Future Choices—Project Report*, 2nd ed.; Foresight Programme of the Government Office for Science: London, UK, 2007. Available online: www.bis.gov.uk/assets/bispartners/foresight/docs/obesity/17.pdf (accessed on 28 September 2018).
6. Cummins, S.; Macintyre, S. Food environments and obesity—Neighbourhood or nation? *Int. J. Epidemiol.* **2006**, *35*, 100–104. [CrossRef] [PubMed]
7. Jeffery, R.W.; Utter, J. The changing environment and population obesity in the United States. *Obes. Res.* **2003**, *11*, 12S–22S. [CrossRef] [PubMed]
8. Levitsky, D.A. The non-regulation of food intake in humans: Hope for reversing the epidemic of obesity. *Physiol. Behav.* **2005**, *86*, 623–632. [CrossRef] [PubMed]

9. Schulte, E.M.; Avena, N.M.; Gearhardt, A.N. Which foods may be addictive? The roles of processing, fat content, and glycemic load. *PLoS ONE* **2015**, *10*, e0117959. [CrossRef] [PubMed]
10. Blundell, J.E.; Burley, V.J.; Cotton, J.R.; Lawton, C.L. Dietary fat and the control of energy intake: Evaluating the effects of fat on meal size and postmeal satiety. *Am. J. Clin. Nutr.* **1993**, *57*, 772S–778S. [CrossRef] [PubMed]
11. Drewnowski, A.; Brunzell, J.; Sande, K.; Iverius, P.; Greenwood, M. Sweet tooth reconsidered: Taste responsiveness in human obesity. *Physiol. Behav.* **1985**, *35*, 617–622. [CrossRef]
12. Drewnowski, A.; Kurth, C.; Holden-Wiltse, J.; Saari, J. Food preferences in human obesity: Carbohydrates versus fats. *Appetite* **1992**, *18*, 207–221. [CrossRef]
13. Davis, C. An introduction to the Special Issue on 'food addiction'. *Appetite* **2017**, *115*, 1–2. [CrossRef] [PubMed]
14. Avena, N.M.; Gearhardt, A.N.; Gold, M.S.; Wang, G.-J.; Potenza, M.N. Tossing the baby out with the bathwater after a brief rinse? The potential downside of dismissing food addiction based on limited data. *Nat. Rev. Neurosci.* **2012**, *13*, 514. [CrossRef]
15. Corsica, J.A.; Pelchat, M.L. Food addiction: True or false? *Curr. Opin. Gastroenterol.* **2010**, *26*, 165–169. [CrossRef] [PubMed]
16. Meule, A.; Gearhardt, A. Food addiction in the light of DSM-5. *Nutrients* **2014**, *6*, 3653–3671. [CrossRef] [PubMed]
17. Rogers, P.J.; Smit, H.J. Food craving and food "addiction": A critical review of the evidence from a biopsychosocial perspective. *Pharmacol. Biochem. Behav.* **2000**, *66*, 3–14. [CrossRef]
18. Volkow, N.D.; Wise, R.A.; Baler, R. The dopamine motive system: Implications for drug and food addiction. *Nat. Rev. Neurosci.* **2017**, *18*, 741–752. [CrossRef] [PubMed]
19. Ziauddeen, H.; Farooqi, I.S.; Fletcher, P.C. Food addiction: is there a baby in the bathwater? *Nat. Rev. Neurosci.* **2012**, *13*, 514. [CrossRef]
20. Ziauddeen, H.; Farooqi, I.S.; Fletcher, P.C. Obesity and the brain: how convincing is the addiction model? *Nat. Rev. Neurosci.* **2012**, *13*, 279–286. [CrossRef] [PubMed]
21. Frascella, J.; Potenza, M.N.; Brown, L.L.; Childress, A.R. Carving addiction at a new joint? Shared brain vulnerabilities open the way for non-substance addictions. *Ann. N. Y. Acad. Sci.* **2010**, *1187*, 294–315. [CrossRef]
22. Orford, J. Addiction as excessive appetite. *Addiction* **2001**, *96*, 15–31. [CrossRef]
23. American Psychiatric Association. *Diagnostic and Statistical Manual of Mental Disorders*, 5th ed.; American Psychiatric Association: Washington, DC, USA, 2013.
24. Potenza, M.N. Biological contributions to addictions in adolescents and adults: Prevention, treatment and policy implications. *J. Adolesc. Health* **2013**, *52*, S22–S32. [CrossRef]
25. Worhunsky, P.D.; Malison, R.T.; Rogers, R.D.; Potenza, M.N. Altered neural correlates of reward and loss processing during simulated slot-machine fMRI in pathological gambling and cocaine dependence. *Drug Alcohol Depend.* **2014**, *145*, 77–86. [CrossRef]
26. Barry, D.; Clarke, M.; Petry, N.M. Obesity and its relationship to addictions: Is overeating a form of addictive behavior? *Am. J. Addict.* **2009**, *18*, 439–451. [CrossRef] [PubMed]
27. Cassin, S.E.; Buchman, D.Z.; Leung, S.E.; Kantarovich, K.; Hawa, A.; Carter, A.; Sockalingam, S. Ethical, Stigma, and policy implications of food addiction: A scoping review. *Nutrients* **2019**, *11*, 710. [CrossRef] [PubMed]
28. Gearhardt, A.N.; Grilo, C.M.; Dileone, R.J.; Brownell, K.D.; Potenza, M.N. Can food be addictive? Public health and policy implications. *Addiction* **2011**, *106*, 1208–1212. [CrossRef] [PubMed]
29. Volkow, N.D.; Wise, R.A. How can drug addiction help us understand obesity? *Nat. Neurosci.* **2005**, *8*, 555–560. [CrossRef] [PubMed]
30. Wilson, G.T. Eating disorders, obesity and addiction. *Eur. Eat. Disord. Rev.* **2010**, *18*, 341–351. [CrossRef] [PubMed]
31. Davis, C.; Carter, J.C. Compulsive overeating as an addiction disorder. A review of theory and evidence. *Appetite* **2009**, *53*, 1–8. [CrossRef] [PubMed]
32. Meule, A.; von Rezori, V.; Blechert, J. Food addiction and bulimia nervosa. *Eur. Eat. Dis. Rev.* **2014**, *22*, 331–337. [CrossRef] [PubMed]
33. Shell, A.G.; Firmin, M.W. Binge eating disorder and substance use disorder: A case for food addiction. *Psychol. Stud.* **2017**, *62*, 370–376. [CrossRef]
34. Smith, D.G.; Robbins, T.W. The neurobiological underpinnings of obesity and binge eating: A rationale for adopting the food addiction model. *Biol. Psychiatr.* **2013**, *73*, 804–810. [CrossRef]

35. Cassin, S.E.; Von Ranson, K.M. Is binge eating experienced as an addiction? *Appetite* **2007**, *49*, 687–690. [CrossRef]
36. Gearhardt, A.N.; White, M.A.; Masheb, R.M.; Morgan, P.T.; Crosby, R.D.; Grilo, C.M. An examination of the food addiction construct in obese patients with binge eating disorder. *Int. J. Eat. Dis.* **2012**, *45*, 657–663. [CrossRef] [PubMed]
37. Gearhardt, A.N.; Corbin, W.R.; Brownell, K.D. Preliminary validation of the Yale food addiction scale. *Appetite* **2009**, *52*, 430–436. [CrossRef] [PubMed]
38. Gearhardt, A.N.; Corbin, W.R.; Brownell, K.D. Development of the Yale food addiction scale version 2.0. *Psychol. Addict. Behav.* **2016**, *30*, 113–121. [CrossRef] [PubMed]
39. Gearhardt, A.N.; Roberto, C.A.; Seamans, M.J.; Corbin, W.R.; Brownell, K.D. Preliminary validation of the Yale food addiction scale for children. *Eat. Behav.* **2013**, *14*, 508–512. [CrossRef]
40. Murphy, C.M.; Stojek, M.K.; MacKillop, J. Interrelationships among impulsive personality traits, food addiction, and body mass index. *Appetite* **2014**, *73*, 45–50. [CrossRef]
41. Curtis, C.; Davis, C. A qualitative study of binge eating and obesity from an addiction perspective. *Eat Dis.* **2014**, *22*, 19–32. [CrossRef]
42. Pretlow, R.A. Addiction to highly pleasurable food as a cause of the childhood obesity epidemic: A qualitative internet study. *Eat. Disord.* **2011**, *19*, 295–307. [CrossRef]
43. French, S.; Jeffery, R.; Sherwood, N.; Neumark-Sztainer, D. Prevalence and correlates of binge eating in a nonclinical sample of women enrolled in a weight gain prevention program. *Int. J. Obes.* **1999**, *23*, 576–585. [CrossRef]
44. Lu, H.K.; Mannan, H.; Hay, P. Exploring relationships between recurrent binge eating and illicit substance use in a non-clinical sample of women over two years. *Behav. Sci.* **2017**, *7*, 46. [CrossRef]
45. MacDiarmid, J.; Loe, J.; Kyle, J.; McNeill, G. "It was an education in portion size". Experience of eating a healthy diet and barriers to long term dietary change. *Appetite* **2013**, *71*, 411–419. [CrossRef]
46. Bacon, L.; Aphramor, L. Weight science: Evaluating the evidence for a paradigm shift. *Nutr. J.* **2011**, *10*, 9. [CrossRef] [PubMed]
47. Dansinger, M.L.; Gleason, J.A.; Griffith, J.L.; Selker, H.P.; Schaefer, E.J. Comparison of the Atkins, Ornish, Weight Watchers, and Zone Diets for Weight Loss and Heart Disease Risk Reduction. *JAMA* **2005**, *293*, 43. [CrossRef] [PubMed]
48. Jeffery, R.W.; Epstein, L.H.; Wilson, G.T.; Drewnowski, A.; Stunkard, A.J.; Wing, R.R. Long-term maintenance of weight loss: Current status. *Health Psychol.* **2000**, *19*, 5–16. [CrossRef] [PubMed]
49. Lowe, M.R.; Annunziato, R.A.; Markowitz, J.T.; Didie, E.; Bellace, D.L.; Riddell, L.; Maille, C.; McKinney, S.; Stice, E. Multiple types of dieting prospectively predict weight gain during the freshman year of college. *Appetite* **2006**, *47*, 83–90. [CrossRef] [PubMed]
50. Mann, T.; Tomiyama, A.J.; Westling, E.; Lew, A.-M.; Samuels, B.; Chatman, J. Medicare's search for effective obesity treatments: Diets are not the answer. *Am. Psychol.* **2007**, *62*, 220–233. [CrossRef] [PubMed]
51. Pietiläinen, K.H.; Saarni, S.E.; Kaprio, J.; Rissanen, A. Does dieting make you fat? A twin study. *Int. J. Obes.* **2012**, *36*, 456–464. [CrossRef] [PubMed]
52. Ifland, J.; Preuss, H.; Marcus, M.; Rourke, K.; Taylor, W.; Burau, K.; Jacobs, W.; Kadish, W.; Manso, G. Refined food addiction: A classic substance use disorder. *Med. Hypotheses* **2009**, *72*, 518–526. [CrossRef] [PubMed]
53. Russell-Mayhew, S.; Von Ranson, K.M.; Masson, P.C. How does overeaters anonymous help its members? A qualitative analysis. *Eur. Eat. Disord. Rev.* **2010**, *18*, 33–42. [CrossRef] [PubMed]
54. Dawe, S.; Loxton, N.J. The role of impulsivity in the development of substance use and eating disorders. *Neurosci. Biobehav. Rev.* **2004**, *28*, 343–351. [CrossRef]
55. Gullo, M.; Dawe, S. Impulsivity and adolescent substance use: Rashly dismissed as "all-bad"? *Neurosci. Biobehav. Rev.* **2008**, *32*, 1507–1518. [CrossRef] [PubMed]
56. Schulte, E.M.; Grilo, C.M.; Gearhardt, A.N. Shared and unique mechanisms underlying binge eating disorder and addictive disorders. *Clin. Psychol. Rev.* **2016**, *44*, 125–139. [CrossRef] [PubMed]
57. De Wit, H. Impulsivity as a determinant and consequence of drug use: A review of underlying processes. *Addict. Biol.* **2008**, *14*, 22–31. [CrossRef] [PubMed]
58. Hershberger, A.R.; Um, M.; Cyders, M.A. The relationship between the UPPS-P impulsive personality traits and substance use psychotherapy outcomes: A meta-analysis. *Drug Alcohol Depend.* **2017**, *178*, 408–416. [CrossRef] [PubMed]

59. Iacono, W.G.; Malone, S.M.; McGue, M. Behavioral disinhibition and the development of early-onset addiction: Common and specific influences. *Annu. Rev. Clin. Psychol.* **2008**, *4*, 325–348. [CrossRef]
60. Verdejo-García, A.; Lawrence, A.J.; Clark, L. Impulsivity as a vulnerability marker for substance-use disorders: Review of findings from high-risk research, problem gamblers and genetic association studies. *Neurosci. Biobehav. Rev.* **2008**, *32*, 777–810. [CrossRef]
61. Billieux, J.; Gay, P.; Rochat, L.; Khazaal, Y.; Zullino, D.; Van Der Linden, M. Lack of inhibitory control predicts cigarette smoking dependence: Evidence from a non-deprived sample of light to moderate smokers. *Drug Alcohol Depend.* **2010**, *112*, 164–167. [CrossRef] [PubMed]
62. Butler, G.; Montgomery, A. Impulsivity, risk taking and recreational 'ecstasy' (MDMA) use. *Drug Alcohol Depend.* **2004**, *76*, 55–62. [CrossRef]
63. García-Marchena, N.; De Guevara-Miranda, D.L.; Pedraz, M.; Araos, P.F.; Rubio, G.; Ruiz, J.J.; Pavón, F.J.; Serrano, A.; Castilla-Ortega, E.; Santín, L.J.; et al. Higher impulsivity as a distinctive trait of severe cocaine addiction among individuals treated for cocaine or alcohol use disorders. *Front. Psychol.* **2018**, *9*, 26. [CrossRef] [PubMed]
64. Lawrence, A.J.; Luty, J.; Bogdan, N.A.; Sahakian, B.J.; Clark, L. Impulsivity and response inhibition in alcohol dependence and problem gambling. *Psychopharmacology* **2009**, *207*, 163–172. [CrossRef]
65. Monterosso, J.R.; Aron, A.R.; Cordova, X.; Xu, J.; London, E.D. Deficits in response inhibition associated with chronic methamphetamine abuse. *Drug Alcohol Depend.* **2005**, *79*, 273–277. [CrossRef]
66. Verdejo-García, A.J.; Perales, J.C.; Pérez-García, M. Cognitive impulsivity in cocaine and heroin polysubstance abusers. *Addict. Behav.* **2007**, *32*, 950–966. [CrossRef] [PubMed]
67. Noël, X.; Van Der Linden, M.; Brevers, D.; Campanella, S.; Verbanck, P.; Hanak, C.; Kornreich, C.; Verbruggen, F. Separating intentional inhibition of prepotent responses and resistance to proactive interference in alcohol-dependent individuals. *Drug Alcohol Depend.* **2013**, *128*, 200–205. [CrossRef] [PubMed]
68. Emery, R.L.; Levine, M.D. Questionnaire and behavioral task measures of impulsivity are differentially associated with body mass index: A comprehensive meta-analysis. *Psychol. Bull.* **2017**, *143*, 868–902. [CrossRef] [PubMed]
69. Guerrieri, R.; Nederkoorn, C.; Jansen, A. The interaction between impulsivity and a varied food environment: Its influence on food intake and overweight. *Int. J. Obes.* **2008**, *32*, 708–714. [CrossRef] [PubMed]
70. Jentsch, J.D.; Pennington, Z.T. Reward, interrupted: Inhibitory control and its relevance to addictions. *Neuropharmacology* **2014**, *76*, 479–486. [CrossRef] [PubMed]
71. Meule, A. Impulsivity and overeating: A closer look at the subscales of the Barratt impulsiveness scale. *Front. Psychol.* **2013**, *4*, 177. [CrossRef]
72. Chalmers, D.K.; Bowyer, C.A.; Olenick, N.L. Problem drinking and obesity: A comparison in personality patterns and lifestyle. *Int. J. Addict.* **1990**, *25*, 803–817. [CrossRef]
73. Davis, C.; Levitan, R.D.; Carter, J.; Kaplan, A.S.; Reid, C.; Curtis, C.; Patte, K.; Kennedy, J.L. Personality and eating behaviors: A case-control study of binge eating disorder. *Int. J. Eat. Disord.* **2008**, *41*, 243–250. [CrossRef]
74. Rydén, A.; Sullivan, M.; Torgerson, J.S.; Karlsson, J.; Lindroos, A.-K.; Taft, C. Severe obesity and personality: A comparative controlled study of personality traits. *Int. J. Obes.* **2003**, *27*, 1534–1540. [CrossRef]
75. Batterink, L.; Yokum, S.; Stice, E. Body mass correlates inversely with inhibitory control in response to food among adolescent girls: An fMRI study. *NeuroImage* **2010**, *52*, 1696–1703. [CrossRef]
76. Davis, C.; Patte, K.; Curtis, C.; Reid, C. Immediate pleasures and future consequences. A neuropsychological study of binge eating and obesity. *Appetite* **2010**, *54*, 208–213. [CrossRef] [PubMed]
77. Nederkoorn, C.; Braet, C.; Van Eijs, Y.; Tanghe, A.; Jansen, A. Why obese children cannot resist food: The role of impulsivity. *Eat. Behav.* **2006**, *7*, 315–322. [CrossRef] [PubMed]
78. Friese, M.; Hofmann, W. Control me or I will control you: Impulses, trait self-control, and the guidance of behavior. *J. Res. Personal.* **2009**, *43*, 795–805. [CrossRef]
79. Friese, M.; Hofmann, W.; Wänke, M. When impulses take over: Moderated predictive validity of explicit and implicit attitude measures in predicting food choice and consumption behaviour. *Br. J. Soc. Psychol.* **2008**, *47*, 397–419. [CrossRef] [PubMed]
80. Hofmann, W.; Friese, M.; Roefs, A. Three ways to resist temptation: The independent contributions of executive attention, inhibitory control, and affect regulation to the impulse control of eating behavior. *J. Exp. Soc. Psychol.* **2009**, *45*, 431–435. [CrossRef]

81. Crescioni, A.W.; Ehrlinger, J.; Alquist, J.L.; Conlon, K.E.; Baumeister, R.F.; Schatschneider, C.; Dutton, G.R. High trait self-control predicts positive health behaviors and success in weight loss. *J. Health Psychol.* **2011**, *16*, 750–759. [CrossRef] [PubMed]
82. De Boer, B.J.; van Hooft, E.A.J.; Bakker, A.B. Stop and start control: A distinction within self-control. *Eur. J. Personal.* **2011**, *25*, 349–362. [CrossRef]
83. Gerrits, J.H.; O'Hara, R.E.; Piko, B.F.; Gibbons, F.X.; De Ridder, D.T.D.; Keresztes, N.; Kamble, S.V.; De Wit, J.B.F. Self-control, diet concerns and eater prototypes influence fatty foods consumption of adolescents in three countries. *Health Educ. Res.* **2010**, *25*, 1031–1041. [CrossRef]
84. Jasinska, A.J.; Yasuda, M.; Burant, C.F.; Gregor, N.; Khatri, S.; Sweet, M.; Falk, E.B. Impulsivity and inhibitory control deficits are associated with unhealthy eating in young adults. *Appetite* **2012**, *59*, 738–747. [CrossRef]
85. Allan, J.L.; Johnston, M.; Campbell, N. Unintentional eating. What determines goal-incongruent chocolate consumption? *Appetite* **2010**, *54*, 422–425. [CrossRef]
86. Allom, V.; Mullan, B. Individual differences in executive function predict distinct eating behaviours. *Appetite* **2014**, *80*, 123–130. [CrossRef] [PubMed]
87. Guerrieri, R.; Nederkoorn, C.; Jansen, A. How impulsiveness and variety influence food intake in a sample of healthy women. *Appetite* **2007**, *48*, 119–122. [CrossRef] [PubMed]
88. Churchill, S.; Jessop, D.C. Reflective and non-reflective antecedents of health-related behaviour: Exploring the relative contributions of impulsivity and implicit self-control to the prediction of dietary behaviour. *Br. J. Health Psychol.* **2011**, *16*, 257–272. [CrossRef] [PubMed]
89. Meule, A.; De Zwaan, M.; Müller, A. Attentional and motor impulsivity interactively predict 'food addiction' in obese individuals. *Compr. Psychiatr.* **2017**, *72*, 83–87. [CrossRef] [PubMed]
90. VanderBroek-Stice, L.; Stojek, M.K.; Beach, S.R.H.; Vandellen, M.R.; MacKillop, J. Multidimensional assessment of impulsivity in relation to obesity and food addiction. *Appetite* **2017**, *112*, 59–68. [CrossRef] [PubMed]
91. Appelhans, B.M.; Woolf, K.; Pagoto, S.L.; Schneider, K.L.; Whited, M.C.; Liebman, R. Inhibiting food reward: Delay discounting, food reward sensitivity, and palatable food intake in overweight and obese women. *Obesity* **2011**, *19*, 2175–2182. [CrossRef] [PubMed]
92. Dissabandara, L.O.; Loxton, N.J.; Dias, S.R.; Dodd, P.R.; Daglish, M.; Stadlin, A. Dependent heroin use and associated risky behaviour: The role of rash impulsiveness and reward sensitivity. *Addict. Behav.* **2014**, *39*, 71–76. [CrossRef]
93. Franken, I.H.; Muris, P. Individual differences in reward sensitivity are related to food craving and relative body weight in healthy women. *Appetite* **2005**, *45*, 198–201. [CrossRef]
94. Davis, C.; Patte, K.; Levitan, R.; Reid, C.; Tweed, S.; Curtis, C. From motivation to behaviour: A model of reward sensitivity, overeating, and food preferences in the risk profile for obesity. *Appetite* **2007**, *48*, 12–19. [CrossRef]
95. De Cock, N.; Van Lippevelde, W.; Goossens, L.; De Clercq, B.; Vangeel, J.; Lachat, C.; Beullens, K.; Huybregts, L.; Vervoort, L.; Eggermont, S.; et al. Sensitivity to reward and adolescents' unhealthy snacking and drinking behavior: The role of hedonic eating styles and availability. *Int. J. Behav. Nutr. Phys. Act.* **2016**, *13*, 17. [CrossRef]
96. Lawrence, N.S.; Hinton, E.C.; Parkinson, J.A.; Lawrence, A.D. Nucleus accumbens response to food cues predicts subsequent snack consumption in women and increased body mass index in those with reduced self-control. *NeuroImage* **2012**, *63*, 415–422. [CrossRef] [PubMed]
97. Davis, C.; Strachan, S.; Berkson, M. Sensitivity to reward: Implications for overeating and overweight. *Appetite* **2004**, *42*, 131–138. [CrossRef] [PubMed]
98. Volkow, N.D.; Fowler, J.S.; Wang, G.-J. The addicted human brain: Insights from imaging studies. *J. Clin. Investig.* **2003**, *111*, 1444–1451. [CrossRef] [PubMed]
99. Volkow, N.D.; Wang, G.-J.; Telang, F.; Fowler, J.S.; Thanos, P.K.; Logan, J.; Alexoff, D.; Ding, Y.-S.; Wong, C.; Ma, Y.; et al. Low dopamine striatal D2 receptors are associated with prefrontal metabolism in obese subjects: Possible contributing factors. *NeuroImage* **2008**, *42*, 1537–1543. [CrossRef] [PubMed]
100. Wang, G.J.; Volkow, N.D.; Logan, J.; Pappas, N.R.; Wong, C.T.; Zhu, W.; Netusil, N.; Fowler, J.S. Brain dopamine and obesity. *Lancet* **2001**, *357*, 354–357. [CrossRef]
101. Blum, K.; Braverman, E.R.; Holder, J.M.; Lubar, J.F.; Monastra, V.J.; Miller, D.; Lubar, J.O.; Chen, T.J.; Comings, D.E. The reward deficiency syndrome: A biogenetic model for the diagnosis and treatment of impulsive, addictive and compulsive behaviors. *J. Psychoact. Drugs* **2000**, *32*, 1–112. [CrossRef]

102. Bowirrat, A.; Oscar-Berman, M. Relationship between dopaminergic neurotransmission, alcoholism, and reward deficiency syndrome. *Am. J. Med. Genet. Part B Neuropsychiatr. Genet.* **2005**, *132*, 29–37. [CrossRef]
103. Burger, K.S.; Stice, E. Variability in reward responsivity and obesity: Evidence from brain imaging studies. *Curr. Drug Abus. Rev.* **2011**, *4*, 182–189. [CrossRef]
104. Robinson, T. The neural basis of drug craving: An incentive-sensitization theory of addiction. *Brain Res. Rev.* **1993**, *18*, 247–291. [CrossRef]
105. Robinson, T.E.; Berridge, K.C. Incentive-sensitization and addiction. *Addiction* **2001**, *96*, 103–114. [CrossRef]
106. Robinson, T.E.; Berridge, K.C. Addiction. *Annu. Rev. Psychol.* **2003**, *54*, 25–53. [CrossRef] [PubMed]
107. Berridge, K.C. 'Liking' and 'wanting' food rewards: Brain substrates and roles in eating disorders. *Physiol. Behav.* **2009**, *97*, 537–550. [CrossRef] [PubMed]
108. Berridge, K.C.; Ho, C.-Y.; Richard, J.M.; DiFeliceantonio, A.G. The tempted brain eats: Pleasure and desire circuits in obesity and eating disorders. *Brain Res.* **2010**, *1350*, 43–64. [CrossRef] [PubMed]
109. Hetherington, M.; Pirie, L.; Nabb, S.; Hetherington, M. Stimulus satiation: Effects of repeated exposure to foods on pleasantness and intake. *Appetite* **2002**, *38*, 19–28. [CrossRef] [PubMed]
110. Ersche, K.D.; Jones, P.S.; Williams, G.B.; Turton, A.J.; Robbins, T.W.; Bullmore, E.T. Abnormal brain structure implicated in stimulant drug addiction. *Science* **2012**, *335*, 601–604. [CrossRef] [PubMed]
111. Everitt, B.J.; Belin, D.; Economidou, D.; Pelloux, Y.; Dalley, J.W.; Robbins, T.W. Neural mechanisms underlying the vulnerability to develop compulsive drug-seeking habits and addiction. *Philos. Trans. R. Soc. B Biol. Sci.* **2008**, *363*, 3125–3135. [CrossRef] [PubMed]
112. Flagel, S.B.; Akil, H.; Robinson, T.E. Individual differences in the attribution of incentive salience to reward-related cues: Implications for addiction. *Neuropharmacology* **2009**, *56*, 139–148. [CrossRef]
113. Saunders, B.T.; Robinson, T.E. Individual variation in resisting temptation: Implications for addiction. *Neurosci. Biobehav. Rev.* **2013**, *37*, 1955–1975. [CrossRef] [PubMed]
114. Kekic, M.; McClelland, J.; Campbell, I.; Nestler, S.; Rubia, K.; David, A.S.; Schmidt, U. The effects of prefrontal cortex transcranial direct current stimulation (tDCS) on food craving and temporal discounting in women with frequent food cravings. *Appetite* **2014**, *78*, 55–62. [CrossRef]
115. Weingarten, H.P.; Elston, D. The phenomenology of food cravings. *Appetite* **1990**, *15*, 231–246. [CrossRef]
116. Pelchat, M.L. Food cravings in young and elderly adults. *Appetite* **1997**, *28*, 103–113. [CrossRef] [PubMed]
117. Lopez, R.B.; Hofmann, W.; Wagner, D.D.; Kelley, W.M.; Heatherton, T.F. Neural predictors of giving in to temptation in daily life. *Psychol. Sci.* **2014**, *25*, 1337–1344. [CrossRef] [PubMed]
118. Cocores, J.A.; Gold, M.S. The Salted Food Addiction Hypothesis may explain overeating and the obesity epidemic. *Med. Hypotheses* **2009**, *73*, 892–899. [CrossRef] [PubMed]
119. Corsica, J.A.; Spring, B.J. Carbohydrate craving: A double-blind, placebo controlled test of the self-medication hypothesis. *Eat. Behav.* **2008**, *9*, 447–454. [CrossRef] [PubMed]
120. Hill, A.J.; Heaton-Brown, L. The experience of food craving: A prospective investigation in healthy women. *J. Psychosom. Res.* **1994**, *38*, 801–814. [CrossRef]
121. Massey, A.; Hill, A.J. Dieting and food craving. A descriptive, quasi-prospective study. *Appetite* **2012**, *58*, 781–785. [CrossRef]
122. Rozin, P.; Levine, E.; Stoess, C. Chocolate craving and liking. *Appetite* **1991**, *17*, 199–212. [CrossRef]
123. Cepeda-Benito, A.; Gleaves, D.H.; Williams, T.L.; Erath, S.A. The development and validation of the state and trait food-cravings questionnaires. *Behav. Ther.* **2000**, *31*, 151–173. [CrossRef]
124. Hill, A.J.; Weaver, C.F.; Blundell, J.E. Food craving, dietary restraint and mood. *Appetite* **1991**, *17*, 187–197. [CrossRef]
125. Nicholls, W.; Hulbert-Williams, L. British english translation of the food craving inventory (FCI-UK). *Appetite* **2013**, *67*, 37–43. [CrossRef]
126. Nijs, I.M.; Franken, I.H.; Muris, P. The modified trait and state food-cravings questionnaires: Development and validation of a general index of food craving. *Appetite* **2007**, *49*, 38–46. [CrossRef] [PubMed]
127. White, M.A.; Whisenhunt, B.L.; Williamson, D.A.; Greenway, F.L.; Netemeyer, R.G. Development and validation of the food-craving inventory. *Obes. Res.* **2002**, *10*, 107–114. [CrossRef] [PubMed]
128. Benton, D.; Greenfield, K.; Morgan, M. The development of the attitudes to chocolate questionnaire. *Personal. Individ. Differ.* **1998**, *24*, 513–520. [CrossRef]
129. Boswell, R.G.; Kober, H. Food cue reactivity and craving predict eating and weight gain: A meta-analytic review. *Obes. Rev.* **2016**, *17*, 159–177. [CrossRef] [PubMed]

130. Burton, P.; Smit, H.J.; Lightowler, H.J. The influence of restrained and external eating patterns on overeating. *Appetite* **2007**, *49*, 191–197. [CrossRef] [PubMed]
131. Dalton, M.; Blundell, J.; Finlayson, G.S. Examination of food reward and energy intake under laboratory and free-living conditions in a trait binge eating subtype of obesity. *Front. Psychol.* **2013**, *4*, 757. [CrossRef] [PubMed]
132. Lafay, L.; Mennen, L.; Charles, M.A.; Eschwège, E.; Borys, J.-M. Gender differences in the relation between food cravings and mood in an adult community: Results from the Fleurbaix Laventie Ville Sant study. *Int. J. Eat. Disord.* **2001**, *29*, 195–204. [CrossRef]
133. Davis, C.; Curtis, C.; Levitan, R.D.; Carter, J.C.; Kaplan, A.S.; Kennedy, J.L. Evidence that 'food addiction' is a valid phenotype of obesity. *Appetite* **2011**, *57*, 711–717. [CrossRef]
134. Meule, A. Food cravings in food addiction: Exploring a potential cut-off value of the food cravings questionnaire-trait-reduced. *Eat. Weight Dis. Stud. Anorex. Bulim. Obes.* **2018**, *23*, 39–43. [CrossRef]
135. Meule, A.; Kübler, A. Food cravings in food addiction: The distinct role of positive reinforcement. *Eat. Behav.* **2012**, *13*, 252–255. [CrossRef]
136. Mussell, M.P.; Mitchell, J.E.; De Zwaan, M.; Crosby, R.D.; Seim, H.C.; Crow, S.J. Clinical characteristics associated with binge eating in obese females: A descriptive study. *Int. J. Obes. Relat. Metab. Disord. J. Int. Assoc. Stud. Obes.* **1996**, *20*, 324–331.
137. Ng, L.; Davis, C. Cravings and food consumption in binge eating disorder. *Eat. Behav.* **2013**, *14*, 472–475. [CrossRef]
138. Van den Eynde, F.; Koskina, A.; Syrad, H.; Guillaume, S.; Broadbent, H.; Campbell, I.C.; Schmidt, U. State and trait food craving in people with bulimic eating disorders. *Eat. Behav.* **2012**, *13*, 414–417. [CrossRef] [PubMed]
139. Bottlender, M.; Soyka, M. Impact of craving on alcohol relapse during, and 12 months following, outpatient treatment. *Alcohol Alcohol.* **2004**, *39*, 357–361. [CrossRef] [PubMed]
140. Litt, M.D.; Cooney, N.L.; Morse, P. Reactivity to alcohol-related stimuli in the laboratory and in the field: Predictors of craving in treated alcoholics. *Addiction* **2000**, *95*, 889–900. [CrossRef] [PubMed]
141. Paliwal, P.; Hyman, S.M.; Sinha, R. Craving predicts time to cocaine relapse: Further validation of the now and brief versions of the cocaine craving questionnaire. *Drug Alcohol Depend.* **2008**, *93*, 252–259. [CrossRef] [PubMed]
142. Gendall, K.A.; Sullivan, P.F.; Joyce, P.R.; Fear, J.L.; Bulik, C.M. Psychopathology and personality of young women who experience food cravings. *Addict. Behav.* **1997**, *22*, 545–555. [CrossRef]
143. Meule, A.; Lutz, A.; Vögele, C.; Kübler, A. Food cravings discriminate differentially between successful and unsuccessful dieters and non-dieters. Validation of the food cravings questionnaires in German. *Appetite* **2012**, *58*, 88–97. [CrossRef]
144. Meule, A.; Westenhöfer, J.; Kübler, A. Food cravings mediate the relationship between rigid, but not flexible control of eating behavior and dieting success. *Appetite* **2011**, *57*, 582–584. [CrossRef]
145. Carter, B.L.; Tiffany, S.T. Meta-analysis of cue-reactivity in addiction research. *Addiction* **1999**, *94*, 327–340. [CrossRef]
146. Davidson, D.; Tiffany, S.T.; Johnston, W.; Flury, L.; Li, T.-K. Using the cue-availability paradigm to assess cue reactivity. *Alcohol. Clin. Exp. Res.* **2003**, *27*, 1251–1256. [CrossRef]
147. Mahler, S.V.; De Wit, H. Cue-Reactors: Individual differences in cue-induced craving after food or smoking abstinence. *PLoS ONE* **2010**, *5*, e15475. [CrossRef] [PubMed]
148. Styn, M.A.; Bovbjerg, D.H.; Lipsky, S.; Erblich, J. Cue-induced cigarette and food craving: A common effect? *Addict. Behav.* **2013**, *38*, 1840–1843. [CrossRef] [PubMed]
149. Cornell, C.E.; Rodin, J.; Weingarten, H. Stimulus-induced eating when satiated. *Physiol. Behav.* **1989**, *45*, 695–704. [CrossRef]
150. Nederkoorn, C.; Smulders, F.; Jansen, A. Cephalic phase responses, craving and food intake in normal subjects. *Appetite* **2000**, *35*, 45–55. [CrossRef] [PubMed]
151. Meule, A.; Küppers, C.; Harms, L.; Friederich, H.-C.; Schmidt, U.; Blechert, J.; Brockmeyer, T. Food cue-induced craving in individuals with bulimia nervosa and binge-eating disorder. *PLoS ONE* **2018**, *13*, e0204151. [CrossRef] [PubMed]
152. Karhunen, L.J.; Lappalainen, R.I.; Tammela, L.; Turpeinen, A.K.; Uusitupa, M.I.J. Subjective and physiological cephalic phase responses to food in obese binge-eating women. *Int. J. Eat. Disord.* **1997**, *21*, 321–328. [CrossRef]

153. Sobik, L.; Hutchison, K.; Craighead, L. Cue-elicited craving for food: A fresh approach to the study of binge eating. *Appetite* **2005**, *44*, 253–261. [CrossRef]
154. Kjelgaard, H.H.; Holstein, B.E.; Due, P.; Brixval, C.S.; Rasmussen, M. Adolescent weight status: Associations with structural and functional dimensions of social relations. *J. Adolesc. Health* **2017**, *60*, 460–468. [CrossRef]
155. Schwimmer, J.B.; Burwinkle, T.M.; Varni, J.W. Health-Related quality of life of severely obese children and adolescents. *JAMA* **2003**, *289*, 1813. [CrossRef]
156. Williams, J.; Wake, M.; Hesketh, K.; Maher, E.; Waters, E. Health-Related quality of life of overweight and obese children. *JAMA* **2005**, *293*, 70–76. [CrossRef] [PubMed]
157. Griffiths, L.J.; Wolke, D.; Page, A.S.; Horwood, J.P. Obesity and bullying: Different effects for boys and girls. *Arch. Dis. Child.* **2006**, *91*, 121–125. [CrossRef] [PubMed]
158. Pearce, M.J.; Boergers, J.; Prinstein, M.J. Adolescent obesity, overt and relational peer victimization, and romantic relationships. *Obes. Res.* **2002**, *10*, 386–393. [CrossRef] [PubMed]
159. Hayden-Wade, H.A.; Stein, R.I.; Ghaderi, A.; Saelens, B.E.; Zabinski, M.F.; Wilfley, D.E. Prevalence, characteristics, and correlates of teasing experiences among overweight children vs. non-overweight peers. *Obes. Res.* **2005**, *13*, 1381–1392. [CrossRef] [PubMed]
160. Neumark-Sztainer, D.; Falkner, N.; Story, M.; Perry, C.; Hannan, P.J.; Mulert, S. Weight-teasing among adolescents: Correlations with weight status and disordered eating behaviors. *Int. J. Obes.* **2002**, *26*, 123–131. [CrossRef] [PubMed]
161. Puhl, R.; Brownell, K.D. Bias, Discrimination, and Obesity. *Obes. Res.* **2001**, *9*, 788–805. [CrossRef] [PubMed]
162. Puhl, R.M.; Heuer, C.A. The stigma of obesity: A review and update. *Obesity* **2009**, *17*, 941–964. [CrossRef] [PubMed]
163. Puhl, R.M.; Latner, J.D. Stigma, obesity, and the health of the nation's children. *Psychol. Bull.* **2007**, *133*, 557–580. [CrossRef]
164. Chen, E.Y.; Brown, M. Obesity stigma in sexual relationships. *Obes. Res.* **2005**, *13*, 1393–1397. [CrossRef]
165. Chao, A.M.; Shaw, J.A.; Pearl, R.L.; Alamuddin, N.; Hopkins, C.M.; Bakizada, Z.M.; Wadden, T.A. Prevalence and psychosocial correlates of food addiction in persons with obesity seeking weight reduction. *Compr. Psychiatr.* **2017**, *73*, 97–104. [CrossRef]
166. Blomquist, K.K.; Ansell, E.B.; White, M.A.; Masheb, R.M.; Grilo, C.M. Interpersonal problems and developmental trajectories of binge eating disorder. *Compr. Psychiatr.* **2012**, *53*, 1088–1095. [CrossRef] [PubMed]
167. Lo Coco, G.; Gullo, S.; Salerno, L.; Iacoponelli, R. The association among interpersonal problems, binge behaviors, and self-esteem, in the assessment of obese individuals. *Compr. Psychiatr.* **2011**, *52*, 164–170. [CrossRef] [PubMed]
168. Jia, H.; Lubetkin, E.I. Trends in quality-adjusted life-years lost contributed by smoking and obesity. *Am. J. Prev. Med.* **2010**, *38*, 138–144. [CrossRef] [PubMed]
169. Bray, G.A. Medical consequences of obesity. *J. Clin. Endocrinol. Metab.* **2004**, *89*, 2583–2589. [CrossRef] [PubMed]
170. Carpenter, K.M.; Hasin, D.S.; Allison, D.B.; Faith, M.S. Relationships between obesity and DSM-IV major depressive disorder, suicide ideation, and suicide attempts: Results from a general population study. *Am. J. Public Health* **2000**, *90*, 251–257. [PubMed]
171. Haslam, D.W.; James, W.P.T. Obesity. *Lancet* **2005**, *366*, 1197–1209. [CrossRef]
172. Kahn, S.E.; Hull, R.L.; Utzschneider, K.M. Mechanisms linking obesity to insulin resistance and type 2 diabetes. *Nature* **2006**, *444*, 840–846. [CrossRef] [PubMed]
173. Lopresti, A.L.; Drummond, P.D. Obesity and psychiatric disorders: Commonalities in dysregulated biological pathways and their implications for treatment. *Prog. Neuro Psychopharmacol. Biol. Psychiatr.* **2013**, *45*, 92–99. [CrossRef]
174. Luppino, F.S.; de Wit, L.M.; Bouvy, P.F.; Stijnen, T.; Cuijpers, P.; Penninx, B.W.J.H.; Zitman, F.G. Overweight, obesity, and depression: A systematic review and meta-analysis of longitudinal studies. *Arch. Gen. Psychiatr.* **2010**, *67*, 220–229. [CrossRef]
175. Mannan, M.; Mamun, A.; Doi, S.; Clavarino, A. Prospective associations between depression and obesity for adolescent males and females—A systematic review and meta-analysis of longitudinal studies. *PLoS ONE* **2016**, *11*, 0157240. [CrossRef]
176. Mokdad, A.H.; Ford, E.S.; Bowman, B.A.; Dietz, W.H.; Vinicor, F.; Bales, V.S.; Marks, J.S. Prevalence of obesity, diabetes, and obesity-related health risk factors, 2001. *JAMA* **2003**, *289*, 76–79. [CrossRef] [PubMed]

177. Van Gaal, L.F.; Mertens, I.L.; De Block, C.E. Mechanisms linking obesity with cardiovascular disease. *Nature* **2006**, *444*, 875–880. [CrossRef] [PubMed]
178. Elkins, G.; Whitfield, P.; Marcus, J.; Symmonds, R.; Rodriguez, J.; Cook, T. Noncompliance with behavioral recommendations following bariatric surgery. *Obes. Surg.* **2005**, *15*, 546–551. [CrossRef]
179. Toussi, R.; Fujioka, K.; Coleman, K.J. Pre- and postsurgery behavioral compliance, patient health, and postbariatric surgical weight loss. *Obesity* **2009**, *17*, 996–1002. [CrossRef] [PubMed]
180. Colantuoni, C.; Schwenker, J.; McCarthy, J.; Rada, P.; Ladenheim, B.; Cadet, J.-L.; Schwartz, G.J.; Moran, T.H.; Hoebel, B.G. Excessive sugar intake alters binding to dopamine and mu-opioid receptors in the brain. *NeuroReport* **2001**, *12*, 3549–3552. [CrossRef]
181. Rada, P.; Avena, N.; Hoebel, B. Daily bingeing on sugar repeatedly releases dopamine in the accumbens shell. *Neuroscience* **2005**, *134*, 737–744. [CrossRef] [PubMed]
182. Harrison, D.M. Oral sucrose for pain management in infants: Myths and misconceptions. *J. Neonatal Nurs.* **2008**, *14*, 39–46. [CrossRef]
183. King, J.M. Patterns of sugar consumption in early infancy. *Community Dent. Oral Epidemiol.* **1978**, *6*, 47–52. [CrossRef]
184. Rossow, I.; Kjaernes, U.; Holst, D. Patterns of sugar consumption in early childhood. *Community Dent. Oral Epidemiol.* **1990**, *18*, 12–16. [CrossRef]
185. Slater, R.; Cornelissen, L.; Fabrizi, L.; Patten, D.; Yoxen, J.; Worley, A.; Boyd, S.; Meek, J.; Fitzgerald, M. Oral sucrose as an analgesic drug for procedural pain in newborn infants: A randomised controlled trial. *Lancet* **2010**, *376*, 1225–1232. [CrossRef]
186. Nielsen, S.J.; Popkin, B.M. Patterns and trends in food portion sizes, 1977–1998. *JAMA* **2003**, *289*, 450–453. [CrossRef] [PubMed]
187. Wideman, C.H.; Nadzam, G.R.; Murphy, H.M. Implications of an animal model of sugar addiction, withdrawal and relapse for human health. *Nutr. Neurosci.* **2005**, *8*, 269–276. [CrossRef] [PubMed]
188. Avena, N.M.; Rada, P.; Hoebel, B.G. Evidence for sugar addiction: Behavioral and neurochemical effects of intermittent, excessive sugar intake. *Neurosci. Biobehav. Rev.* **2008**, *32*, 20–39. [CrossRef] [PubMed]
189. Galic, M.A.; Persinger, M.A. Voluminous sucrose consumption in female rats: Increased "nippiness" during periods of sucrose removal and possible oestrus periodicity. *Psychol. Rep.* **2002**, *90*, 58–60. [CrossRef] [PubMed]
190. Avena, N.M.; Bocarsly, M.E.; Rada, P.; Kim, A.; Hoebel, B.G. After daily bingeing on a sucrose solution, food deprivation induces anxiety and accumbens dopamine/acetylcholine imbalance. *Physiol. Behav.* **2008**, *94*, 309–315. [CrossRef] [PubMed]
191. Hetherington, M.M.; MacDiarmid, J.I. "Chocolate addiction": A preliminary study of its description and its relationship to problem eating. *Appetite* **1993**, *21*, 233–246. [CrossRef] [PubMed]
192. Carnell, S.; Gibson, C.; Benson, L.; Ochner, C.N.; Geliebter, A. Neuroimaging and obesity: Current knowledge and future directions. *Obes. Rev.* **2012**, *13*, 43–56. [CrossRef]
193. Parvaz, M.A.; Alia-Klein, N.; Woicik, P.A.; Volkow, N.D.; Goldstein, R.Z. Neuroimaging for drug addiction and related behaviors. *Rev. Neurosci.* **2011**, *22*, 609–624. [CrossRef] [PubMed]
194. Zhang, Y.; von Deneen, K.M.; Tian, J.; Gold, M.S.; Liu, Y. Food addiction and neuroimaging. *Curr. Pharm. Des.* **2011**, *17*, 1149–1157. [CrossRef] [PubMed]
195. Goldstein, R.Z.; Volkow, N.D. Drug addiction and its underlying neurobiological basis: Neuroimaging evidence for the involvement of the frontal cortex. *Am. J. Psychiatr.* **2002**, *159*, 1642–1652. [CrossRef]
196. Goldstein, R.Z.; Volkow, N.D. Dysfunction of the prefrontal cortex in addiction: Neuroimaging findings and clinical implications. *Nat. Rev. Neurosci.* **2011**, *12*, 652–669. [CrossRef] [PubMed]
197. Koob, G.F.; Volkow, N.D. Neurobiology of addiction: A neurocircuitry analysis. *Lancet Psychiatr.* **2016**, *3*, 760–773. [CrossRef]
198. Volkow, N.D.; Wang, G.-J.; Fowler, J.S.; Telang, F. Overlapping neuronal circuits in addiction and obesity: Evidence of systems pathology. *Philos. Trans. R. Soc. B Biol. Sci.* **2008**, *363*, 3191–3200. [CrossRef] [PubMed]
199. Volkow, N.D.; Wang, G.-J.; Tomasi, D.; Baler, R.D. The addictive dimensionality of obesity. *Biol. Psychiatr.* **2013**, *73*, 811–818. [CrossRef] [PubMed]
200. Pierce, R.C.; Kumaresan, V. The mesolimbic dopamine system: The final common pathway for the reinforcing effect of drugs of abuse? *Neurosci. Biobehav. Rev.* **2006**, *30*, 215–238. [CrossRef] [PubMed]

201. Hernández, L.; Hoebel, B.G. Food reward and cocaine increase extracellular dopamine in the nucleus accumbens as measured by microdialysis. *Life Sci.* **1988**, *42*, 1705–1712. [CrossRef]
202. Radhakishun, F.S.; Van Ree, J.M.; Westerink, B.H. Scheduled eating increases dopamine release in the nucleus accumbens of food-deprived rats as assessed with on-line brain dialysis. *Neurosci. Lett.* **1988**, *85*, 351–356. [CrossRef]
203. Yoshida, M.; Yokoo, H.; Mizoguchi, K.; Kawahara, H.; Tsuda, A.; Nishikawa, T.; Tanaka, M. Eating and drinking cause increased dopamine release in the nucleus accumbens and ventral tegmental area in the rat: Measurement by in vivo microdialysis. *Neurosci. Lett.* **1992**, *139*, 73–76. [CrossRef]
204. Martel, P.; Fantino, M. Influence of the amount of food ingested on mesolimbic dopaminergic system activity: A microdialysis study. *Pharmacol. Biochem. Behav.* **1996**, *55*, 297–302. [CrossRef]
205. Martel, P.; Fantino, M. Mesolimbic dopaminergic system activity as a function of food reward: A microdialysis study. *Pharmacol. Biochem. Behav.* **1996**, *53*, 221–226. [CrossRef]
206. Carelli, R.M.; Ijames, S.G.; Crumling, A.J. Evidence that separate neural circuits in the nucleus accumbens encode cocaine versus "natural" (water and food) reward. *J. Neurosci.* **2000**, *20*, 4255–4266. [CrossRef] [PubMed]
207. Caine, S.B.; Koob, G.F. Effects of mesolimbic dopamine depletion on responding maintained by cocaine and food. *J. Exp. Anal. Behav.* **1994**, *61*, 213–221. [CrossRef] [PubMed]
208. Rothemund, Y.; Preuschhof, C.; Bohner, G.; Bauknecht, H.-C.; Klingebiel, R.; Flor, H.; Klapp, B.F. Differential activation of the dorsal striatum by high-calorie visual food stimuli in obese individuals. *NeuroImage* **2007**, *37*, 410–421. [CrossRef]
209. Small, D.M.; Jones-Gotman, M.; Dagher, A. Feeding-induced dopamine release in dorsal striatum correlates with meal pleasantness ratings in healthy human volunteers. *NeuroImage* **2003**, *19*, 1709–1715. [CrossRef]
210. Volkow, N.D.; Wang, G.-J.; Fowler, J.S.; Logan, J.; Jayne, M.; Franceschi, D.; Wong, C.; Gatley, S.J.; Gifford, A.N.; Ding, Y.-S.; et al. Nonhedonic food motivation in humans involves dopamine in the dorsal striatum and methylphenidate amplifies this effect. *Synapse* **2002**, *44*, 175–180. [CrossRef] [PubMed]
211. Volkow, N.D.; Wang, G.-J.; Maynard, L.; Jayne, M.; Fowler, J.S.; Zhu, W.; Logan, J.; Gatley, S.J.; Ding, Y.-S.; Wong, C.; et al. Brain dopamine is associated with eating behaviors in humans. *Int. J. Eat. Disord.* **2003**, *33*, 136–142. [CrossRef] [PubMed]
212. Stoeckel, L.E.; Weller, R.E.; Cook, E.W.; Twieg, D.B.; Knowlton, R.C.; Cox, J.E.; Iii, E.W.C. Widespread reward-system activation in obese women in response to pictures of high-calorie foods. *NeuroImage* **2008**, *41*, 636–647. [CrossRef]
213. Stice, E.; Spoor, S.; Bohon, C.; Small, D.M. Relation between obesity and blunted striatal response to food is moderated by TaqIA A1 allele. *Science* **2008**, *322*, 449–452. [CrossRef]
214. Wang, G.-J.; Volkow, N.D.; Felder, C.; Fowler, J.S.; Levy, A.V.; Pappas, N.R.; Wong, C.T.; Zhu, W.; Netusil, N. Enhanced resting activity of the oral somatosensory cortex in obese subjects. *NeuroReport* **2002**, *13*, 1151–1155. [CrossRef] [PubMed]
215. Bello, N.T.; Lucas, L.R.; Hajnal, A. Repeated sucrose access influences dopamine D2 receptor density in the striatum. *NeuroReport* **2002**, *13*, 1575–1578. [CrossRef] [PubMed]
216. Johnson, P.M.; Kenny, P.J. Dopamine D2 receptors in addiction-like reward dysfunction and compulsive eating in obese rats. *Nat. Neurosci.* **2010**, *13*, 635–641. [CrossRef] [PubMed]
217. Volkow, N.D.; Wang, G.-J.; Fowler, J.S.; Tomasi, D.; Telang, F.; Baler, R. Addiction: Decreased reward sensitivity and increased expectation sensitivity conspire to overwhelm the brain's control circuit. *BioEssays* **2010**, *32*, 748–755. [CrossRef] [PubMed]
218. Fehr, C.; Yakushev, I.; Hohmann, N.; Buchholz, H.G.; Landvogt, C.; Deckers, H.; Eberhardt, A.; Smolka, M.N.; Scheurich, A.; Dielentheis, T.; et al. Association of low striatal dopamine D 2 receptor availability with nicotine dependence similar to that seen with other drugs of abuse. *Am. J. Psychiatr.* **2008**, *165*, 507–514. [CrossRef] [PubMed]
219. Heinz, A. Correlation Between Dopamine D2 Receptors in the Ventral Striatum and Central Processing of Alcohol Cues and Craving. *Am. J. Psychiatr.* **2004**, *161*, 1783–1789. [CrossRef] [PubMed]
220. Volkow, N.D.; Fowler, J.S.; Wang, G.-J.; Hitzemann, R.; Logan, J.; Schlyer, D.J.; Dewey, S.L.; Wolf, A.P. Decreased dopamine D2 receptor availability is associated with reduced frontal metabolism in cocaine abusers. *Synapse* **1993**, *14*, 169–177. [CrossRef]

221. Volkow, N.D.; Wang, G.-J.; Fowler, J.S.; Logan, J.; Hitzemann, R.; Ding, Y.-S.; Pappas, N.; Shea, C.; Piscani, K. Decreases in dopamine receptors but not in dopamine transporters in alcoholics. *Alcohol. Clin. Exp. Res.* **1996**, *20*, 1594–1598. [CrossRef] [PubMed]
222. Volkow, N.D.; Chang, L.; Wang, G.-J.; Fowler, J.S.; Ding, Y.-S.; Sedler, M.; Logan, J.; Franceschi, D.; Gatley, J.; Hitzemann, R.; et al. Low level of brain dopamine D2 receptors in methamphetamine abusers: Association with metabolism in the orbitofrontal cortex. *Am. J. Psychiatr.* **2001**, *158*, 2015–2021. [CrossRef]
223. Thanos, P.K.; Volkow, N.D.; Freimuth, P.; Umegaki, H.; Ikari, H.; Roth, G.; Ingram, D.K.; Hitzemann, R. Overexpression of dopamine D2 receptors reduces alcohol. *J. Neurochem.* **2001**, *78*, 1094–1103. [CrossRef]
224. Volkow, N.D.; Wang, G.-J.; Begleiter, H.; Porjesz, B.; Fowler, J.S.; Telang, F.; Wong, C.; Ma, Y.; Logan, J.; Goldstein, R.; et al. High levels of dopamine D2 receptors in unaffected members of alcoholic families. *Arch. Gen. Psychiatr.* **2006**, *63*, 999–1008. [CrossRef]
225. Reinholz, J.; Skopp, O.; Breitenstein, C.; Bohr, I.; Winterhoff, H.; Knecht, S. Compensatory weight gain due to dopaminergic hypofunction: New evidence and own incidental observations. *Nutr. Metab.* **2008**, *5*, 35. [CrossRef]
226. Wang, G.; Volkow, N.D.; Thanos, P.K.; Fowler, J.S. Similarity between obesity and drug addiction as assessed by neurofunctional imaging: A concept review. *J. Addict. Dis.* **2004**, *23*, 39–53. [CrossRef] [PubMed]
227. Hardman, C.A.; Herbert, V.M.; Brunstrom, J.M.; Munafò, M.R.; Rogers, P.J.; Brunstrom, J. Dopamine and food reward: Effects of acute tyrosine/phenylalanine depletion on appetite. *Physiol. Behav.* **2012**, *105*, 1202–1207. [CrossRef] [PubMed]
228. Geiger, B.M.; Haburcak, M.; Avena, N.M.; Moyer, M.C.; Hoebel, B.G.; Pothos, E.N. Deficits of mesolimbic dopamine neurotransmission in rat dietary obesity. *Neuroscience* **2009**, *159*, 1193–1199. [CrossRef] [PubMed]
229. Baptista, T.; Parada, M.; Hernández, L. Long term administration of some antipsychotic drugs increases body weight and feeding in rats. Are D2 dopamine receptors involved? *Pharmacol. Biochem. Behav.* **1987**, *27*, 399–405. [CrossRef]
230. Clifton, P.G.; Rusk, I.N.; Cooper, S.J. Effects of dopamine D1 and dopamine D2 antagonists on the free feeding and drinking patterns of rats. *Behav. Neurosci.* **1991**, *105*, 272–281. [CrossRef]
231. Scislowski, P.; Tozzo, E.; Zhang, Y.; Phaneuf, S.; Prevelige, R.; Cincotta, A. Biochemical mechanisms responsible for the attenuation of diabetic and obese conditions in ob/ob mice treated with dopaminergic agonists. *Int. J. Obes.* **1999**, *23*, 425–431. [CrossRef]
232. Goudie, A.J.; Cooper, G.D.; Halford, J.C.G. Antipsychotic-induced weight gain. *Diabetes Obes. Metab.* **2005**, *7*, 478–487. [CrossRef]
233. Cincotta, A.H.; Meier, A.H. Bromocriptine (Ergoset) reduces body weight and improves glucose tolerance in obese subjects. *Diabetes Care* **1996**, *19*, 667–670. [CrossRef]
234. Manning, P.J.; Grattan, D.; Merriman, T.; Manning, T.; Williams, S.; Sutherland, W. Pharmaceutical interventions for weight-loss maintenance: No effect from cabergoline. *Int. J. Obes.* **2018**, *42*, 1871–1879. [CrossRef]
235. Gibson, C.D.; Karmally, W.; McMahon, D.J.; Wardlaw, S.L.; Korner, J. Randomized pilot study of cabergoline, a dopamine receptor agonist: Effects on body weight and glucose tolerance in obese adults. *Diabetes Obes. Metab.* **2012**, *14*, 335–340. [CrossRef]
236. Frank, G.K.W.; Shott, M.E.; Hagman, J.O.; Schiel, M.A.; DeGuzman, M.C.; Rossi, B. The Partial Dopamine D2 Receptor Agonist Aripiprazole is Associated With Weight Gain in Adolescent Anorexia Nervosa. *Int. J Eat. Disord.* **2017**, *50*, 447–450. [CrossRef] [PubMed]
237. Steele, K.E.; Prokopowicz, G.P.; Schweitzer, M.A.; Magunsuon, T.H.; Lidor, A.O.; Kuwabawa, H.; Kuma, A.; Brasic, J.; Wong, D.F. Alterations of central dopamine receptors before and after gastric bypass surgery. *Obes. Surg.* **2010**, *20*, 369–374. [CrossRef] [PubMed]
238. Van Der Zwaal, E.M.; De Weijer, B.A.; Van De Giessen, E.M.; Janssen, I.; Berends, F.J.; Van De Laar, A.; Ackermans, M.T.; Fliers, E.; La Fleur, S.E.; Booij, J.; et al. Striatal dopamine D2/3 receptor availability increases after long-term bariatric surgery-induced weight loss. *Eur. Neuropsychopharmacol.* **2016**, *26*, 1190–1200. [CrossRef] [PubMed]
239. Ochner, C.N.; Kwok, Y.; Conceição, E.; Pantazatos, S.P.; Puma, L.M.; Carnell, S.; Teixeira, J.; Hirsch, J.; Geliebter, A. Selective reduction in neural responses to high calorie foods following gastric bypass surgery. *Ann. Surg.* **2011**, *253*, 502–507. [CrossRef] [PubMed]

240. Ghahremani, D.G.; Lee, B.; Robertson, C.L.; Tabibnia, G.; Morgan, A.T.; De Shetler, N.; Brown, A.K.; Monterosso, J.R.; Aron, A.A.; Mandelkern, M.A.; et al. Striatal dopamine D_2/D_3 receptors mediate response inhibition and related activity in frontostriatal neural circuitry in humans. *J. Neurosci.* **2012**, *32*, 7316–7324. [CrossRef] [PubMed]
241. Lavagnino, L.; Arnone, D.; Cao, B.; Soares, J.C.; Selvaraj, S. Inhibitory control in obesity and binge eating disorder: A systematic review and meta-analysis of neurocognitive and neuroimaging studies. *Neurosci. Biobehav. Rev.* **2016**, *68*, 714–726. [CrossRef] [PubMed]
242. Volkow, N.D.; Wang, G.-J.; Telang, F.; Fowler, J.S.; Goldstein, R.Z.; Alia-Klein, N.; Logan, J.; Wong, C.; Thanos, P.K.; Ma, Y.; et al. Inverse association between BMI and prefrontal metabolic activity in healthy adults. *Obesity* **2009**, *17*, 60–65. [CrossRef] [PubMed]
243. Le, D.S.N.T.; Pannacciulli, N.; Chen, K.; Del Parigi, A.; Salbe, A.D.; Reiman, E.M.; Krakoff, J. Less activation of the left dorsolateral prefrontal cortex in response to a meal: A feature of obesity. *Am. J. Clin. Nutr.* **2006**, *84*, 725–731. [CrossRef]
244. Le, D.S.N.; Pannacciulli, N.; Chen, K.; Salbe, A.D.; Hill, O.J.; Wing, R.R.; Reiman, E.M.; Krakoff, J. Less activation in the left dorsolateral prefrontal cortex in the reanalysis of the response to a meal in obese than in lean women and its association with successful weight loss. *Am. J. Clin. Nutr.* **2007**, *86*, 573–579. [CrossRef]
245. Gautier, J.F.; Chen, K.; Salbe, A.D.; Bandy, D.; Pratley, R.E.; Heiman, M.; Ravussin, E.; Reiman, E.M.; Tataranni, P.A. Differential brain responses to satiation in obese and lean men. *Diabetes* **2000**, *49*, 838–846. [CrossRef]
246. Del Parigi, A.; Chen, K.; Salbe, A.D.; Hill, J.O.; Wing, R.R.; Reiman, E.M.; Tataranni, P.A. Successful dieters have increased neural activity in cortical areas involved in the control of behavior. *Int. J. Obes.* **2007**, *31*, 440–448. [CrossRef] [PubMed]
247. Hollmann, M.; Hellrung, L.; Pleger, B.; Schlögl, H.; Kabisch, S.; Stumvoll, M.; Villringer, A.; Horstmann, A. Neural correlates of the volitional regulation of the desire for food. *Int. J. Obes.* **2012**, *36*, 648–655. [CrossRef] [PubMed]
248. McCaffery, J.M.; Haley, A.P.; Sweet, L.H.; Phelan, S.; Raynor, H.A.; Del Parigi, A.; Cohen, R.; Wing, R.R. Differential functional magnetic resonance imaging response to food pictures in successful weight-loss maintainers relative to normal-weight and obese controls. *Am. J. Clin. Nutr.* **2009**, *90*, 928–934. [CrossRef] [PubMed]
249. Weygandt, M.; Mai, K.; Dommes, E.; Leupelt, V.; Hackmack, K.; Kahnt, T.; Rothemund, Y.; Spranger, J.; Haynes, J.-D. The role of neural impulse control mechanisms for dietary success in obesity. *NeuroImage* **2013**, *83*, 669–678. [CrossRef] [PubMed]
250. Nederkoorn, C.; Houben, K.; Hofmann, W.; Roefs, A.; Jansen, A. Control yourself or just eat what you like? Weight gain over a year is predicted by an interactive effect of response inhibition and implicit preference for snack foods. *Health Psychol.* **2010**, *29*, 389–393. [CrossRef] [PubMed]
251. Cohen, J.; Yates, K.F.; Duong, M.; Convit, A. Obesity, orbitofrontal structure and function are associated with food choice: A cross-sectional study. *BMJ Open* **2011**, *1*, e000175. [CrossRef] [PubMed]
252. Maayan, L.; Hoogendoorn, C.; Sweat, V.; Convit, A. Disinhibited eating in obese adolescents is associated with orbitofrontal volume reductions and executive dysfunction. *Obesity* **2011**, *19*, 1382–1387. [CrossRef] [PubMed]
253. Masouleh, S.K.; Arelin, K.; Horstmann, A.; Lampe, L.; Kipping, J.A.; Luck, T.; Riedel-Heller, S.G.; Schroeter, M.L.; Stumvoll, M.; Villringer, A.; et al. Higher body mass index in older adults is associated with lower gray matter volume: Implications for memory performance. *Neurobiol. Aging* **2016**, *40*, 1–10. [CrossRef]
254. Pannacciulli, N.; Del Parigi, A.; Chen, K.; Le, D.S.N.; Reiman, E.M.; Tataranni, P.A. Brain abnormalities in human obesity: A voxel-based morphometric study. *NeuroImage* **2006**, *31*, 1419–1425. [CrossRef]
255. Taki, Y.; Kinomura, S.; Sato, K.; Inoue, K.; Goto, R.; Okada, K.; Uchida, S.; Kawashima, R.; Fukuda, H. Relationship between body mass index and gray matter volume in 1428 healthy individuals. *Obesity* **2008**, *16*, 119–124. [CrossRef]
256. Walther, K.; Birdsill, A.C.; Glisky, E.L.; Ryan, L. Structural brain differences and cognitive functioning related to body mass index in older females. *Hum. Brain Mapp.* **2010**, *31*, 1052–1064. [CrossRef] [PubMed]
257. Yao, L.; Li, W.; Dai, Z.; Dong, C. Eating behavior associated with gray matter volume alternations: A voxel based morphometry study. *Appetite* **2016**, *96*, 572–579. [CrossRef] [PubMed]

258. Yokum, S.; Stice, E. Cognitive regulation of food craving: Effects of three cognitive reappraisal strategies on neural response to palatable foods. *Int. J. Obes.* **2013**, *37*, 1565–1570. [CrossRef] [PubMed]
259. Bechara, A. Decision making, impulse control and loss of willpower to resist drugs: A neurocognitive perspective. *Nat. Neurosci.* **2005**, *8*, 1458–1463. [CrossRef] [PubMed]
260. Davis, C.; Levitan, R.D.; Muglia, P.; Bewell, C.; Kennedy, J.L. Decision-making deficits and overeating: A risk model for obesity. *Obes. Res.* **2004**, *12*, 929–935. [CrossRef] [PubMed]
261. Elias, M.F.; Elias, P.K.; Sullivan, L.M.; Wolf, P.A.; D'Agostino, R.B. Obesity, diabetes and cognitive deficit: The Framingham heart study. *Neurobiol. Aging* **2005**, *26*, 11–16. [CrossRef] [PubMed]
262. Gunstad, J.; Paul, R.H.; Cohen, R.A.; Tate, D.F.; Spitznagel, M.B.; Gordon, E. Elevated body mass index is associated with executive dysfunction in otherwise healthy adults. *Compr. Psychiatr.* **2007**, *48*, 57–61. [CrossRef] [PubMed]
263. Lowe, C.J.; Reichelt, A.C.; Hall, P.A. The prefrontal cortex and obesity: A health neuroscience perspective. *Trends Cogn. Sci.* **2019**, *23*, 349–361. [CrossRef] [PubMed]
264. Blum, K.; Braverman, E.R.; Wood, R.C.; Gill, J.; Li, C.; Chen, T.J.; Taub, M.; Montgomery, A.R.; Sheridan, P.J.; Cull, J.G. Increased prevalence of the Taq I A1allele of the dopamine receptor gene (DRD2) in obesity with comorbid substance use disorder: A preliminary report. *Pharmacogenetics* **1996**, *6*, 297–305. [CrossRef] [PubMed]
265. Comings, D.E.; Muhleman, D.; Ahn, C.; Gysin, R.; Flanagan, S.D. The dopamine D2 receptor gene: A genetic risk factor in substance abuse. *Drug Alcohol Depend.* **1994**, *34*, 175–180. [CrossRef]
266. Han, D.H.; Yoon, S.J.; Sung, Y.H.; Lee, Y.S.; Kee, B.S.; Lyoo, I.K.; Renshaw, P.F.; Cho, S.C. A preliminary study: Novelty seeking, frontal executive function, and dopamine receptor (D2) TaqI A gene polymorphism in patients with methamphetamine dependence. *Compr. Psychiatr.* **2008**, *49*, 387–392. [CrossRef] [PubMed]
267. Jönsson, E.G.; Nöthen, M.M.; Grünhage, F.; Farde, L.; Nakashima, Y.; Propping, P.; Sedvall, G.C.; Nöthen, M.; Jönsson, E. Polymorphisms in the dopamine D2 receptor gene and their relationships to striatal dopamine receptor density of healthy volunteers. *Mol. Psychiatr.* **1999**, *4*, 290–296. [CrossRef]
268. Noble, E. Addiction and its reward process through polymorphisms of the D2 dopamine receptor gene: A review. *Eur. Psychiatr.* **2000**, *15*, 79–89. [CrossRef]
269. Spitz, M.R.; Detry, M.A.; Pillow, P.; Hu, Y.; Amos, C.I.; Hong, W.K.; Wu, X. Variant alleles of the D2 dopamine receptor gene and obesity. *Nutr. Res.* **2000**, *20*, 371–380. [CrossRef]
270. Eisenberg, D.T.; MacKillop, J.; Modi, M.; Beauchemin, J.; Dang, D.; Lisman, S.A.; Lum, J.K.; Wilson, D.S. Examining impulsivity as an endophenotype using a behavioral approach: A DRD2 TaqI A and DRD4 48-bp VNTR association study. *Behav. Brain Funct.* **2007**, *3*, 2. [CrossRef] [PubMed]
271. Kirsch, P.; Reuter, M.; Mier, D.; Lonsdorf, T.; Stark, R.; Gallhofer, B.; Vaitl, D.; Hennig, J. Imaging gene–substance interactions: The effect of the DRD2 TaqIA polymorphism and the dopamine agonist bromocriptine on the brain activation during the anticipation of reward. *Neurosci. Lett.* **2006**, *405*, 196–201. [CrossRef]
272. Klein, T.A.; Neumann, J.; Reuter, M.; Hennig, J.; Von Cramon, D.Y.; Ullsperger, M. Genetically determined differences in learning from errors. *Science* **2007**, *318*, 1642–1645. [CrossRef]
273. Montag, C.; Weber, B.; Jentgens, E.; Elger, C.; Reuter, M. An epistasis effect of functional variants on the BDNF and DRD2 genes modulates gray matter volume of the anterior cingulate cortex in healthy humans. *Neuropsychology* **2010**, *48*, 1016–1021. [CrossRef]
274. Gasquoine, P.G. Localization of function in anterior cingulate cortex: From psychosurgery to functional neuroimaging. *Neurosci. Biobehav. Rev.* **2013**, *37*, 340–348. [CrossRef]
275. Peoples, L.L. Will, anterior cingulate cortex, and addiction. *Science* **2002**, *296*, 1623–1624. [CrossRef]
276. Brody, A.L.; Mandelkern, M.A.; Olmstead, R.E.; Jou, J.; Tiongson, E.; Allen, V.; Scheibal, D.; London, E.D.; Monterosso, J.R.; Tiffany, S.T.; et al. Neural substrates of resisting craving during cigarette cue exposure. *Biol. Psychiatr.* **2007**, *62*, 642–651. [CrossRef] [PubMed]
277. Pursey, K.M.; Stanwell, P.; Gearhardt, A.N.; Collins, C.E.; Burrows, T.L. The prevalence of food addiction as assessed by the Yale food addiction scale: A systematic review. *Nutrients* **2014**, *6*, 4552–4590. [CrossRef] [PubMed]
278. Carter, J.C.; Van Wijk, M.; Rowsell, M. Symptoms of 'food addiction' in binge eating disorder using the Yale food addiction scale version 2.0. *Appetite* **2019**, *133*, 362–369. [CrossRef] [PubMed]

279. De Vries, S.-K.; Meule, A. Food addiction and bulimia nervosa: New data based on the Yale food addiction scale 2.0. *Eur. Eat. Disord. Rev.* **2016**, *24*, 518–522. [CrossRef] [PubMed]
280. Hall, P.A.; Fong, G.T.; Epp, L.J.; Elias, L.J. Executive function moderates the intention-behavior link for physical activity and dietary behavior. *Psychol. Health* **2008**, *23*, 309–326. [CrossRef] [PubMed]
281. Hall, P.A. Executive control resources and frequency of fatty food consumption: Findings from an age-stratified community sample. *Health Psychol.* **2012**, *31*, 235–241. [CrossRef] [PubMed]
282. Alcorn, J.L.; Marks, K.R.; Stoops, W.W.; Rush, C.R.; Lile, J.A. Attentional bias to cannabis cues in cannabis users but not cocaine users. *Addict. Behav.* **2019**, *88*, 129–136. [CrossRef] [PubMed]
283. Bradley, B.; Field, M.; Mogg, K.; De Houwer, J. Attentional and evaluative biases for smoking cues in nicotine dependence: Component processes of biases in visual orienting. *Behav. Pharmacol.* **2004**, *15*, 29–36. [CrossRef]
284. Field, M.; Eastwood, B.; Bradley, B.P.; Mogg, K. Selective processing of cannabis cues in regular cannabis users. *Drug Alcohol Depend.* **2006**, *85*, 75–82. [CrossRef]
285. Qureshi, A.; Monk, R.L.; Pennington, C.R.; Wilcockson, T.D.; Heim, D. Alcohol-related attentional bias in a gaze contingency task: Comparing appetitive and non-appetitive cues. *Addict. Behav.* **2019**, *90*, 312–317. [CrossRef]
286. Field, M.; Cox, W. Attentional bias in addictive behaviors: A review of its development, causes, and consequences. *Drug Alcohol Depend.* **2008**, *97*, 1–20. [CrossRef] [PubMed]
287. Franken, I.H. Drug craving and addiction: Integrating psychological and neuropsychopharmacological approaches. *Prog. Neuro Psychopharmacol. Biol. Psychiatr.* **2003**, *27*, 563–579. [CrossRef]
288. Detandt, S.; Verbanck, P.; Bazan, A.; Quertemont, E. Smoking addiction: The shift from head to hands: Approach bias towards smoking-related cues in low-dependent versus dependent smokers. *J. Psychopharmacol.* **2017**, *31*, 819–829. [CrossRef] [PubMed]
289. Weckler, H.; Kong, G.; Larsen, H.; Cousijn, J.; Wiers, R.W.; Krishnan-Sarin, S. Impulsivity and approach tendencies towards cigarette stimuli: Implications for cigarette smoking and cessation behaviors among youth. *Exp. Clin. Psychopharmacol.* **2017**, *25*, 363–372. [CrossRef] [PubMed]
290. Watson, P.; De Wit, S.; Hommel, B.; Wiers, R.W. Motivational mechanisms and outcome expectancies underlying the approach bias toward addictive substances. *Front. Psychol.* **2012**, *3*, 440. [CrossRef] [PubMed]
291. Beraha, E.M.; Cousijn, J.; Hermanides, E.; Goudriaan, A.E.; Wiers, R.W. Implicit Associations and Explicit Expectancies toward Cannabis in Heavy Cannabis Users and Controls. *Front. Psychol.* **2013**, *4*, 1–9. [CrossRef]
292. De Houwer, J.; Custers, R.; De Clercq, A. Do smokers have a negative implicit attitude toward smoking? *Cogn. Emot.* **2006**, *20*, 1274–1284. [CrossRef]
293. Houben, K.; Wiers, R.W. Are drinkers implicitly positive about drinking alcohol? Personalizing the alcohol-IAT to reduce negative extrapersonal contamination. *Alcohol Alcohol.* **2007**, *42*, 301–307. [CrossRef]
294. McCarthy, D.M.; Thompsen, D.M. Implicit and explicit measures of alcohol and smoking cognitions. *Psychol. Addict. Behav.* **2006**, *20*, 436–444. [CrossRef]
295. Brignell, C.; Griffiths, T.; Bradley, B.P.; Mogg, K. Attentional and approach biases for pictorial food cues. Influence of external eating. *Appetite* **2009**, *52*, 299–306. [CrossRef]
296. Deluchi, M.; Costa, F.S.; Friedman, R.; Gonçalves, R.; Bizarro, L. Attentional bias to unhealthy food in individuals with severe obesity and binge eating. *Appetite* **2017**, *108*, 471–476. [CrossRef] [PubMed]
297. Hardman, C.A.; Scott, J.; Field, M.; Jones, A. To eat or not to eat. The effects of expectancy on reactivity to food cues. *Appetite* **2014**, *76*, 153–160. [CrossRef] [PubMed]
298. Popien, A.; Frayn, M.; Von Ranson, K.M.; Sears, C.R. Eye gaze tracking reveals heightened attention to food in adults with binge eating when viewing images of real-world scenes. *Appetite* **2015**, *91*, 233–240. [CrossRef] [PubMed]
299. Field, M.; Werthmann, J.; Franken, I.; Hofmann, W.; Hogarth, L.; Roefs, A. The role of attentional bias in obesity and addiction. *Health Psychol.* **2016**, *35*, 767–780. [CrossRef] [PubMed]
300. Stojek, M.; Shank, L.M.; Vannucci, A.; Bongiorno, D.M.; Nelson, E.E.; Waters, A.J.; Engel, S.G.; Boutelle, K.N.; Pine, D.S.; Yanovski, J.A.; et al. A systematic review of attentional biases in disorders involving binge eating. *Appetite* **2018**, *123*, 367–389. [CrossRef]
301. Castellanos, E.H.; Charboneau, E.; Dietrich, M.S.; Park, S.; Bradley, B.P.; Mogg, K.; Cowan, R.L.; Bradley, B. Obese adults have visual attention bias for food cue images: Evidence for altered reward system function. *Int. J. Obes.* **2009**, *33*, 1063–1073. [CrossRef]

302. Iceta, S.; Benoit, J.; Cristini, P.; Lambert-Porcheron, S.; Segrestin, B.; Laville, M.; Poulet, E.; Disse, E. Attentional bias and response inhibition in severe obesity with food disinhibition: A study of P300 and N200 event-related potential. *Int. J. Obes.* **2019**, *1*. [CrossRef]
303. Nijs, I.M.; Muris, P.; Euser, A.S.; Franken, I.H. Differences in attention to food and food intake between overweight/obese and normal-weight females under conditions of hunger and satiety. *Appetite* **2010**, *54*, 243–254. [CrossRef]
304. Werthmann, J.; Roefs, A.; Nederkoorn, C.; Mogg, K.; Bradley, B.P.; Jansen, A. Can (not) take my eyes off it: Attention bias for food in overweight participants. *Health Psychol.* **2011**, *30*, 561–569. [CrossRef]
305. Hardman, C.A.; Rogers, P.J.; Etchells, K.A.; Houstoun, K.V.E.; Munafò, M.R. The effects of food-related attentional bias training on appetite and food intake. *Appetite* **2013**, *71*, 295–300. [CrossRef]
306. Kemps, E.; Tiggemann, M.; Orr, J.; Grear, J. Attentional retraining can reduce chocolate consumption. *J. Exp. Psychol. Appl.* **2013**, *20*, 94–102. [CrossRef] [PubMed]
307. Schmitz, F.; Svaldi, J. Effects of bias modification training in binge eating disorder. *Behav. Ther.* **2017**, *48*, 707–717. [CrossRef] [PubMed]
308. Werthmann, J.; Field, M.; Roefs, A.; Nederkoorn, C.; Jansen, A. Attention bias for chocolate increases chocolate consumption—An attention bias modification study. *J. Behav. Ther. Exp. Psychiatr.* **2014**, *45*, 136–143. [CrossRef] [PubMed]
309. Zhang, S.; Cui, L.; Sun, X.; Zhang, Q. The effect of attentional bias modification on eating behavior among women craving high-calorie food. *Appetite* **2018**, *129*, 135–142. [CrossRef] [PubMed]
310. Smith, E.; Treffiletti, A.; Bailey, E.P.; Moustafa, A. The effect of attentional bias modification training on food intake in overweight and obese women. *J. Health Psychol.* **2018**, 1–11. [CrossRef] [PubMed]
311. Kakoschke, N.L.; Kemps, E.; Tiggemann, M. Attentional bias modification encourages healthy eating. *Eat. Behav.* **2014**, *15*, 120–124. [CrossRef] [PubMed]
312. Kemps, E.; Tiggemann, M.; Elford, J. Sustained effects of attentional re-training on chocolate consumption. *J. Behav. Ther. Exp. Psychiatr.* **2015**, *49*, 94–100. [CrossRef] [PubMed]
313. Veenstra, E.M.; De Jong, P.J. Restrained eaters show enhanced automatic approach tendencies towards food. *Appetite* **2010**, *55*, 30–36. [CrossRef]
314. Becker, D.; Jostmann, N.B.; Wiers, R.W.; Holland, R.W. Approach avoidance training in the eating domain: Testing the effectiveness across three single session studies. *Appetite* **2015**, *85*, 58–65. [CrossRef]
315. Brockmeyer, T.; Hahn, C.; Reetz, C.; Schmidt, U.; Friederich, H.-C. Approach bias and cue reactivity towards food in people with high versus low levels of food craving. *Appetite* **2015**, *95*, 197–202. [CrossRef]
316. Havermans, R.C.; Giesen, J.C.; Houben, K.; Jansen, A. Weight, gender, and snack appeal. *Eat. Behav.* **2011**, *12*, 126–130. [CrossRef] [PubMed]
317. Mehl, N.; Mueller-Wieland, L.; Mathar, D.; Horstmann, A. Retraining automatic action tendencies in obesity. *Physiol. Behav.* **2018**, *192*, 50–58. [CrossRef] [PubMed]
318. Mogg, K.; Bradley, B.P.; O'Neill, B.; Bani, M.; Merlo-Pich, E.; Koch, A.; Bullmore, E.T.; Nathan, P.J. Effect of dopamine D_3 receptor antagonism on approach responses to food cues in overweight and obese individuals. *Behav. Pharmacol.* **2012**, *23*, 603–608. [CrossRef] [PubMed]
319. Kemps, E.; Tiggemann, M.; Martin, R.; Elliott, M. Implicit approach–avoidance associations for craved food cues. *J. Exp. Psychol. Appl.* **2013**, *19*, 30–38. [CrossRef] [PubMed]
320. Schumacher, S.E.; Kemps, E.; Tiggemann, M. Bias modification training can alter approach bias and chocolate consumption. *Appetite* **2016**, *96*, 219–224. [CrossRef] [PubMed]
321. Brockmeyer, T.; Hahn, C.; Reetz, C.; Schmidt, U.; Friederich, H.-C. Approach Bias Modification in Food Craving-A Proof-of-Concept Study. *Eur. Eat. Disord. Rev.* **2015**, *23*, 352–360. [CrossRef] [PubMed]
322. De Houwer, J. A conceptual and theoretical analysis of evaluative conditioning. *Span. J. Psychol.* **2007**, *10*, 230–241. [CrossRef] [PubMed]
323. De Houwer, J.; Thomas, S.; Baeyens, F. Association learning of likes and dislikes: A review of 25 years of research on human evaluative conditioning. *Psychol. Bull.* **2001**, *127*, 853–869. [CrossRef] [PubMed]
324. Dwyer, D.M.; Jarratt, F.; Dick, K. Evaluative conditioning with foods as CSs and body shapes as USs: No evidence for sex differences, extinction, or overshadowing. *Cogn. Emot.* **2007**, *21*, 281–299. [CrossRef]
325. Hollands, G.J.; Prestwich, A.; Marteau, T.M. Using aversive images to enhance healthy food choices and implicit attitudes: An experimental test of evaluative conditioning. *Health Psychol.* **2011**, *30*, 195–203. [CrossRef]

326. Lascelles, K.R.; Field, A.P.; Davey, G.C. Using foods as CSs and body shapes as UCSs: A putative role for associative learning in the development of eating disorders. *Behav. Ther.* **2003**, *34*, 213–235. [CrossRef]
327. Lebens, H.; Roefs, A.; Martijn, C.; Houben, K.; Nederkoorn, C.; Jansen, A. Making implicit measures of associations with snack foods more negative through evaluative conditioning. *Eat. Behav.* **2011**, *12*, 249–253. [CrossRef] [PubMed]
328. Wang, Y.; Wang, G.; Zhang, D.; Wang, L.; Cui, X.; Zhu, J.; Fang, Y. Learning to dislike chocolate: Conditioning negative attitudes toward chocolate and its effect on chocolate consumption. *Front. Psychol.* **2017**, *8*, 1–7. [CrossRef] [PubMed]
329. Walsh, E.M.; Kiviniemi, M.T. Changing how I feel about the food: Experimentally manipulated affective associations with fruits change fruit choice behaviors. *J. Behav. Med.* **2014**, *37*, 322–331. [CrossRef] [PubMed]
330. Shaw, J.A.; Forman, E.M.; Espel, H.M.; Butryn, M.L.; Herbert, J.D.; Lowe, M.R.; Nederkoorn, C. Can evaluative conditioning decrease soft drink consumption? *Appetite* **2016**, *105*, 60–70. [CrossRef]
331. Houben, K.; Havermans, R.C.; Wiers, R.W. Learning to dislike alcohol: Conditioning negative implicit attitudes toward alcohol and its effect on drinking behavior. *Psychopharmacology* **2010**, *211*, 79–86. [CrossRef] [PubMed]
332. Zerhouni, O.; Houben, K.; El Methni, J.; Rutte, N.; Werkman, E.; Wiers, R.W. I didn't feel like drinking, but I guess why: Evaluative conditioning changes on explicit attitudes toward alcohol and healthy foods depends on contingency awareness. *Learn. Motiv.* **2019**, *66*, 1–12. [CrossRef]
333. Hofmann, W.; De Houwer, J.; Perugini, M.; Baeyens, F.; Crombez, G. Evaluative conditioning in humans: A meta-analysis. *Psychol. Bull.* **2010**, *136*, 390–421. [CrossRef]
334. Benedict, T.; Richter, J.; Gast, A. The influence of misinformation manipulations on evaluative conditioning. *Acta Psychol.* **2019**, *194*, 28–36. [CrossRef]
335. Aron, A.R.; Robbins, T.W.; Poldrack, R.A. Inhibition and the right inferior frontal cortex. *Trends Cogn. Sci.* **2004**, *8*, 170–177. [CrossRef]
336. Aron, A.R.; Robbins, T.W.; Poldrack, R.A. Inhibition and the right inferior frontal cortex: One decade on. *Trends Cogn. Sci.* **2014**, *18*, 177–185. [CrossRef] [PubMed]
337. Logan, G.D. Executive control of thought and action. *Acta Psychol.* **1985**, *60*, 193–210. [CrossRef]
338. Logan, G.D.; Cowan, W.B. On the ability to inhibit thought and action: A theory of an act of control. *Psychol. Rev.* **1984**, *91*, 295–327. [CrossRef]
339. Logan, G.D.; Schachar, R.J.; Tannock, R. Impulsivity and inhibitory control. *Psychol. Sci.* **1997**, *8*, 60–64. [CrossRef]
340. Miyake, A.; Friedman, N.P.; Emerson, M.J.; Witzki, A.H.; Howerter, A.; Wager, T.D. The unity and diversity of executive functions and their contributions to complex "frontal lobe" tasks: A latent variable analysis. *Cogn. Psychol.* **2000**, *41*, 49–100. [CrossRef] [PubMed]
341. Nigg, J.T.; Wong, M.M.; Martel, M.M.; Jester, J.M.; Puttler, L.I.; Glass, J.M.; Adams, K.M.; Fitzgerald, H.E.; Zucker, R.A. Poor response inhibition as a predictor of problem drinking and illicit drug use in adolescents at risk for alcoholism and other substance use disorders. *J. Am. Acad. Child Adolesc. Psychiatr.* **2006**, *45*, 468–475. [CrossRef]
342. Krishnan-Sarin, S.; Reynolds, B.; Duhig, A.M.; Smith, A.; Liss, T.; McFetridge, A.; Cavallo, D.A.; Carroll, K.M.; Potenza, M.N. Behavioral impulsivity predicts treatment outcome in a smoking cessation program for adolescents. *Drug Alcohol Depend.* **2007**, *88*, 79–82. [CrossRef]
343. Czapla, M.; Simon, J.J.; Richter, B.; Kluge, M.; Friederich, H.C.; Herpertz, S.; Loeber, S. The impact of cognitive impairment and impulsivity on relapse of alcohol-dependent patients: Implications for psychotherapeutic treatment. *Addict. Biol.* **2016**, *21*, 873–884. [CrossRef]
344. Houben, K.; Wiers, R.W. Response inhibition moderates the relationship between implicit associations and drinking behavior. *Alcohol. Clin. Exp. Res.* **2009**, *33*, 626–633. [CrossRef]
345. Nederkoorn, C.; Smulders, F.T.; Havermans, R.C.; Roefs, A.; Jansen, A. Impulsivity in obese women. *Appetite* **2006**, *47*, 253–256. [CrossRef]
346. Price, M.; Lee, M.; Higgs, S. Food-specific response inhibition, dietary restraint and snack intake in lean and overweight/obese adults: A moderated-mediation model. *Int. J. Obes.* **2016**, *40*, 877. [CrossRef] [PubMed]
347. Houben, K. Overcoming the urge to splurge: Influencing eating behavior by manipulating inhibitory control. *J. Behav. Ther. Exp. Psychiatr.* **2011**, *42*, 384–388. [CrossRef] [PubMed]
348. Nederkoorn, C.; Dassen, F.C.; Franken, L.; Resch, C.; Houben, K. Impulsivity and overeating in children in the absence and presence of hunger. *Appetite* **2015**, *93*, 57–61. [CrossRef] [PubMed]

349. Houben, K.; Nederkoorn, C.; Jansen, A. Eating on impulse: The relation between overweight and food-specific inhibitory control. *Obesity* **2014**, *22*, e6–e8. [CrossRef] [PubMed]
350. Mühlberg, C.; Mathar, D.; Villringer, A.; Horstmann, A.; Neumann, J. Stopping at the sight of food—How gender and obesity impact on response inhibition. *Appetite* **2016**, *107*, 663–676. [CrossRef] [PubMed]
351. Meule, A.; Lutz, A.P.; Vögele, C.; Kübler, A. Impulsive reactions to food-cues predict subsequent food craving. *Eat. Behav.* **2014**, *15*, 99–105. [CrossRef] [PubMed]
352. Allan, J.L.; Johnston, M.; Campbell, N. Missed by an inch or a mile? Predicting the size of intention–behaviour gap from measures of executive control. *Psychol. Health* **2011**, *26*, 635–650. [CrossRef]
353. Rosval, L.; Steiger, H.; Bruce, K.; Israël, M.; Richardson, J.; Aubut, M. Impulsivity in women with eating disorders: Problem of response inhibition, planning, or attention? *Int. J. Eat. Disord.* **2006**, *39*, 590–593. [CrossRef]
354. Bowley, C.; Faricy, C.; Hegarty, B.; Johnstone, S.J.; Smith, J.L.; Kelly, P.J.; Rushby, J.A. The effects of inhibitory control training on alcohol consumption, implicit alcohol-related cognitions and brain electrical activity. *Int. J. Psychophysiol.* **2013**, *89*, 342–348. [CrossRef]
355. Di Lemma, L.C.G.; Field, M. Cue avoidance training and inhibitory control training for the reduction of alcohol consumption: A comparison of effectiveness and investigation of their mechanisms of action. *Psychopharmacology* **2017**, *234*, 2489–2498. [CrossRef]
356. Houben, K.; Havermans, R.C.; Nederkoorn, C.; Jansen, A. Beer à no-go: Learning to stop responding to alcohol cues reduces alcohol intake via reduced affective associations rather than increased response inhibition. *Addiction* **2012**, *107*, 1280–1287. [CrossRef] [PubMed]
357. Houben, K.; Nederkoorn, C.; Wiers, R.W.; Jansen, A. Resisting temptation: Decreasing alcohol-related affect and drinking behaviour by training response inhibition. *Drug Alcohol Depend.* **2011**, *116*, 132–136. [CrossRef] [PubMed]
358. Kilwein, T.M.; Bernhardt, K.A.; Stryker, M.L.; Looby, A. Decreased alcohol consumption after pairing alcohol-related cues with an inhibitory response. *J. Subst. Use* **2017**, *23*, 154–161. [CrossRef]
359. Verbruggen, F.; Adams, R.; Chambers, C.D. Proactive motor control reduces monetary risk taking in gambling. *Psychol. Sci.* **2012**, *23*, 805–815. [CrossRef] [PubMed]
360. Verbruggen, F.; Adams, R.C.; Van't Wout, F.; Stevens, T.; McLaren, I.P.L.; Chambers, C.D. Are the effects of response inhibition on gambling long-lasting? *PLoS ONE* **2013**, *7*, e70155. [CrossRef] [PubMed]
361. Adams, R.C.; Lawrence, N.S.; Verbruggen, F.; Chambers, C.D. Training response inhibition to reduce food consumption: Mechanisms, stimulus specificity and appropriate training protocols. *Appetite* **2017**, *109*, 11–23. [CrossRef] [PubMed]
362. Houben, K.; Jansen, A. Training inhibitory control. A recipe for resisting sweet temptations. *Appetite* **2011**, *56*, 345–349. [CrossRef]
363. Houben, K.; Jansen, A. Chocolate equals stop. Chocolate-specific inhibition training reduces chocolate intake and go associations with chocolate. *Appetite* **2015**, *87*, 318–323. [CrossRef]
364. Lawrence, N.S.; Verbruggen, F.; Morrison, S.; Adams, R.C.; Chambers, C.D. Stopping to food pictures reduces food intake: Effects of cue specificity, control conditions and individual differences. *Appetite* **2015**, *85*, 91–103. [CrossRef]
365. Oomen, D.; Grol, M.; Spronk, D.; Booth, C.; Fox, E. Beating uncontrolled eating: Training inhibitory control to reduce food intake and food cue sensitivity. *Appetite* **2018**, *131*, 73–83. [CrossRef]
366. Camp, B.; Lawrence, N.S. Giving pork the chop: Response inhibition training to reduce meat intake. *Appetite* **2019**, *141*, 104315. [CrossRef] [PubMed]
367. Chen, Z.; Holland, R.W.; Quandt, J.; Dijksterhuis, A.; Veling, H. When mere action versus inaction leads to robust preference change. *J. Personal. Soc. Psychol.* **2019**. [CrossRef] [PubMed]
368. Porter, L.; Bailey-Jones, C.; Priudokaite, G.; Allen, S.; Wood, K.; Stiles, K.; Lawrence, N.S. From cookies to carrots; the effect of inhibitory control training on children's snack selections. *Appetite* **2018**, *124*, 111–123. [CrossRef] [PubMed]
369. Van Koningsbruggen, G.M.; Veling, H.; Stroebe, W.; Aarts, H. Comparing two psychological interventions in reducing impulsive processes of eating behaviour: Effects on self-selected portion size. *Br. J. Health Psychol.* **2013**, *19*, 767–782. [CrossRef] [PubMed]
370. Veling, H.; Aarts, H.; Stroebe, W. Using stop signals to reduce impulsive choices for palatable unhealthy foods. *Br. J. Health Psychol.* **2013**, *18*, 354–368. [CrossRef] [PubMed]

371. Veling, H.; Aarts, H.; Stroebe, W. Stop signals decrease choices for palatable foods through decreased food evaluation. *Front. Psychol.* **2013**, *4*, 875. [CrossRef] [PubMed]
372. Gish, M.Y. Addiction to food: How go/no-go tasks affect appetite. *Biomed. Health Sci. Res.* **2015**, *6*, 221–231. [CrossRef]
373. Veling, H.; van Koningsbruggen, G.M.; Aarts, H.; Stroebe, W. Targeting impulsive processes of eating behavior via the internet. Effects on body weight. *Appetite* **2014**, *78*, 102–109. [CrossRef]
374. Allom, V.; Mullan, B.; Hagger, M. Does inhibitory control training improve health behaviour? A meta-analysis. *Health Psychol. Rev.* **2015**, *10*, 1–38. [CrossRef]
375. Jones, A.; Di Lemma, L.C.; Robinson, E.; Christiansen, P.; Nolan, S.; Tudur-Smith, C.; Field, M. Inhibitory control training for appetitive behaviour change: A meta-analytic investigation of mechanisms of action and moderators of effectiveness. *Appetite* **2016**, *97*, 16–28. [CrossRef]
376. Lawrence, N.S.; O'Sullivan, J.; Parslow, D.; Javaid, M.; Adams, R.C.; Chambers, C.D.; Kos, K.; Verbruggen, F. Training response inhibition to food is associated with weight loss and reduced energy intake. *Appetite* **2015**, *95*, 17–28. [CrossRef] [PubMed]
377. Kakoschke, N.L.; Kemps, E.; Tiggemann, M. The effect of combined avoidance and control training on implicit food evaluation and choice. *J. Behav. Ther. Exp. Psychiatr.* **2017**, *55*, 99–105. [CrossRef] [PubMed]
378. Schakel, L.; Veldhuijzen, D.S.; Van Middendorp, H.; Van Dessel, P.; De Houwer, J.; Bidarra, R.; Evers, A.W.M. The effects of a gamified approach avoidance training and verbal suggestions on food outcomes. *PLoS ONE* **2018**, *13*, e0201309. [CrossRef] [PubMed]
379. Chen, Z.; Veling, H.; Dijksterhuis, A.; Holland, R.W. Do impulsive individuals benefit more from food go/no-go training? Testing the role of inhibition capacity in the no-go devaluation effect. *Appetite* **2018**, *124*, 99–110. [CrossRef] [PubMed]
380. McLaren, I.P.L.; Verbruggen, F. Association and inhibition. In *The Wiley Blackwell Handbook on the Cognitive Neuroscience of Learning*; Murphy, R.A., Honey, R.C., Eds.; John Wiley and Sons: Chichester, UK, 2016.
381. Veling, H.; Holland, R.W.; Van Knippenberg, A. When approach motivation and behavioral inhibition collide: Behavior regulation through stimulus devaluation. *J. Exp. Soc. Psychol.* **2008**, *44*, 1013–1019. [CrossRef]
382. Verbruggen, F.; Best, M.; Bowditch, W.A.; Stevens, T.; McLaren, I.P. The inhibitory control reflex. *Neuropsychology* **2014**, *65*, 263–278. [CrossRef] [PubMed]
383. Alonso-Alonso, M. Translating tDCS into the field of obesity: Mechanism-driven approaches. *Front. Hum. Neurosci.* **2013**, *7*, 512.
384. Feil, J.; Zangen, A. Brain stimulation in the study and treatment of addiction. *Neurosci. Biobehav. Rev.* **2010**, *34*, 559–574. [CrossRef] [PubMed]
385. Hall, P.A.; Vincent, C.M.; Burhan, A.M. Non-invasive brain stimulation for food cravings, consumption, and disorders of eating: A review of methods, findings and controversies. *Appetite* **2018**, *124*, 78–88. [CrossRef]
386. Lowe, C.J.; Vincent, C.; Hall, P.A. Effects of noninvasive brain stimulation on food cravings and consumption: A meta-analytic review. *Psychosom. Med.* **2017**, *79*, 2–13. [CrossRef]
387. Nardone, R.; Bergmann, J.; Christova, M.; Lochner, P.; Tezzon, F.; Golaszewski, S.; Trinka, E.; Brigo, F. Non-invasive brain stimulation in the functional evaluation of alcohol effects and in the treatment of alcohol craving: A review. *Neurosci. Res.* **2012**, *74*, 169–176. [CrossRef] [PubMed]
388. Song, S.; Zilverstand, A.; Gui, W.; Li, H.J.; Zhou, X. Effects of single-session versus multi-session non-invasive brain stimulation on craving and consumption in individuals with drug addiction, eating disorders or obesity: A meta-analysis. *Brain Stimul.* **2018**, *12*, 606–618. [CrossRef] [PubMed]
389. Maizey, L.; Allen, C.P.; Dervinis, M.; Verbruggen, F.; Varnava, A.; Kozlov, M.; Adams, R.C.; Stokes, M.; Klemen, J.; Bungert, A.; et al. Comparative incidence rates of mild adverse effects to transcranial magnetic stimulation. *Clin. Neurophysiol.* **2013**, *124*, 536–544. [CrossRef] [PubMed]
390. Matsumoto, H.; Ugawa, Y. Adverse effects of tDCS and tACS: A review. *Clin. Neurophysiol. Pract.* **2017**, *2*, 19–25. [CrossRef] [PubMed]
391. Pascual-Leone, A.; Houser, C.; Reese, K.; Shotland, L.; Grafman, J.; Sato, S.; Valls-Sole, J.; Brasil-Neto, J.; Wassermann, E.; Cohen, L.; et al. Safety of rapid-rate transcranial magnetic stimulation in normal volunteers. *Electroencephalogr. Clin. Neurophysiol. Potentials Sect.* **1993**, *89*, 120–130. [CrossRef]
392. Poreisz, C.; Boros, K.; Antal, A.; Paulus, W. Safety aspects of transcranial direct current stimulation concerning healthy subjects and patients. *Brain Res. Bull.* **2007**, *72*, 208–214. [CrossRef] [PubMed]

393. Rossi, S.; Hallett, M.; Rossini, P.M.; Pascual-Leone, A. Safety, ethical considerations, and application guidelines for the use of transcranial magnetic stimulation in clinical practice and research. *Clin. Neurophysiol.* **2009**, *120*, 2008–2039. [CrossRef] [PubMed]
394. Taylor, R.; Gálvez, V.; Loo, C. Transcranial magnetic stimulation (TMS) safety: A practical guide for psychiatrists. *Australas. Psychiatr.* **2018**, *26*, 189–192. [CrossRef]
395. Fitzgerald, P.; Fountain, S.; Daskalakis, Z. A comprehensive review of the effects of rTMS on motor cortical excitability and inhibition. *Clin. Neurophysiol.* **2006**, *117*, 2584–2596. [CrossRef]
396. Amiaz, R.; Levy, D.; Vainiger, D.; Grunhaus, L.; Zangen, A. Repeated high-frequency transcranial magnetic stimulation over the dorsolateral prefrontal cortex reduces cigarette craving and consumption. *Addiction* **2009**, *104*, 653–660. [CrossRef]
397. Bolloni, C.; Panella, R.; Pedetti, M.; Frascella, A.G.; Gambelunghe, C.; Piccoli, T.; Maniaci, G.; Brancato, A.; Cannizzaro, C.; Diana, M. Bilateral transcranial magnetic stimulation of the prefrontal cortex reduces cocaine intake: A pilot study. *Front. Psychol.* **2016**, *7*, 101. [CrossRef]
398. Mishra, B.R.; Praharaj, S.K.; Katshu, M.Z.U.H.; Sarkar, S.; Nizamie, S.H. Comparison of anticraving efficacy of right and left repetitive transcranial magnetic stimulation in alcohol dependence: A randomized double-blind study. *J. Neuropsychiatry Clin. Neurosci.* **2015**, *27*, 54. [CrossRef] [PubMed]
399. Su, H.; Zhong, N.; Gan, H.; Wang, J.; Han, H.; Chen, T.; Li, X.; Ruan, X.; Zhu, Y.; Jiang, H.; et al. High frequency repetitive transcranial magnetic stimulation of the left dorsolateral prefrontal cortex for methamphetamine use disorders: A randomised clinical trial. *Drug Alcohol Depend.* **2017**, *175*, 84–91. [CrossRef] [PubMed]
400. Terraneo, A.; Leggio, L.; Saladini, M.; Ermani, M.; Bonci, A.; Gallimberti, L.; Information, P.E.K.F.C. Transcranial magnetic stimulation of dorsolateral prefrontal cortex reduces cocaine use: A pilot study. *Eur. Neuropsychopharmacol.* **2016**, *26*, 37–44. [CrossRef] [PubMed]
401. Beeli, G.; Casutt, G.; Baumgartner, T.; Jäncke, L. Modulating presence and impulsiveness by external stimulation of the brain. *Behav. Brain Funct.* **2008**, *4*, 33. [CrossRef] [PubMed]
402. Garavan, H.; Hester, R.; Murphy, K.; Fassbender, C.; Kelly, C. Individual differences in the functional neuroanatomy of inhibitory control. *Brain Res.* **2006**, *1105*, 130–142. [CrossRef] [PubMed]
403. Oldrati, V.; Patricelli, J.; Colombo, B.; Antonietti, A. The role of dorsolateral prefrontal cortex in inhibition mechanism: A study on cognitive reflection test and similar tasks through neuromodulation. *Neuropsychology* **2016**, *91*, 499–508. [CrossRef] [PubMed]
404. Zheng, D.; Oka, T.; Bokura, H.; Yamaguchi, S. The key locus of common response inhibition network for no-go and stop signals. *J. Cogn. Neurosci.* **2008**, *20*, 1434–1442. [CrossRef] [PubMed]
405. Diana, M. The dopamine hypothesis of drug addiction and its potential therapeutic value. *Front. Psychol.* **2011**, *2*, 64. [CrossRef] [PubMed]
406. Strafella, A.P.; Paus, T.; Barrett, J.; Dagher, A. Repetitive Transcranial Magnetic Stimulation of the Human Prefrontal Cortex Induces Dopamine Release in the Caudate Nucleus. *J. Neurosci.* **2001**, *21*, RC157. [CrossRef] [PubMed]
407. Batista, E.K.; Klauss, J.; Fregni, F.; Nitsche, M.A.; Nakamura-Palacios, E.M. A Randomized Placebo-Controlled Trial of Targeted Prefrontal Cortex Modulation with Bilateral tDCS in Patients with Crack-Cocaine Dependence. *Int. J. Neuropsychopharmacol.* **2015**, *18*, 066. [CrossRef] [PubMed]
408. Boggio, P.S.; Liguori, P.; Sultani, N.; Rezende, L.; Fecteau, S.; Fregni, F. Cumulative priming effects of cortical stimulation on smoking cue-induced craving. *Neurosci. Lett.* **2009**, *463*, 82–86. [CrossRef] [PubMed]
409. Boggio, P.S.; Sultani, N.; Fecteau, S.; Merabet, L.; Mecca, T.; Pascual-Leone, A.; Basaglia, A.; Fregni, F. Prefrontal cortex modulation using transcranial DC stimulation reduces alcohol craving: A double-blind, sham-controlled study. *Drug Alcohol Depend.* **2008**, *92*, 55–60. [CrossRef] [PubMed]
410. Boggio, P.S.; Zaghi, S.; Villani, A.B.; Fecteau, S.; Pascual-Leone, A.; Fregni, F. Modulation of risk-taking in marijuana users by transcranial direct current stimulation (tDCS) of the dorsolateral prefrontal cortex (DLPFC). *Drug Alcohol Depend.* **2010**, *112*, 220–225. [CrossRef] [PubMed]
411. Fecteau, S.; Agosta, S.; Hone-Blanchet, A.; Fregni, F.; Boggio, P.; Ciraulo, D.; Pascual-Leone, A. Modulation of smoking and decision-making behaviors with transcranial direct current stimulation in tobacco smokers: A preliminary study. *Drug Alcohol Depend.* **2014**, *140*, 78–84. [CrossRef] [PubMed]
412. Klauss, J.; Anders, Q.S.; Felippe, L.V.; Nitsche, M.A.; Nakamura-Palacios, E.M. Multiple sessions of transcranial direct current stimulation (tDCS) reduced craving and relapses for alcohol use: A randomized placebo-controlled trial in alcohol use disorder. *Front. Pharmacol.* **2018**, *9*, 716. [CrossRef] [PubMed]

413. Antal, A.; Terney, D.; Poreisz, C.; Paulus, W. Towards unravelling task-related modulations of neuroplastic changes induced in the human motor cortex. *Eur. J. Neurosci.* **2007**, *26*, 2687–2691. [CrossRef] [PubMed]
414. Liebetanz, D.; Nitsche, M.A.; Tergau, F.; Paulus, W. Pharmacological approach to the mechanisms of transcranial DC-stimulation-induced after-effects of human motor cortex excitability. *Brain* **2002**, *125*, 2238–2247. [CrossRef] [PubMed]
415. Nitsche, M.A.; Paulus, W. Excitability changes induced in the human motor cortex by weak transcranial direct current stimulation. *J. Physiol.* **2000**, *527*, 633–639. [CrossRef]
416. Nitsche, M.; Schauenburg, A.; Lang, N.; Liebetanz, D.; Exner, C.; Paulus, W.; Tergau, F. Facilitation of implicit motor learning by weak transcranial direct current stimulation of the primary motor cortex in the human. *J. Cogn. Neurosci.* **2003**, *15*, 619–626. [CrossRef]
417. Nitsche, M.A.; Seeber, A.; Frommann, K.; Klein, C.C.; Rochford, C.; Nitsche, M.S.; Fricke, K.; Liebetanz, D.; Lang, N.; Antal, A.; et al. Modulating parameters of excitability during and after transcranial direct current stimulation of the human motor cortex. *J. Physiol.* **2005**, *568*, 291–303. [CrossRef] [PubMed]
418. Priori, A. Brain polarization in humans: A reappraisal of an old tool for prolonged non-invasive modulation of brain excitability. *Clin. Neurophysiol.* **2003**, *114*, 589–595. [CrossRef]
419. Nitsche, M.A.; Paulus, W. Sustained excitability elevations induced by transcranial DC motor cortex stimulation in humans. *Neurology* **2001**, *57*, 1899–1901. [CrossRef]
420. Gandiga, P.C.; Hummel, F.C.; Cohen, L.G. Transcranial DC stimulation (tDCS): A tool for double-blind sham-controlled clinical studies in brain stimulation. *Clin. Neurophysiol.* **2006**, *117*, 845–850. [CrossRef] [PubMed]
421. Nitsche, M.A.; Cohen, L.G.; Wassermann, E.M.; Priori, A.; Lang, N.; Antal, A.; Paulus, W.; Hummel, F.; Boggio, P.S.; Fregni, F.; et al. Transcranial direct current stimulation: State of the art 2008. *Brain Stimul.* **2008**, *1*, 206–223. [CrossRef] [PubMed]
422. Stagg, C.J.; Nitsche, M.A. Physiological basis of transcranial direct current stimulation. *Neuroscience* **2011**, *17*, 37–53. [CrossRef]
423. Wiers, R.W.; Gladwin, T.E.; Hofmann, W.; Salemink, E.; Ridderinkhof, K.R. Cognitive bias modification and cognitive control training in addiction and related psychopathology: Mechanisms, clinical perspectives, and ways forward. *Clin. Psychol. Sci.* **2013**, *1*, 192–212. [CrossRef]
424. Goldman, R.L.; Borckardt, J.J.; Frohman, H.A.; O'Neil, P.M.; Madan, A.; Campbell, L.K.; Budak, A.; George, M.S. Prefrontal cortex transcranial direct current stimulation (tDCS) temporarily reduces food cravings and increases the self-reported ability to resist food in adults with frequent food craving. *Appetite* **2011**, *56*, 741–746. [CrossRef]
425. Kim, S.H.; Chung, J.H.; Kim, T.H.; Lim, S.H.; Kim, Y.; Lee, Y.A.; Song, S.W. The effects of repetitive transcranial magnetic stimulation on eating behaviors and body weight in obesity: A randomized controlled study. *Brain Stimul.* **2018**, *11*, 528–535. [CrossRef]
426. Lapenta, O.M.; Di Sierve, K.; De Macedo, E.C.; Fregni, F.; Boggio, P.S. Transcranial direct current stimulation modulates ERP-indexed inhibitory control and reduces food consumption. *Appetite* **2014**, *83*, 42–48. [CrossRef]
427. Ljubisavljevic, M.; Maxood, K.; Bjekic, J.; Oommen, J.; Nagelkerke, N. Long-Term effects of repeated prefrontal cortex transcranial direct current stimulation (tDCS) on food craving in normal and overweight young adults. *Brain Stimul.* **2016**, *9*, 826–833. [CrossRef] [PubMed]
428. Montenegro, R.A.; Okano, A.H.; Cunha, F.A.; Gurgel, J.L.; Fontes, E.B.; Farinatti, P.T. Prefrontal cortex transcranial direct current stimulation associated with aerobic exercise change aspects of appetite sensation in overweight adults. *Appetite* **2012**, *58*, 333–338. [CrossRef] [PubMed]
429. Ray, M.K.; Sylvester, M.D.; Osborn, L.; Helms, J.; Turan, B.; Burgess, E.E.; Boggiano, M.M. The critical role of cognitive-based trait differences in transcranial direct current stimulation (tDCS) suppression of food craving and eating in frank obesity. *Appetite* **2017**, *116*, 568–574. [CrossRef]
430. Uher, R.; Yoganathan, D.; Mogg, A.; Eranti, S.V.; Treasure, J.; Campbell, I.C.; McLoughlin, D.M.; Schmidt, U. Effect of left prefrontal repetitive transcranial magnetic stimulation on food craving. *Biol. Psychiatr.* **2005**, *58*, 840–842. [CrossRef]
431. Van den Eynde, F.; Claudino, A.M.; Mogg, A.; Horrell, L.; Stahl, D.; Ribeiro, W.; Uher, R.; Campbell, I.; Schmidt, U. Repetitive transcranial magnetic stimulation reduces cue-induced food craving in bulimic disorders. *Biol. Psychiatr.* **2010**, *67*, 793–795. [CrossRef]

432. Barth, K.S.; Rydin-Gray, S.; Kose, S.; Borckardt, J.J.; O'Neil, P.M.; Shaw, D.; Madan, A.; Budak, A.; George, M.S. Food Cravings and the Effects of Left Prefrontal Repetitive Transcranial Magnetic Stimulation Using an Improved Sham Condition. *Front. Psychol.* **2011**, *2*, 9. [CrossRef]
433. Fregni, F.; Orsati, F.; Pedrosa, W.; Fecteau, S.; Tome, F.A.; Nitsche, M.A.; Mecca, T.; Macedo, E.C.; Pascual-Leone, A.; Boggio, P.S. Transcranial direct current stimulation of the prefrontal cortex modulates the desire for specific foods. *Appetite* **2008**, *51*, 34–41. [CrossRef] [PubMed]
434. Jansen, J.M.; Daams, J.G.; Koeter, M.W.; Veltman, D.J.; Brink, W.V.D.; Goudriaan, A.E. Effects of non-invasive neurostimulation on craving: A meta-analysis. *Neurosci. Biobehav. Rev.* **2013**, *37*, 2472–2480. [CrossRef] [PubMed]
435. Ray, M.K.; Sylvester, M.D.; Helton, A.; Pittman, B.R.; Wagstaff, L.E.; McRae, T.R.; Turan, B.; Fontaine, K.R.; Amthor, F.R.; Boggiano, M.M.; et al. The effect of expectation on transcranial direct current stimulation (tDCS) to suppress food craving and eating in individuals with overweight and obesity. *Appetite* **2019**, *136*, 1–7. [CrossRef] [PubMed]
436. Sedgmond, J.; Lawrence, N.S.; Verbruggen, F.; Morrison, S.; Chambers, C.D.; Adams, R.C. Prefrontal brain stimulation during food-related inhibition training: Effects on food craving, food consumption and inhibitory control. *R. Soc. Open Sci.* **2019**, *6*, 181186. [CrossRef]
437. Hanlon, C.A.; Hartwell, K.J.; Canterberry, M.; Li, X.; Owens, M.; LeMatty, T.; Prisciandaro, J.J.; Borckardt, J.; Brady, K.T.; George, M.S. Reduction of cue-induced craving through realtime neurofeedback in nicotine users: The role of region of interest selection and multiple visits. *Psychiatr. Res.* **2013**, *213*, 79–81. [CrossRef] [PubMed]
438. Hartwell, K.J.; Prisciandaro, J.J.; Borckardt, J.; Li, X.; George, M.S.; Brady, K.T. Real-time fMRI in the treatment of nicotine dependence: A conceptual review and pilot studies. *Psychol. Addict. Behav.* **2013**, *27*, 501–509. [CrossRef] [PubMed]
439. Li, X.; Hartwell, K.J.; Borckardt, J.; Prisciandaro, J.J.; Saladin, M.E.; Morgan, P.S.; Johnson, K.A.; LeMatty, T.; Brady, K.T.; George, M.S. Volitional reduction of anterior cingulate cortex activity produces decreased cue craving in smoking cessation: A preliminary real-time fMRI study. *Addict. Biol.* **2012**, *18*, 739–748. [CrossRef] [PubMed]
440. Dehghani-Arani, F.; Rostami, R.; Nadali, H. Neurofeedback training for opiate addiction: Improvement of mental health and craving. *Appl. Psychophysiol. Biofeedback* **2013**, *38*, 133–141. [CrossRef] [PubMed]
441. Dehghani-Arani, F.; Rostami, R.; Nostratabadi, M. Effectiveness of neurofeedback training as a treatment for opioid-dependent patients. *Clin. EEG Neurosci.* **2010**, *41*, 170–177. [CrossRef] [PubMed]
442. Horrell, T.; El-Baz, A.; Baruth, J.; Tasman, A.; Sokhadze, G.; Stewart, C.; Sokhadze, E. Neurofeedback Effects on Evoked and Induced EEG Gamma Band Reactivity to Drug-related Cues in Cocaine Addiction. *J. Neurother.* **2010**, *14*, 195–216. [CrossRef] [PubMed]
443. Scott, W.C.; Kaiser, D.; Othmer, S.; Sideroff, S.I. Effects of an EEG biofeedback protocol on a mixed substance abusing population. *Am. J. Drug Alcohol Abus.* **2005**, *31*, 455–469. [CrossRef]
444. Dewiputri, W.I.; Auer, T. Functional Magnetic Resonance Imaging (fMRI) neurofeedback: Implementations and applications. *Malays. J. Med Sci.* **2013**, *20*, 5–15.
445. Frank, S.; Lee, S.; Preissl, H.; Schultes, B.; Birbaumer, N.; Veit, R. The obese brain athlete: Self-regulation of the anterior insula in adiposity. *PLoS ONE* **2012**, *7*, e42570. [CrossRef]
446. Frank, S.; Kullmann, S.; Veit, R. Food related processes in the insular cortex. *Front. Hum. Neurosci.* **2013**, *7*, 499. [CrossRef]
447. Ihssen, N.; Sokunbi, M.O.; Lawrence, A.D.; Lawrence, N.S.; Linden, D.E. Neurofeedback of visual food cue reactivity: A potential avenue to alter incentive sensitization and craving. *Brain Imaging Behav.* **2017**, *11*, 915–924. [CrossRef]
448. Kohl, S.H.; Veit, R.; Spetter, M.S.; Günther, A.; Rina, A.; Lührs, M.; Birbaumer, N.; Preissl, H.; Hallschmid, M. Real-time fMRI neurofeedback training to improve eating behavior by self-regulation of the dorsolateral prefrontal cortex: A randomized controlled trial in overweight and obese subjects. *NeuroImage* **2019**, *191*, 596–609. [CrossRef] [PubMed]
449. Spetter, M.S.; Malekshahi, R.; Birbaumer, N.; Lührs, M.; Van Der Veer, A.H.; Scheffler, K.; Spuckti, S.; Preissl, H.; Veit, R.; Hallschmid, M. Volitional regulation of brain responses to food stimuli in overweight and obese subjects: A real-time fMRI feedback study. *Appetite* **2017**, *112*, 188–195. [CrossRef] [PubMed]

450. Weiner, S. The addiction of overeating: Self-help groups as treatment models. *J. Clin. Psychol.* **1998**, *54*, 163–167. [CrossRef]
451. Magill, M.; Ray, L.A. Cognitive-behavioural treatment with alcohol and illicit drug users: A meta-analysis of randomized controlled trials. *J. Stud. Alcohol Drugs* **2009**, *70*, 516–527. [CrossRef] [PubMed]
452. Göhner, W.; Schlatterer, M.; Seelig, H.; Frey, I.; Berg, A.; Fuchs, R. Two-Year follow-up of an interdisciplinary cognitive-behavioral intervention program for obese adults. *J. Psychol.* **2012**, *146*, 371–391. [CrossRef] [PubMed]
453. Marchesini, G.; Natale, S.; Chierici, S.; Manini, R.; Besteghi, L.; Di Domizio, S.; Sartini, A.; Pasqui, F.; Baraldi, L.; Forlani, G.; et al. Effects of cognitive–behavioural therapy on health-related quality of life in obese subjects with and without binge eating disorder. *Int. J. Obes.* **2002**, *26*, 1261–1267. [CrossRef] [PubMed]
454. Linardon, J.; Wade, T.D.; Garcia, X.D.L.P.; Brennan, L. The efficacy of cognitive-behavioral therapy for eating disorders: A systematic review and meta-analysis. *J. Consult. Clin. Psychol.* **2017**, *85*, 1080–1094. [CrossRef] [PubMed]
455. Vanderlinden, J.; Adriaensen, A.; Vancampfort, D.; Pieters, G.; Probst, M.; Vansteelandt, K. A Cognitive-behavioral therapeutic program for patients with obesity and binge eating disorder: Short- and long- term follow-up data of a prospective study. *Behav. Modif.* **2012**, *36*, 670–686. [CrossRef] [PubMed]
456. Wilfley, D.E.; Welch, R.R.; Stein, R.I.; Spurrell, E.B.; Cohen, L.R.; Saelens, B.E.; Dounchis, J.Z.; Frank, M.A.; Wiseman, C.V.; Matt, G.E. A randomized comparison of group cognitive-behavioral therapy and group interpersonal psychotherapy fort the treatment of overweight individuals with binge-eating disorder. *Arch. Gen. Psychiatr.* **2002**, *59*, 713721. [CrossRef] [PubMed]
457. Wilson, G.T.; Fairburn, C.C.; Agras, W.S.; Walsh, B.T.; Kraemer, H. Cognitive-behavioral therapy for bulimia nervosa: Time course and mechanisms of change. *J. Consult. Clin. Psychol.* **2002**, *70*, 267–274. [CrossRef] [PubMed]
458. Gearhardt, A.N.; Corbin, W.R.; Brownell, K.D. Food addiction: An examination of the diagnostic criteria for dependence. *J. Addict. Med.* **2009**, *3*, 1–8. [CrossRef] [PubMed]
459. Earnshaw, V.; Smith, L.; Copenhaver, M. Drug addiction stigma in the context of methadone maintenance therapy: An investigation into understudied sources of stigma. *Int. J. Ment. Healthy Addict.* **2013**, *11*, 110–122. [CrossRef] [PubMed]
460. DePierre, J.A.; Puhl, R.M.; Luedicke, J. A new stigmatized identity? Comparisons of a "food addict" label with other stigmatized health conditions. *Basic Appl. Soc. Psychol.* **2013**, *35*, 10–21. [CrossRef]
461. Rasmussen, N. Stigma and the addiction paradigm for obesity: Lessons from 1950s America. *Addiction* **2015**, *110*, 217–225. [CrossRef] [PubMed]
462. Hardman, C.A.; Rogers, P.J.; Dallas, R.; Scott, J.; Ruddock, H.K.; Robinson, E. "Food addiction is real". The effects of exposure to this message on self-diagnosed food addiction and eating behaviour. *Appetite* **2015**, *91*, 179–184. [CrossRef] [PubMed]
463. Latner, J.D.; Puhl, R.M.; Murakami, J.M.; O'Brien, K.S. Food addiction as a causal model of obesity. Effects on stigma, blame, and perceived psychopathology. *Appetite* **2014**, *77*, 79–84. [CrossRef]
464. Reid, J.; O'Brien, K.S.; Puhl, R.; Hardman, C.A.; Carter, A. Food addiction and its potential links with weight stigma. *Curr. Addict. Rep.* **2018**, *5*, 192–201. [CrossRef]

© 2019 by the authors. Licensee MDPI, Basel, Switzerland. This article is an open access article distributed under the terms and conditions of the Creative Commons Attribution (CC BY) license (http://creativecommons.org/licenses/by/4.0/).

Article

Food Addiction in Eating Disorders and Obesity: Analysis of Clusters and Implications for Treatment

Susana Jiménez-Murcia [1,2,3,*], Zaida Agüera [1,2,4], Georgios Paslakis [5,6], Lucero Munguia [3], Roser Granero [2,7], Jéssica Sánchez-González [1], Isabel Sánchez [1,2], Nadine Riesco [1,2], Ashley N Gearhardt [8], Carlos Dieguez [2,9], Gilda Fazia [10,11], Cristina Segura-García [11,12], Isabel Baenas [1], José M Menchón [1,3,13] and Fernando Fernández-Aranda [1,2,3,*]

1. Department of Psychiatry, University Hospital of Bellvitge-IDIBELL, L'Hospitalet de Llobregat, 08907 Barcelona, Spain; zaguera@bellvitgehospital.cat (Z.A.); jsanchezg@bellvitgehospital.cat (J.S.-G.); isasanchez@bellvitgehospital.cat (I.S.); nriesco@bellvitgehospital.cat (N.R.); ibaenas@bellvitgehospital.cat (I.B.); jmenchon@bellvitgehospital.cat (J.M.M.)
2. CIBER Fisiopatología Obesidad y Nutrición (CIBERobn), Instituto de Salud Carlos III, L'Hospitalet de Llobregat, 08907 Barcelona, Spain; roser.granero@uab.cat (R.G.); carlos.dieguez@usc.es (C.D.)
3. Clinical Sciences Department, School of Medicine, University of Barcelona, 08907 L'Hospitalet de Llobregat, Spain; laarcreed_lm@hotmail.com
4. Department of Public Health, Mental Health and Maternal-Child Nursing, School of Nursing, University of Barcelona, 08907 L'Hospitalet de Llobregat, Spain
5. Toronto General Hospital, University Health Network, Toronto, ON M5G 2C4, Canada; paslakis@outlook.de
6. Department of Psychiatry, University of Toronto, Toronto, ON M5T 1R8, Canada
7. Department of Psychobiology and Methodology, Autonomous University of Barcelona, 08193 Barcelona, Spain
8. Department of Psychology, University of Michigan, Ann Arbor, MI 48109, USA; agearhar@umich.edu
9. Department Physiology, Center for Research in Molecular Medicine and Chronic Diseases (CIMUS), Health Research Institute of Santiago USC-IDIS, University Santiago de Compostela, 15705 Santiago de Compostela, Spain
10. Department of Health Sciences, University "Magna Graecia" of Catanzaro, Viale Europa, 88100 Catanzaro, Italy; gildafazia@gmail.com
11. Center for Clinical Research and Treatment of Eating Disorders, Azienda Ospedaliera Universitaria Mater Domini, 88100 Catanzaro, Italy; segura@unicz.it
12. Department of Medical and Surgical Sciences, University "Magna Graecia" of Catanzaro, 88100 Catanzaro, Italy
13. CIBER de Salud Mental (CIBERSAM), Instituto de Salud Carlos III, L'Hospitalet de Llobregat, 08907 Barcelona, Spain
* Correspondence: sjimenez@bellvitgehospital.cat (S.J.-M.); ffernandez@bellvitgehospital.cat (F.F.-A.); Tel.: +34 93 260 79 88 (S.J.-M.); +34-932-60-72-27 (F.F.-A.)

Received: 31 July 2019; Accepted: 14 October 2019; Published: 3 November 2019

Abstract: Food addiction (FA) has been associated with greater psychopathology in individuals with eating disorders (ED) and obesity (OBE). The current study aims to provide a better phenotypic characterization of the FA construct by conducting a clustering analysis of FA in both conditions (ED and OBE). The total sample was comprised of 234 participants that scored positive on the Yale Food Addiction Scale 2.0. (YFAS-2) (119 bulimia nervosa (BN), 50 binge eating disorder (BED), 49 other specified feeding or eating disorder (OSFED) and 16 OBE). All participants completed a comprehensive battery of questionnaires. Three clusters of FA participants were identified. Cluster 1 (dysfunctional) was characterized by the highest prevalence of OSFED and BN, the highest ED severity and psychopathology, and more dysfunctional personality traits. Cluster 2 (moderate) showed a high prevalence of BN and BED and moderate levels of ED psychopathology. Finally, cluster 3 (adaptive) was characterized by a high prevalence of OBE and BED, low levels of ED psychopathology, and more functional personality traits. In conclusion, this study identified three distinct clusters of ED-OBE patients with FA and provides some insight into a better phenotypic characterization of the FA construct when considering psychopathology, personality and ED pathology. Future studies should address whether these three food addiction categories are indicative of therapy outcome.

Keywords: food addiction; eating disorders; bulimia nervosa; binge eating disorders; obesity; other specified feeding or eating disorders; cluster analysis

1. Introduction

Food addiction (FA) is a concept that has been of increasing scientific interest and debate. An immense body of literature within the field of eating disorder (ED) research has emerged, with whole special issues of scientific journals being dedicated to its characterization [1,2]. In addition to obvious phenomenological similarities between addiction and ED (e.g., loss of control, continued use despite negative consequences, cravings), a great number of neurobiological findings have emerged to additionally support the new concept, not only in preclinical studies, but also in humans [3–6].

Still, not all controversies have yet been resolved. Starting off as a concept to explain a potential subtype of obesity (OBE) [7–10], FA has also been associated with ED, such as bulimia nervosa (BN) [11–13], binge eating disorder (BED) [14–16], and even anorexia nervosa (AN) [17]. In a systematic review of studies on FA in non-clinical and clinical cohorts, it was especially BED that was associated with the most severe FA symptoms [15]. FA also seems to be prevalent in individuals with OBE waiting for bariatric surgery [18,19] and predicts less effective weight reduction throughout dietary intervention before surgery [20]. Interestingly, surgery-induced weight loss may lead to remission of FA [21,22].

The Yale Food Addiction Scale (YFAS) is the main instrument for the assessment of FA and has been developed to assess FA based on the known criteria used for the assessment of substance dependence, but applied for high palatable foods [23,24]. Higher scores on the YFAS have been associated with higher body mass index (BMI), binge eating episodes, impulsivity, and cravings for highly palatable food [25], as well as with neural responses similar to those found in substance use disorders [26–29].

There is scarce evidence with regard to the identification of key determinants of FA based on personality traits or ED-related symptoms and most of what is known is derived from non-clinical cohorts. In a non-clinical sample, negative urgency (irrational acting in aversive affective states) and low levels of task persistence (lack of perseverance) were shown to be significantly and directly associated with FA and FA mediated their association to BMI [30]. In another study in undergraduates, negative urgency, impulsivity when under distress, and emotion dysregulation positively predicted symptom count on the YFAS [31]. Similar findings were shown in a clinical ED cohort, although negative urgency appeared to be the only independent predictor for FA, while self-directedness and emotion dysregulation predicted negative urgency and were highly related to ED-related symptomatology, but not to food addiction itself [32,33]. In individuals with OBE awaiting bariatric surgery, FA was associated with personality traits such as neuroticism, impulsivity, and alexithymia [34], but also more emotion dysregulation, more harm avoidance, and less self-directedness [35].

FA has also been associated with a positive screen for more severe variants of ED-related psychopathology [7] as well as with a positive screen for major mental health symptoms, major depressive episode, anxiety, early life adversities, such as psychological and sexual abuse, and an overall reduced quality of life [36–38]. Finally, female gender was a predictor of severe food addiction [36] and high reward sensitivity was significantly associated with more severe FA symptoms in females [39].

Due to these meaningful interrelationships in the literature between FA and personality traits, as well as psychopathology, it is important to consider the association of FA with these factors in an integrated way (rather than looking at each construct in isolation). Evaluating how personality and psychopathology cluster within the FA construct could lead to a better understanding of potential distinct phenotypes within FA that could have different clinical profiles.

Based on this premise, the aim of the present paper was to explore empirically the severity of clusters of FA-positive (FA+) participants based on psychopathological symptomatology (namely ED-related psychopathology and general psychopathology) and personality traits, and to investigate how the clinical features and diagnosis of ED and OBE were distributed among them. This is the first study that attempts this type of analysis in order to identify subgroups among participants with positive FA in order to provide a better phenotypic characterization of the FA construct. We hypothesized that, despite overlaps, patients with ED would predominantly fit into a cluster characterized by a more severe ED-related psychopathology that would be different than the cluster predominantly found among individuals with OBE. We then evaluated whether there were differences by ED subtypes, with a subgroup of patients (basically BED, but also BN) who are more similar to participants with OBE. We also hypothesized that, among others, ED-related severity, general psychopathology, and personality traits would be important determinants of the FA phenotypes. The rationale for performing this kind of analysis was to gain insight into the possibility of different FA clusters that would ideally translate into future symptom-targeted treatments for FA phenotypes in individuals with OBE and for those suffering from ED.

2. Materials and Methods

2.1. Participants

From an initial sample of 395, the final participants of the current study were 234 females who scored positive for the FA (based on YFAS 2.) and who also had a diagnosis of ED or OBE (119 with BN, 50 with BED, 49 with other specified feeding or eating disorder (OSFED), and 16 with OBE). All participants in the OBE group ($n = 16$) as well as $n = 42$ participants in the BED group (84.0%), $n = 27$ in the BN group (22.7%), and $n = 2$ in the OSFED group (4.1%) had a BMI higher than 30. All participants included in the study were consecutively referred for assessment and treatment at the Unit of Eating Disorders of the Department of Psychiatry of the University Hospital of Bellvitge in Barcelona, between May of 2016 and November 2018, diagnosed according to the DSM-5 [40] criteria, and were between 18 and 60 years old.

Male participants referred for the Unit in that time period were excluded from the study, as the number of them in our sample was too small for meaningful statistical comparisons ($n = 21$; four with BN, eight with BED, two with OSFED, and seven with OBE). Considering the controversial results of the presence of FA in AN [17], as well as the characteristic fears and cognitive distortions about food and weight involved in AN that may influence patients understanding of what is considered excessive food intake or abnormal eating behavior [41], patients diagnosed with AN were excluded from the study as well. Likewise, due to the different reported prevalence of FA in OBE patients [20] compared to OBE with a comorbid ED [14,15,17], the homogeneity of the sample was preserved by only inducing patients with OBE without ED ($n = 53$), and only those with positive FA were included ($n = 16$).

According to the Declaration of Helsinki, the present study was approved by the Clinical Research Ethics Committee (CEIC) of Bellvitge University Hospital (ethic approval code: PR205/17), and written informed consent was obtained from all participants. All the assessments were conducted by experienced psychologist and psychiatrists.

2.2. Assessment

For the assessment, anthropometric measures such as weight and height (without the participants wearing clothes or shoes) were taken to calculate the BMI (i.e., weight (kg)/height (m^2)). In addition to clinically relevant variables (like age of onset or duration of the disorder) and sociodemographical characteristics, a battery of the Spanish-validated versions of the following instruments was used.

2.2.1. Eating Disorders Inventory 2 (EDI-2)

The Eating Disorders Inventory 2 (EDI-2) [42] is a 91-item multidimensional self-report questionnaire answered on a 6-point Likert scale that assesses different cognitive and behavioral characteristics typical for ED: Drive for Thinness, Bulimia, Body Dissatisfaction, Ineffectiveness, Perfectionism, Interpersonal Distrust, Interoceptive Awareness, Maturity Fears, Asceticism, Impulse Regulation, and Social Insecurity. The validated version for the Spanish population was developed by Garner, 1998 [43], with a mean internal consistency of 0.63 (coefficient alpha).

2.2.2. Symptom Checklist-90-Revised (SCL-90-R)

The Symptom Checklist-90-Revised (SCL-90-R) [44] is used to evaluate a broad range of psychological problems and symptoms of psychopathology. It consist of a 90-item questionnaire that measures nine primary symptom dimensions: Somatization, Obsession-Compulsion, Interpersonal Sensitivity, Depression, Anxiety, Hostility, Phobic Anxiety, Paranoid Ideation, and Psychoticism; and includes three global indices: global severity index (overall psychological distress), positive symptom distress index (the intensity of symptoms), and a positive symptom total (self-reported symptoms). The global severity index can be used as a summary of the test. The validation of the scale in a Spanish population [45], obtained a mean internal consistency of 0.75 (coefficient alpha).

2.2.3. Yale Food Addiction Scale 2.0 (YFAS-2)

The Yale Food Addiction Scale 2.0 (YFAS-2) [24] is a 35-item self-report questionnaire for measuring FA during the previous 12 months. This original instrument (YFAS) was based on the Diagnostic and Statistical Manual of Mental Disorders (DSM-IV-TR) criteria for substance dependence and was adapted to the context of food consumption. The newer version of the instrument, YFAS-2, is based on DSM-5 Criteria and evaluates 11 symptoms. The scale produces two measurements: (a) a continuous symptom count score that reflects the number of fulfilled diagnostic criteria (ranging from 0 to 11), and (b) a food addiction threshold based on the number of symptoms (at least 2) and self-reported clinically significant impairment or distress. This final measurement allows for the binary classification of food addiction (present versus absent). Additionally, based on the revised DSM-5 taxonomy, is possible to establish severity cut-offs: mild (2–3 symptoms), moderate (4–5 symptoms), and severe (6–11 symptoms). The translation and validation of the YFAS-2 for Spanish speaking samples with ED was carried out by Granero et al. (2018) [13], showing excellent internal reliability coefficient ($\alpha = 0.94$), as well as an excellent accuracy in discriminating between healthy controls and ED subsamples ($\kappa = 0.75$).

2.2.4. Temperament and Character Inventory-Revised (TCI-R)

The Temperament and Character Inventory-Revised (TCI-R) [46] is a 240-item, 5-point Likert scale, questionnaire that measures four temperament dimensions (Harm Avoidance, Novelty Seeking, Reward Dependence, and Persistence) and three character dimensions (Self-Directedness, Cooperativeness, and Self-Transcendence) of personality. The Spanish validation in an adult population was carried out by Gutiérrez-Zotes et al. [47]; the internal consistency (coefficient alpha) of the scales was 0.87.

2.3. Statistical Analysis

Statistical analysis was carried out with SPSS23 for Windows. Empirical clusters were explored with the two-step-cluster procedure, using as indicators the ED severity level (EDI-2 scores), the global psychopathological state (SCL-90-R scores), the personality profile (TCI-R scores), and the DSM-5 diagnostic subtype. Two-step-cluster used the log-likelihood distance measure through a multinomial probability mass function for categorical variables and a normal density function for continuous variables. This clustering technique has desirable features which makes it different from traditional grouping and latent class techniques [48,49]: (a) scalability, which allows analysis of large data files by constructing a cluster-features-tree which is used as a summary of the records; (b) automatic selection of the number of clusters-classes; and (c) handling of categorical and quantitative variables. Criteria for the final model selected in this study were adequate goodness-of-fit (based on a cohesion and separation index) and adequate clinical interpretability [50]. In this study, the Silhouettes coefficient was used as a measure of the goodness of the final cluster solution. This coefficient estimates the cohesion of the elements within a cluster and the separation between the clusters, and it ranges from −1 to +1 (being the values lower than 0.30 interpreted as poor fitting, between 0.30 and 0.50 as fair fitting, and higher than 0.50 as good fitting [51]).

Comparison between clusters was based on chi-square tests (χ^2) for categorical variables and analysis of variance (ANOVA) for quantitative criteria. Effect size was measured through Cohen's-d coefficient for mean and proportion comparisons (low effect size was considered for $|d| > 0.2$, moderate for $|d| > 0.5$, and large for $|d| > 0.8$; [52]). The Finner's method was used to control Type-I error due to multiple statistical comparisons (this procedure is included into the Familywise error rate stepwise procedures and offers a more powerful test than the classical Bonferroni correction) [53].

3. Results

3.1. Characteristics of the Sample

All the participants in the analyses were women who met criteria for FA positive screening score based on the YFAS-2 questionnaire $n = 234$. The distribution of the whole sample according to diagnostic group was 16 OBE (without ED) (6.8%), 50 BED (21.4%), 119 BN (50.9%), and 49 OSFED (20.9%). Most of the participants were single (62.0%, compared to 24.8% married or living with a stable partner and 13.2% separated or divorced) and achieved primary (40.6%) or secondary (44.9%) education levels. Table A1 (Appendix A) includes the distribution of the clinical profiles stratified and compared by the diagnostic subtype.

3.2. Cluster Composition

The best grouping structure selected for the sample of the study was the three-cluster solution, which coincided with the most optimal solution chosen by the two-step-cluster procedure. This solution obtained a Silhouettes index (0.30) into the fair range, suggesting mild-moderate evidence of the cluster structure in the data. The comparison between the largest cluster size ($n = 90$, 38.5%) and the smallest ($n = 60$, 25.6%) yielded a ratio of 1.50.

Figure 1 summarizes the results of the clustering procedure: the bar-graph with the indicator relevance for each variable (which reports how good each variable was for the grouping and can differentiate between the derived clusters: the higher the importance of the measure, the less likely it is that the variation for the variable between clusters is due to chance and the more likely it is due to underlying differences) and the centroids (which summarizes the cluster patterns for the set of variables and allows clinical interpretation of the empirical clusters). The indicator variables with the highest contribution into the clustering were SCL-90-R scales measuring the symptom levels (concretely psychotic, depressive, interpersonal sensitivity, anxiety, paranoia), followed by the EDI-2 scales measuring ED severity (impulse regulation, social insecurity, and ineffectiveness). The personality traits measured with the TCI-R obtained low relevance for the clustering, except for harm avoidance (which achieved moderate-low capacity), as well as the diagnostic subtype (which achieved poor capacity for the differentiation between the groups).

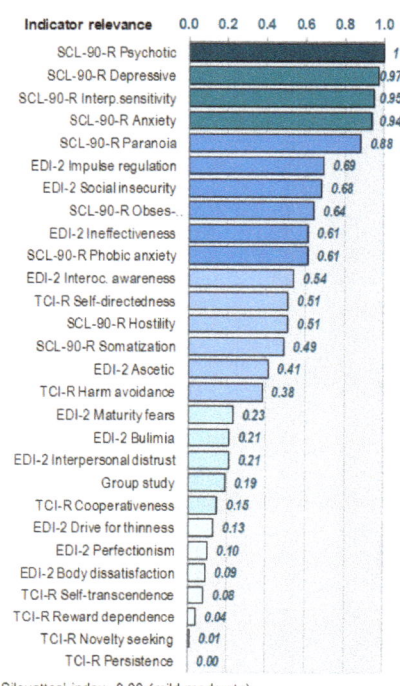

Figure 1. Results of the clustering procedure. Note. EDI-2: Eating disorders Inventory 2. SCL-90-R: Symptom Checklist-90-Revised. SCL-90-R Obsess-comp.: Obsession-Compulsion SCL-90-R subscale. TCI-R: Temperament and Character Inventory-Revised.

3.3. Comparison between the Empirical Clusters

The first part of Table 1 contains the comparison between the three empirical clusters for the civil status and the education levels. Differences between the groups only were found for the civil status: compared with the other two groups, cluster 1 had the highest proportion of single participants.

The second block of Table 1 contains the comparison between the clusters for the clinical profile. As a whole, clusters were ordered by the psychopathological state and personality profiles. Cluster 1 ($n = 60$, 25.6%) included the youngest participants (mean 28.8 years-old), with the lowest age of onset (mean 18.5 years-old), the shortest duration of the eating problems (mean 10.5 years), and the lowest BMI (mean 27.6 kg/m^2), but with the highest severity in eating problems (the highest means in the EDI-2), the worst psychopathological state (the highest means in the SCL-90R), the highest levels in the personality traits of harm avoidance and self-transcendence, and the lowest levels in self-directedness and cooperativeness. Cluster 1 was labeled in this study as the "dysfunctional cluster".

Table 1. Comparison between clusters for variables of the study.

	Cluster 1 n = 60 (25.6%)		Cluster 2 n = 90 (38.5%)		Cluster 3 n = 84 (35.9%)		Cl1-Cl2		Pairwise Comparisons Cl1-Cl3		Cl2-Cl3	
	n	%	n	%	n	%	p	|d|	p	|d|	p	|d|
Civil status—Single	46	76.7%	53	58.9%	46	54.8%	0.049 *	0.39	0.026 *	0.51 †	0.620	0.08
Married—Couple	10	16.7%	22	24.4%	26	31.0%		0.19		0.34		0.15
Separated—divorced	4	6.7%	15	16.7%	12	14.3%		0.32		0.25		0.07
Education Primary	29	48.3%	33	36.7%	33	39.3%	0.164	0.24	0.083	0.18	0.688	0.05
Secondary	27	45.0%	43	47.8%	35	41.7%		0.06		0.07		0.12
University	4	6.7%	14	15.6%	16	19.0%		0.29		0.38		0.09
	Mean	SD	Mean	SD	Mean	SD	p	|d|	p	|d|	p	|d|
Age (years-old)	28.77	11.05	35.66	12.99	34.54	12.13	0.001 *	0.57 †	0.006 *	0.50 †	0.546	0.09
Age of onset (years-old)	18.50	9.17	21.97	10.71	21.66	9.10	0.034 *	0.35	0.057	0.35	0.837	0.03
Duration (years)	10.49	9.10	14.30	10.20	12.64	8.72	0.016 *	0.39	0.177	0.24	0.246	0.17
YFAS-2: total criteria	8.22	2.66	9.59	1.80	7.55	2.77	0.001 *	0.61 †	0.102	0.25	0.001 *	0.87 †
BMI (kg/m^2)	27.55	11.13	30.29	9.08	31.48	10.99	0.114	0.27	0.025 *	0.36	0.447	0.12
EDI-2: Drive for thinness	17.67	2.84	15.60	4.82	13.77	4.83	0.005 *	0.52 †	0.001 *	0.98 †	0.007 *	0.38
EDI-2: Body dissatisfaction	21.60	5.61	20.44	6.52	17.18	7.06	0.288	0.19	0.001 *	0.69 †	0.001 *	0.51 †
EDI-2: Interoceptive awar.	18.25	5.77	15.54	5.33	8.19	5.12	0.003 *	0.52 †	0.001 *	1.84 †	0.001 *	1.41 †
EDI-2: Bulimia	8.68	5.30	11.67	3.85	6.92	4.69	0.001 *	0.64 †	0.023	0.35	0.001 *	1.11 †
EDI-2: Interpers. distrust	9.35	5.23	7.08	4.76	4.12	3.89	0.003 *	0.52 †	0.001 *	1.14 †	0.001 *	0.68 †
EDI-2: Ineffectiveness	18.73	6.57	15.27	6.06	7.08	4.50	0.001 *	0.55 †	0.001 *	2.07 †	0.001 *	1.53 †
EDI-2: Maturity fears	12.70	5.69	9.03	5.37	6.19	5.18	0.001 *	0.66 †	0.001 *	1.20 †	0.001 *	0.54 †
EDI-2: Perfectionism	7.67	4.63	6.53	4.21	4.52	3.65	0.101	0.26	0.001 *	0.75 †	0.002 *	0.51 †
EDI-2: Impulse regulation	13.63	5.46	7.50	4.38	3.27	3.41	0.001 *	1.24 †	0.001 *	2.27 †	0.001 *	1.08 †
EDI-2: Ascetism	10.45	3.18	8.88	2.78	5.69	2.91	0.001 *	0.53 †	0.001 *	1.56 †	0.001 *	1.12 †
EDI-2: Social Insecurity	13.40	4.54	9.60	4.10	4.51	3.03	0.001 *	0.88 †	0.001 *	2.30 †	0.001 *	1.41 †
EDI-2: Total	152.13	27.03	127.14	20.64	81.45	22.16	0.001 *	1.04 †	0.001 *	2.86 †	0.001 *	2.13 †
SCL-90-R: somatization	2.75	0.63	2.13	0.71	1.46	0.72	0.001 *	0.93 †	0.001 *	1.92 †	0.001 *	0.94 †
SCL-90-R: obsess-comp.	2.83	0.59	2.34	0.57	1.53	0.64	0.001 *	0.84 †	0.001 *	2.11 †	0.001 *	1.34 †
SCL-90-R: interpers.sens.	3.03	0.53	2.50	0.54	1.38	0.67	0.001 *	1.00 †	0.001 *	2.73 †	0.001 *	1.84 †
SCL-90-R: depressive	3.24	0.43	2.68	0.53	1.67	0.62	0.001 *	1.17 †	0.001 *	2.93 †	0.001 *	1.74 †
SCL-90-R: anxiety	2.76	0.55	1.96	0.64	1.08	0.54	0.001 *	1.36 †	0.001 *	3.09 †	0.001 *	1.49 †
SCL-90-R: hostility	2.32	0.88	1.43	0.73	0.87	0.67	0.001 *	1.10 †	0.001 *	1.86 †	0.001 *	0.81 †
SCL-90-R: phobic anxiety	2.04	0.83	1.32	0.79	0.51	0.52	0.001 *	0.89 †	0.001 *	2.22 †	0.001 *	1.21 †
SCL-90-R: paranoia	2.48	0.62	1.57	0.59	0.87	0.53	0.001 *	1.50 †	0.001 *	2.77 †	0.001 *	1.25 †
SCL-90-R: psychotic	2.31	0.52	1.56	0.45	0.85	0.48	0.001 *	1.55 †	0.001 *	2.92 †	0.001 *	1.53 †
SCL-90-R: GSI index	2.73	0.36	2.06	0.33	1.26	0.38	0.001 *	1.90 †	0.001 *	3.96 †	0.001 *	2.26 †
SCL-90-R: PST index	81.63	6.60	71.93	8.36	55.42	13.26	0.001 *	1.29 †	0.001 *	2.50 †	0.001 *	1.49 †
SCL-90-R: PSDI index	3.01	0.34	2.60	0.36	2.03	0.36	0.001 *	1.20 †	0.001 *	2.81 †	0.001 *	1.57 †
TCI-R: Novelty seeking	103.32	16.27	99.00	16.77	100.79	14.63	0.105	0.26	0.347	0.16	0.460	0.11
TCI-R: Harm avoidance	133.15	13.32	126.94	16.59	109.63	16.15	0.018 *	0.41	0.001 *	1.59 †	0.001 *	1.06 †
TCI-R: Reward-depend.	95.83	17.43	98.68	15.33	103.25	15.83	0.289	0.17	0.007 *	0.45	0.062	0.29
TCI-R: Persistence	103.42	22.25	102.77	20.58	107.33	19.98	0.852	0.03	0.267	0.19	0.149	0.23
TCI-R: Self-directedness	96.03	13.89	104.39	15.30	124.70	17.28	0.002 *	0.57 †	0.001 *	1.83 †	0.001 *	1.24 †
TCI-R: Cooperativeness	124.40	19.00	133.38	16.53	139.31	11.39	0.001 *	0.50 †	0.001 *	0.95 †	0.013 *	0.42
TCI-R: Self-transcendence	73.35	13.87	63.43	15.43	63.73	17.46	0.001 *	0.68 †	0.001 *	0.61 †	0.903	0.02

Note. SD: standard deviation. * Bold: significant comparison (0.05 level). † Bold: effect size into the moderate (|d| > 0.50) to high range (|d| > 0.80). YFAS-2: Yale food addiction scale 2.0. BMI: Body mass index. EDI-2: Eating disorders Inventory 2. SCL-90-R: Symptom Checklist-90-Revised. TCI-R: Temperament and Character Inventory-Revised.

Cluster 2 (n = 90, 38.5%) included the participants with the longest duration of the eating problems (mean 14.3 years) and the highest level of food addiction (mean 9.6). The mean for BMI (30.3 kg/m2) was higher than in cluster 1, but similar to that obtained in cluster 3. The mean scores in the EDI-2 and the SCL-90R were high for participants into cluster 2, but clearly lower than values registered for cluster 1. Cluster 2 was labeled in this study as the "moderate cluster".

Cluster 3 (n = 84, 35.9%) was characterized by similar mean scores in age (34.5 years-old), onset of eating problems (21.7 years) and duration of the disorder (12.6 years) than cluster 2. The most adaptive scores in the clinical profile was registered for cluster 3 (the lowest scores in the EDI-2 and the SCL-90R), as well as the highest means in the personality traits of persistence, self-directedness and cooperativeness. Cluster 3 was labeled in this study as the "functional cluster".

Figure 2 includes the 100%-stacked bar chart with the distribution of the DSM-5 ED diagnostic subtype into each empirical cluster. OBE patients were primary included into cluster 3 (68.85), while BED patients were mostly distributed into clusters 3 and 2 (48.0% and 44.0%, respectively). Approximately half of the BN patients were into cluster 2 (52.1%), and the remaining participants into this group were distributed between cluster 1 (27.7%) and cluster 3 (20.2%). A little more than half of the OSFED patients were in cluster 3 (51%), and almost 41% in cluster 1.

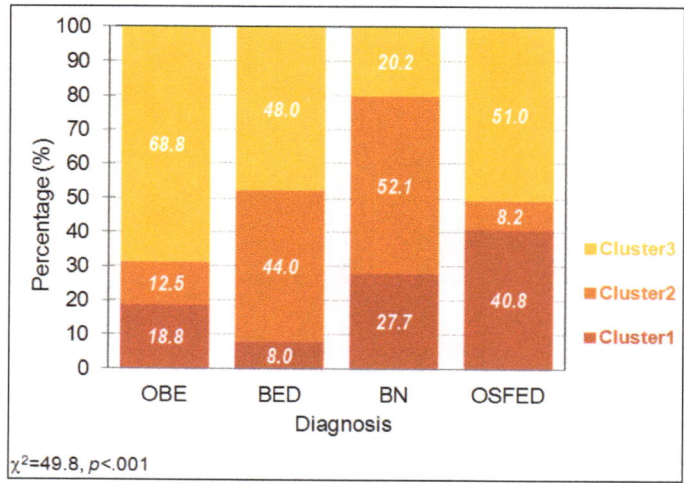

Figure 2. Distribution of the diagnostic subtype and the food addiction (FA) severity group within the empirical clusters. Note. OBE: obesity. BED: binge eating disorder. BN: bulimia nervosa. OSFED: other specified feeding eating disorder.

As a synthesis of the results, Figure 3 contains the radar-chart comparing the empirical clusters for the main clinical variables of the study. To allow adequate interpretability, z-standardized scores were plotted.

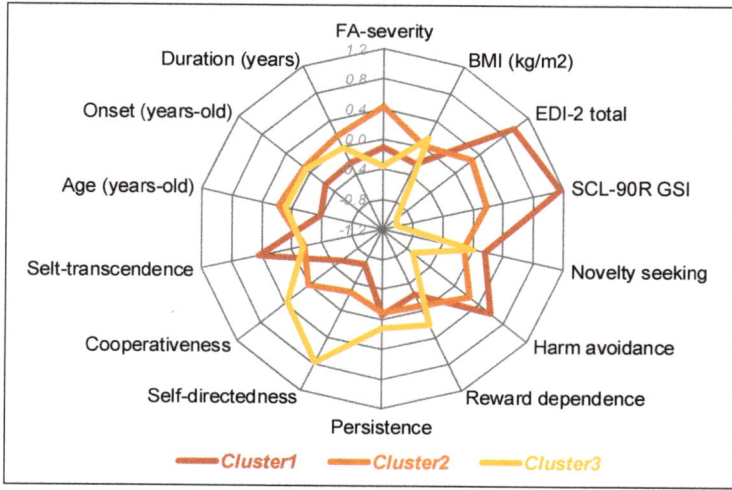

Figure 3. Radar-chart comparing the z-standardized mean scores between the empirical clusters Note. FA: Food addiction. BMI: Body mass index. EDI-2: Eating disorders Inventory 2. SCL-90-R GSI: Global Severity Index of the Symptom Checklist-90-Revised.

4. Discussion

The aim of the present study was to explore empirical severity clusters with FA-positive (FA+) females, and to investigate whether these FA clusters differed by ED and OBE. A three-cluster structure was detected based on general psychopathology, ED severity, and personality traits. These clusters ranged from a more functional cluster, to moderate and highly dysfunctional group.

Although high FA has previously been associated with high eating psychopathology and more dysfunctional personality traits, [32], this is the first study that analyzes a potential heterogeneity within patients with FA. Although all the participants in the current study were FA+, we found that the identified clusters followed a linearity with respect to FA severity with the most dysfunctional clusters (1 and 2) having the highest FA symptoms level, and the most functional one, the lowest. For the "dysfunctional cluster", we found a higher prevalence of OSFED and BN, both ED conditions characterized by more dysfunctional personality traits, greater impulsivity, and more general psychopathology, [54–57], as well by their worse prognosis [56]. Consistent with this literature, the cluster with more OFED and BN had the worst psychopathological state and highest severity in ED symptomatology.

Personality traits that were elevated in the most dysfunction cluster were greater difficulties in establishing goals and objectives to guide their lives (self-directedness), the highest levels of anxiety, worry, fear (harm avoidance), and being a more self-centered person (lower cooperativeness). All of these characteristics have not only been associated with FA [32,35], but also with BN, substance use disorders (SUD), and comorbidity between BN and SUD [58,59]. This cluster was also associated with higher levels of novelty seeking, which is a predictor of risky behaviors and associated with SUD and ED pathologies [59–64]. Therefore, the commonality of these extreme personality features in addictive profiles, suggests it could be important in the etiology and maintenance of FA and could be the focus of intervention efforts. Additionally, the aforementioned characteristics have been related to higher risk of dropout and poor outcome in OSFED patients with purging behaviors [56], suggesting important clinical relevance. Although the association of FA and treatment outcome was not the aim of this study, prior research suggests that higher initial FA scores have been associated with a worse prognosis in BN [65].

In sum, the severity of this dysfunctional cluster may be driven by the comorbid ED provided and characteristics associated with these disorders. Thus, in order to be successful, the treatment aimed at patients who are in this cluster should target its principal psychological and psychiatric characteristics, in addition to factors that may maintain FA. As previous literature has reported, changes in personality traits are related to an overall improvement in eating pathology [66], and interventions that could target the personality factors elevated in this cluster would likely be of benefit. Additionally, due to the severity of the psychopathological state in this cluster, pharmacological approaches to address comorbid psychiatric conditions could be also considered if it is needed.

The second cluster, the moderate cluster, was the one that presented the highest level of FA, although functioning was more adaptive than cluster 1 (i.e., lower ED severity, intermediate psychopathology levels). Therefore, we hypothesize that it is FA which mostly determines the characteristics of this cluster. Regarding the personality traits, high levels of harm avoidance and low self-directedness were present in this cluster, but at moderate levels compared to cluster 1 (the dysfunctional). Related to the ED pathology, there was a high presence of patients with BN followed by patients with BED within this cluster. Higher FA scores have been already associated with bingeing ED-subtype patients [13,14,17,67], given the tendency to consume more high-fat/high-sugar caloric food during binges episodes, which may result in a higher number of FA symptoms [17] given the similar neural responses in reward pathways modulated by dopamine by those types of food and addictive drugs [6,16,29,68–70]. In fact, new maintenance models of BN and BED have emerged, that take into consideration the addictive response to palatable foods [71]. Related studies have also found that similar patterns of neural underpinning of tolerance and dependence observed in SUD appear to be related to binge-type eating disorders as well [72,73]. This is consistent with the high comorbidity between SUD and binge-type ED [74,75]. Finally, another important aspect to mention is that it has been shown that individuals with BN or BED experience more frequent and more intense food cravings than persons without binge eating [76–79] and that both conditions show significantly larger food cue reactivity (self-reported craving) [80]. This intense desire or urge to eat a particular type of food also, at a neural level, resembles responses to drug cues in SUD [26]. Therefore, it could be suggested that the treatment directed to patients in this cluster should target the reward related neural processes that maintain addictive disorders. In this regard, it has been suggested that previously developed interventions for addictions could be applied to binge eating behaviors [71], such as training in the reduction of food cue reactivity in order to reduce craving [81,82] and a reduction in the intake of high-fat/high-sugar caloric food which hyper-activate reward systems.

The "adaptative" cluster presents with more functional personality traits and low levels of general psychopathology, as well as the lowest levels of FA. Thus, the FA in this cluster may be the result of different factors than patients in clusters 1 and 2, which could have important implications in the treatment. First, it is important to mention that within this cluster, there was the highest presence of patients with OBE without ED. This is consistent with prior findings that patients with OBE with a comorbid ED have a higher level of psychopathology than OBE patients without ED [83–86]. Second, this cluster presents the most functional personality traits, the lowest levels of harm avoidance and self-transcendence, and the highest in cooperativeness and self-directedness. Similar results have been found in healthy control groups when comparing them with ED and behavioral addiction patients [87,88]. Thus, BMI could be playing an important role in this cluster, considering that in our sample there are statistically significant differences between cluster 1 (the dysfunctional) and 3 (the adaptative) in BMI. It has been suggested that visceral adiposity levels could be a mediator of the relationship between middle-dorsal insula network connectivity (insula region relevant for eating behavior) and food craving [28], being that visceral fat disrupts insula coding of bodily homeostatic signals, which may boost externally driven food cravings, and also, there are positive associations between food craving and excessive overeating [89,90]. Given the characteristics of this cluster, for which no dysfunctional personality traits of severe ED symptomology or psychopathology must be addressed, it could be hypothesized that nutritional changes that would have a positive impact on a reduction of BMI could be also be beneficial for the reducing FA, through reducing craving episodes. This is consistent with previous studies that find that FA decreases significantly after bariatric surgery [91], and that the induced weight loss by the surgery resulted in the remission of FA in 93% of patients [22]. Finally, there is a moderate representation of patients with BED in this cluster, which may be due to the common co-occurrence between both conditions (OBE-BED). This may be due to the association of features such as grazing [92], craving [93–95], or hedonic [96] and emotional eating [97,98] with OBE and BED. Addressing these factors (and their negative consequences) could be an important focus of treatment in this FA cluster, as grazing is associated with poorer weight loss treatment outcomes in OBE [92] and craving and the use of food to regulate mood are potential triggers for overeating [41,99,100].

The results of this study should take into account the following limitations. First, the sample only included women, so the results cannot be generalized to males; for future studies, it will be important to explore if the FA cluster structure found in this study is replicated in males. Another limitation is that the size of some of the participant groups, divided by ED subtype and OBE, is small. Finally that, due to the fact that the participants were recruited in the same geographical area, the results may not be generalizable to other samples

5. Conclusions

The findings in the present study describe a three-cluster structure of FA-positive (FA+) participants that differ by ED and OBE profile. The clusters range from more to less functional, depending on psychopathology and personality traits. The identification of phenotypes in FA will not only increase knowledge of each cluster's characteristics, but may allow for better individualization of treatment by identifying novel intervention targets and improving treatment outcomes. Likewise, the present study identifies future lines of research, as longitudinal studies that could analyze the predictive validity of this cluster structure on treatment outcome could be of importance.

Author Contributions: The authors had the following roles in this paper: conceptualization, S.J.-M., R.G., A.N.G., C.D., F.F.-A. and R.G.; methodology, F.F.-A, S.J.-M., Z.A. and R.G.; formal statistical analysis, R.G.; data collection and assessment: Z.A., G.P., L.M., J.S.-G., I.S. and N.R.; writing—original draft preparation, Z.A., G.P., L.M., J.S.-G., R.G., A.N.G., F.F.-A. and S.J.-M.; writing—review and editing, Z.A., G.P., L.M., J.S.-G., I.B., J.M.M., A.N.G., C D., G.F., C.S.-G., F.F.-A., S.J.-M. and R.G.; supervision, F.F.-A. and S.J.-M.; funding acquisition, F.F.-A., J.M.M., C.D. and S.J.-M.; Investigation, S.J.-M., Z.A., G.P., L.M., J.S.-G., G.F. and F.F.-A.; Project administration, S.J.-M. and F.F.-A.; Resources, S.J.-M. and F.F.-A.

Funding: We thank CERCA Programme/Generalitat de Catalunya for institutional support. This manuscript and research was supported by grants from the Ministerio de Economía y Competitividad (PSI2015-68701-R), by the Delegación del Gobierno para el Plan Nacional sobre Drogas (2017I067), by the Instituto de Salud Carlos III (ISCIII) (FIS PI14/00290 and PI17/01167), by the SLT006/17/00246 grant, funded by the Department of Health of the Generalitat de Catalunya by the call "Acció instrumental de programes de recerca orientats en l'àmbit de la recerca i la innovació en salut", and co-funded by FEDER funds/European Regional Development Fund (ERDF), a way to build Europe. CIBERObn and CIBERSAM are both initiatives of ISCIII.

Conflicts of Interest: The authors declare no conflict of interest. The funders had no role in the design of the study; in the collection, analyses, or interpretation of data; in the writing of the manuscript, or in the decision to publish the results.

Appendix A

Table A1. Clinical profiles in the ED-subtypes.

	α	Obesity $n = 16$		BED $n = 119$		BN $n = 50$		OSFED $n = 49$		p
		Mean	SD	Mean	SD	Mean	SD	Mean	SD	
Age (years-old)		44.13	11.89	40.90	12.62	31.31	11.23	27.73	9.94	<0.001 *
Age of onset (years-old)		26.06	11.44	25.90	11.18	19.32	9.14	18.29	6.91	<0.001 *
Duration (years)		18.75	9.94	14.79	9.79	12.26	9.24	9.81	8.53	0.003
YFAS2.0: total criteria	0.943	8.19	2.51	9.54	1.68	9.00	2.18	6.35	2.97	<0.001 *
BMI (kg/m^2)		44.64	8.33	40.46	10.06	26.72	6.85	22.59	4.13	<0.001 *
EDI-2: Drive for thinness	0.736	12.31	3.32	12.82	5.19	16.44	4.23	16.88	3.81	<0.001 *
EDI-2: Body dissatisfaction	0.842	20.19	6.47	21.74	5.63	19.40	6.86	17.55	7.06	0.019
EDI-2: Interoceptive awareness	0.785	11.56	7.12	12.30	6.96	14.69	6.66	12.94	6.58	0.078
EDI-2: Bulimia	0.710	6.13	4.67	10.08	3.72	10.91	4.53	5.14	4.63	<0.001 *
EDI-2: Interpersonal distrust	0.802	5.38	4.53	6.52	5.30	7.28	5.09	5.43	4.55	0.121
EDI-2: Ineffectiveness	0.860	12.25	7.84	11.72	6.45	14.78	7.52	11.27	7.41	0.011 *
EDI-2: Maturity fears	0.784	8.06	6.92	7.06	5.95	9.79	5.65	9.14	5.95	0.048 *
EDI-2: Perfectionism	0.697	5.75	4.16	5.84	3.91	6.69	4.45	5.06	4.26	0.146
EDI-2: Impulse regulation	0.757	4.50	4.47	5.74	5.36	8.56	6.15	7.96	5.73	0.005 *
EDI-2: Ascetic	0.717	8.00	3.76	7.24	3.35	8.70	3.58	7.73	3.21	0.071
EDI-2: Social Insecurity	0.770	6.19	5.39	7.76	4.76	9.90	5.38	7.80	4.62	0.004 *
EDI-2: Total	0.928	100.31	34.62	108.82	33.40	127.13	34.96	106.90	38.03	<0.001 *
SCL-90-R: somatization	0.875	2.07	0.93	2.02	0.75	2.13	0.83	1.88	0.98	0.408
SCL-90-R: obses-comp.	0.818	1.88	0.94	2.06	0.80	2.29	0.72	2.10	0.85	0.094
SCL-90-R: interpers.sens.	0.854	1.91	1.13	2.02	0.93	2.40	0.83	2.14	0.87	0.024 *
SCL-90-R: depressive	0.878	2.12	1.06	2.30	0.88	2.60	0.76	2.40	0.84	0.044 *
SCL-90-R: anxiety	0.865	1.36	0.86	1.62	0.87	1.98	0.84	1.92	0.90	0.008 *
SCL-90-R: hostility	0.840	0.99	0.71	1.26	0.81	1.59	0.98	1.49	0.94	0.035 *
SCL-90-R: phobic anxiety	0.800	1.13	0.93	0.97	0.86	1.31	0.94	1.25	0.96	0.179
SCL-90-R: paranoia	0.745	1.63	0.97	1.42	0.85	1.64	0.83	1.46	0.85	0.355
SCL-90-R: psychotic	0.788	1.21	0.67	1.25	0.73	1.61	0.70	1.59	0.80	0.009 *
SCL-90-R: GSI index	0.971	1.69	0.80	1.77	0.67	2.06	0.64	1.91	0.71	0.025 *
SCL-90-R: PST index	0.971	60.81	20.65	65.24	15.58	71.10	12.26	67.98	15.03	0.012 *
SCL-90-R: PSDI index	0.971	2.44	0.59	2.38	0.55	2.57	0.50	2.48	0.53	0.162
TCI-R: Novelty seeking	0.782	94.69	14.06	100.80	15.70	101.85	17.34	100.00	12.76	0.394
TCI-R: Harm avoidance	0.873	117.69	19.22	120.28	17.00	124.04	18.74	121.73	18.85	0.440
TCI-R: Reward-depend.	0.834	108.94	17.06	99.20	16.70	97.74	15.58	101.43	16.48	0.057
TCI-R: Persistence	0.893	103.94	15.64	97.26	20.19	104.36	21.44	112.76	18.92	0.003 *
TCI-R: Self-directedness	0.844	115.31	23.31	108.64	20.65	106.87	18.25	115.06	19.58	0.054
TCI-R: Cooperativeness	0.844	139.88	11.85	131.56	20.63	132.02	16.52	135.59	12.66	0.195
TCI-R: Self-transcendence	0.863	72.38	21.29	66.14	16.53	65.84	14.92	64.55	17.63	0.420

Note. α: Cronbach's alpha in the sample. BED: binge eating disorder; BN: bulimia nervosa; OSFED: other specified feeding eating disorder. SD: standard deviation. * Bold: significant comparison (.05 level). YFAS: Yale food addiction scale 2.0. BMI: Body mass index. EDI-2: Eating disorders Inventory 2. SCL-90-R: Symptom Checklist-90-Revised. TCI-R: Temperament and Character Inventory-Revised.

References

1. Schulte, E.M.; Joyner, M.A.; Potenza, M.N.; Grilo, C.M.; Gearhardt, A.N. Current Considerations Regarding Food Addiction. *Curr. Psychiatry Rep.* **2015**, *17*, 17–19. [CrossRef] [PubMed]
2. Fernandez-Aranda, F.; Karwautz, A.; Treasure, J. Food addiction: A transdiagnostic construct of increasing interest. *Eur. Eat. Disord. Rev.* **2018**, *26*, 536–540. [CrossRef] [PubMed]
3. Avena, N.M.; Rada, P.; Hoebel, B.G. Evidence for sugar addiction: Behavioral and neurochemical effects of intermittent, excessive sugar intake. *Neurosci. Biobehav. Rev.* **2008**, *32*, 20–39. [CrossRef] [PubMed]
4. Volkow, N.D.; Wang, G.J.; Fowler, J.S.; Tomasi, D.; Baler, R. Food and Drug Reward: Overlapping Circuits in Human Obesity and Addiction Inhibitory control. *Curr. Top. Behav. Neurosci.* **2011**, *11*, 1–24.
5. Tang, D.W.; Fellows, L.K.; Small, D.M.; Dagher, A. Food and drug cues activate similar brain regions: A meta-analysis of functional MRI studies. *Physiol. Behav.* **2012**, *106*, 317–324. [CrossRef] [PubMed]
6. Michaud, A.; Vainik, U.; Garcia-Garcia, I.; Dagher, A. Overlapping Neural Endophenotypes in Addiction and Obesity. *Front. Endocrinol.* **2017**, *8*, 127. [CrossRef] [PubMed]
7. Gearhardt, A.N.; Boswell, R.G.; White, M.A. The association of "food addiction" with disordered eating and body mass index. *Eat. Behav.* **2014**, *15*, 427–433. [CrossRef]
8. Ferrario, C.R. Food Addiction and Obesity. *Neuropsychopharmacol. Rev.* **2017**, *42*, 361–362. [CrossRef]
9. Cope, E.C.; Gould, E. New Evidence Linking Obesity and Food Addiction. *Biol. Psychiatry* **2017**, *81*, 734–736. [CrossRef]
10. Sinha, R. Role of addiction and stress neurobiology on food intake and obesity. *Biol. Psychol.* **2018**, *131*, 5–13. [CrossRef]
11. Meule, A.; Von Rezori, V.; Blechert, J. Food Addiction and Bulimia Nervosa. *Eur. Eat. Disord. Rev.* **2014**, *22*, 331–337. [CrossRef] [PubMed]
12. De Vries, S.-K.; Meule, A. Food Addiction and Bulimia Nervosa: New Data Based on the Yale Food Addiction Scale 2.0. *Eur. Eat. Disord. Rev.* **2016**, *24*, 518–522. [CrossRef] [PubMed]
13. Granero, R.; Jiménez-Murcia, S.; Gearhardt, A.N.; Agüera, Z.; Aymamí, N.; Gómez-Peña, M.; Lozano-Madrid, M.; Mallorquí-Bagué, N.; Mestre-Bach, G.; Neto-Antao, M.I.; et al. Validation of the Spanish Version of the Yale Food Addiction Scale 2.0 (YFAS 2.0) and Clinical Correlates in a Sample of Eating Disorder, Gambling Disorder, and Healthy Control Participants. *Front. Psychiatry* **2018**, *9*, 208. [CrossRef] [PubMed]
14. Gearhardt, A.N.; White, M.A.; Masheb, R.M.; Morgan, P.T.; Crosby, R.D.; Grilo, C.M. An Examination of the Food Addiction Construct in Obese Patients with Binge Eating Disorder. *Int. J. Eat. Disord.* **2012**, *45*, 657–663. [CrossRef]
15. Penzenstadler, L.; Soares, C.; Karila, L.; Khazaal, Y. Systematic Review of Food Addiction as Measured with the Yale Food Addiction Scale: Implications for the Food Addiction Construct. *Curr. Neuropharmacol.* **2019**, *17*, 1–13. [CrossRef]
16. Smith, D.G.; Robbins, T.W. The Neurobiological Underpinnings of Obesity and Binge Eating: A Rationale for Adopting the Food Addiction Model. *Biol. Psychiatry* **2013**, *73*, 804–810. [CrossRef]
17. Granero, R.; Hilker, I.; Agüera, Z.; Jiménez-Murcia, S.; Sauchelli, S.; Islam, M.A.; Fagundo, A.B.; Sánchez, I.; Riesco, N.; Dieguez, C.; et al. Food Addiction in a Spanish Sample of Eating Disorders: DSM-5 Diagnostic Subtype Differentiation and Validation Data. *Eur. Eat. Disord. Rev.* **2014**, *22*, 389–396. [CrossRef]
18. Ivezaj, V.; Wiedemann, A.A.; Grilo, C.M. Food Addiction and Bariatric Surgery: A Systematic Review of the Literature. *Obes. Rev.* **2017**, *18*, 1386–1397. [CrossRef]
19. Müller, A.; Leukefeld, C.; Hase, C.; Gruner, K.; Mall, J.W.; Köhler, H.; Zwaan, M. De Food addiction and other addictive behaviours in bariatric surgery candidates. *Eur. Eat. Disord.* **2018**, *26*, 1–12.
20. Guerrero Pérez, F.; Sánchez-González, J.; Sánchez, I.; Jiménez-Murcia, S.; Granero, R.; Simó-Servat, A.; Ruiz, A.; Virgili, N.; López-Urdiales, R.; Montserrat-Gil de Bernabe, M.; et al. Food addiction and preoperative weight loss achievement in patients seeking bariatric surgery. *Eur. Eat. Disord. Rev.* **2018**, *26*, 645–656. [CrossRef]
21. Murray, S.M.; Tweardy, S.; Geliebter, A.; Avena, N.M. A Longitudinal Preliminary Study of Addiction-Like Responses to Food and Alcohol Consumption Among Individuals Undergoing Weight Loss Surgery. *Obes. Surg.* **2019**. [CrossRef] [PubMed]

22. Pepino, M.Y.; Stein, R.I.; Eagon, J.C.; Klein, S. Bariatric surgery-induced weight loss causes remission of food addiction in extreme obesity. *Obesity* **2014**, *22*, 1792–1798. [CrossRef] [PubMed]
23. Gearhardt, A.N.; Corbin, W.R.; Brownell, K.D. Food addiction: an examination of the diagnostic criteria for dependence. *J. Addict. Med.* **2009**, *3*, 1–7. [CrossRef] [PubMed]
24. Gearhardt, A.N.; Corbin, W.R.; Brownell, K.D. Development of the Yale Food Addiction Scale Version 2.0. *Psychol. Addict. Behav.* **2016**, *30*, 113–121. [CrossRef] [PubMed]
25. Meule, A.; Gearhardt, A.N. Food addiction in the light of DSM-5. *Nutrients* **2014**, *6*, 3653–3671. [CrossRef] [PubMed]
26. Gearhardt, A.N.; Yokum, S.; Orr, P.T.; Stice, E.; Corbin, W.R.; Brownell, K.D. The Neural Correlates of "Food Addiction". *Arch. Gen. Psychiatry* **2011**, *68*, 808–816. [CrossRef] [PubMed]
27. Davis, C.; Loxton, N.J.; Levitan, R.D.; Kaplan, A.S.; Carter, J.C.; Kennedy, J.L. "Food addiction" and its association with a dopaminergic multilocus genetic profile. *Physiol. Behav.* **2013**, *118*, 63–69. [CrossRef]
28. Contreras-Rodriguez, O.; Burrows, T.; Pursey, K.M.; Stanwell, P.; Parkes, L.; Soriano-Mas, C.; Verdejo-Garcia, A. Food addiction linked to changes in ventral striatum functional connectivity between fasting and satiety. *Appetite* **2019**, *133*, 18–23. [CrossRef]
29. Schulte, E.M.; Yokum, S.; Potenza, M.N.; Gearhardt, A.N. Neural systems implicated in obesity as an addictive disorder: from biological to behavioral mechanisms. *Prog. Brain Res.* **2016**, *223*, 329–346.
30. Murphy, C.M.; Stojek, M.K.; MacKillop, J. Interrelationships among impulsive personality traits, food addiction, and Body Mass Index. *Appetite* **2014**, *73*, 45–50. [CrossRef]
31. Pivarunas, B.; Conner, B.T. Impulsivity and emotion dysregulation as predictors of food addiction. *Eat. Behav.* **2015**, *19*, 9–14. [CrossRef] [PubMed]
32. Wolz, I.; Hilker, I.; Granero, R.; Jiménez-Murcia, S.; Gearhardt, A.N.; Dieguez, C.; Casanueva, F.F.; Crujeiras, A.B.; Menchón, J.M.; Fernández-Aranda, F. "Food Addiction" in Patients with Eating Disorders is Associated with Negative Urgency and Difficulties to Focus on Long-Term Goals. *Front. Psychol.* **2016**, *7*, 61. [CrossRef] [PubMed]
33. Wolz, I.; Granero, R.; Fernández-aranda, F. A comprehensive model of food addiction in patients with binge-eating symptomatology: The essential role of negative urgency. *Compr. Psychiatry* **2017**, *74*, 118–124. [CrossRef] [PubMed]
34. Brunault, P.; Ducluzeau, P.H.; Courtois, R.; Bourbao-Tournois, C.; Delbachian, I.; Réveillère, C.; Ballon, N. Food Addiction is Associated with Higher Neuroticism, Lower Conscientiousness, Higher Impulsivity, but Lower Extraversion in Obese Patient Candidates for Bariatric Surgery. *Subst. Use Misuse* **2018**, *53*, 1919–1923. [CrossRef] [PubMed]
35. Ouellette, A.-S.; Rodrigue, C.; Lemieux, S.; Tchernof, A.; Biertho, L.; Bégin, C. An examination of the mechanisms and personality traits underlying food addiction among individuals with severe obesity awaiting bariatric surgery. *Eat. Weight Disord. Stud. Anorexia, Bulim. Obes.* **2017**, *22*, 633–640. [CrossRef]
36. Burrows, T.; Skinner, J.; McKenna, R.; Rollo, M. Food Addiction, Binge Eating Disorder, and Obesity: Is There a Relationship? *Behav. Sci.* **2017**, *7*, 54. [CrossRef]
37. Burrows, T.; Kay-Lambkin, F.; Pursey, K.; Skinner, J.; Dayas, C. Food addiction and associations with mental health symptoms: A systematic review with meta-analysis. *J. Hum. Nutr. Diet.* **2018**, *31*, 544–572. [CrossRef]
38. Nunes-neto, P.R.; Köhler, C.A.; Schuch, F.B.; Solmi, M.; Quevedo, J.; Maes, M.; Murru, A.; Vieta, E.; Mcintyre, R.S.; Mcelroy, S.L.; et al. Food addiction: Prevalence, psychopathological correlates and associations with quality of life in a large sample. *J. Psychiatr. Res.* **2018**, *96*, 145–152. [CrossRef]
39. Loxton, N.J.; Tipman, R.J. Reward sensitivity and food addiction in women. *Appetite* **2017**, *115*, 28–35. [CrossRef]
40. American Psychiatric Association. *Diagnostic and Statistical Manual of Mental Disorders (DSM-5)*; American Psychiatric Association: Arlington, VA, USA, 2013; ISBN 0-89042-555-8.
41. Gendall, K.A.; Joyce, P.R.; Sullivan, P.F.; Bulik, C.M. Food cravers: Characteristics of those who binge. *Int. J. Eat. Disord.* **1998**, *23*, 353–360. [CrossRef]
42. Garner, D. *Eating Disorder Inventory-2*; Psychological Assessment Resources: Odessa, Ukraine, 1991.
43. Garner, D. *Inventario de Trastornos de la Conducta Alimentaria (EDI-2)—Manual*; TEA Ediciones: Madrid, Spain, 1998.
44. Derogatis, L. *SCL-90-R. Administration, Scoring and Procedures Manual—II for the Revised Version*. Clinical Psychometric Research; Clinical Psychometric Research: Baltimore, MD, USA, 1990.

45. Derogatis, L. *SCL-90-R. Cuestionario de 90 Síntomas-Manual*; TEA Ediciones: Madrid, Spain, 2002.
46. Cloninger, C.R. *The Temperament and Character Inventory–Revised*; Washington University, Center for Psychobiology of Personality: St. Louis, MO, USA, 1999.
47. Gutiérrez-Zotes, J.A.; Bayón, C.; Montserrat, C.; Valero, J.; Labad, A.; Cloninger, C.R.; Fernández-Aranda, F. Temperament and Character Inventory—Revised (TCI-R). Standardization and normative data in a general population sample. *ACTAS Esp. Psiquiatr.* **2004**, *32*, 8–15. [PubMed]
48. Bacher, J. A Probabilistic Clustering Model for Variables of Mixed Type. *Qual. Quant.* **2000**, *34*, 223–235. [CrossRef]
49. Huang, Z. Extensions to the k-Means Algorithm for Clustering Large Data Sets with Categorical Values. *Data Mining Knowl. Discov.* **1998**, *2*, 283–304. [CrossRef]
50. Nylund, K.L.; Asparouhov, T.; Muthén, B.O. Deciding on the Number of Classes in Latent Class Analysis and Growth Mixture Modeling: A Monte Carlo Simulation Study. *Struct. Equ. Model. A Multidiscip. J.* **2007**, *14*, 535–569. [CrossRef]
51. Rousseeuw, P.J. Silhouettes: A graphical aid to the interpretation and validation of cluster analysis. *J. Comput. Appl. Math.* **1987**, *20*, 53–65. [CrossRef]
52. Kelley, K.; Preacher, K.J. On Effect Size. *Association* **2012**, *17*, 137–152. [CrossRef]
53. Finner, H. On a Monotonicity Problem in Step-Down Multiple Test Procedures. *J. Am. Stat. Assoc.* **1993**, *88*, 920. [CrossRef]
54. Fontenelle, L.F.; Mendlowicz, M.V.; Moreira, R.O.; Appolinario, J.C. An empirical comparison of atypical bulimia nervosa and binge eating disorder. *Braz. J. Med. Biol. Res.* **2005**, *38*, 1663–1667. [CrossRef]
55. Mustelin, L.; Lehtokari, V.-L.; Keski-Rahkonen, A. Other specified and unspecified feeding or eating disorders among women in the community. *Int. J. Eat. Disord.* **2016**, *49*, 1010–1017. [CrossRef]
56. Riesco, N.; Agüera, Z.; Granero, R.; Jiménez-Murcia, S.; Menchón, J.M.; Fernández-Aranda, F. Other Specified Feeding or Eating Disorders (OSFED): Clinical heterogeneity and cognitive-behavioral therapy outcome. *Eur. Psychiatry* **2018**, *54*, 109–116. [CrossRef]
57. Van Hanswijck de Jonge, P.; van Furth, E.F.; Hubert Lacey, J.; Waller, G. The prevalence of DSM-IV personality pathology among individuals with bulimia nervosa, binge eating disorder and obesity. *Psychol. Med.* **2003**, *33*, 1311–1317. [CrossRef] [PubMed]
58. Del Pino-Gutiérrez, A.; Jiménez-Murcia, S.; Fernández-Aranda, F.; Agüera, Z.; Granero, R.; Hakansson, A.; Fagundo, A.B.; Bolao, F.; Valdepérez, A.; Mestre-Bach, G.; et al. The relevance of personality traits in impulsivity-related disorders: From substance use disorders and gambling disorder to bulimia nervosa. *J. Behav. Addict.* **2017**, *6*, 396–405. [CrossRef] [PubMed]
59. Janiri, L.; Martinotti, G.; Dario, T.; Schifano, F.; Bria, P. The gamblers' Temperament and Character Inventory (TCI) personality profile. *Subst. Use Misuse* **2007**, *42*, 975–984. [CrossRef] [PubMed]
60. Wang, Y.; Liu, Y.; Yang, L.; Gu, F.; Li, X.; Zha, R.; Wei, Z.; Pei, Y.; Zhang, P.; Zhou, Y.; et al. Novelty seeking is related to individual risk preference and brain activation associated with risk prediction during decision making. *Sci. Rep.* **2015**, *5*, 10534. [CrossRef] [PubMed]
61. Stansfield, K.H.; Kirstein, C.L. Effects of novelty on behavior in the adolescent and adult rat. *Dev. Psychobiol.* **2006**, *48*, 10–15. [CrossRef] [PubMed]
62. Grucza, R.A.; Robert Cloninger, C.; Bucholz, K.K.; Constantino, J.N.; Schuckit, M.I.; Dick, D.M.; Bierut, L.J. Novelty seeking as a moderator of familial risk for alcohol dependence. *Alcohol. Clin. Exp. Res.* **2006**, *30*, 1176–1183. [CrossRef] [PubMed]
63. Krug, I.; Pinheiro, A.P.; Bulik, C.; Jiménez-Murcia, S.; Granero, R.; Penelo, E.; Masuet, C.; Agüera, Z.; Fernández-Aranda, F. Lifetime substance abuse, family history of alcohol abuse/dependence and novelty seeking in eating disorders: Comparison study of eating disorder subgroups. *Psychiatry Clin. Neurosci.* **2009**, *63*, 82–87. [CrossRef] [PubMed]
64. Angres, D.; Nielsen, A. The role of the TCI-R (Temperament Character Inventory) in individualized treatment plannning in a population of addicted professionals. *J. Addict. Dis.* **2007**, *26* (Suppl. 1), 51–64. [CrossRef]
65. Hilker, I.; Sánchez, I.; Steward, T.; Jiménez-Murcia, S.; Granero, R.; Gearhardt, A.N.; Rodríguez-Muñoz, R.C.; Dieguez, C.; Crujeiras, A.B.; Tolosa-Sola, I.; et al. Food Addiction in Bulimia Nervosa: Clinical Correlates and Association with Response to a Brief Psychoeducational Intervention. *Eur. Eat. Disord. Rev.* **2016**, *24*, 482–488. [CrossRef]

66. Agüera, Z.; Krug, I.; Sánchez, I.; Granero, R.; Penelo, E.; Peñas-Lledó, E.; Jiménez-Murcia, S.; Menchón, J.M.; Fernández-Aranda, F. Personality Changes in Bulimia Nervosa after a Cognitive Behaviour Therapy. *Eur. Eat. Disord. Rev.* **2012**, *20*, 379–385. [CrossRef]
67. Davis, C.; Curtis, C.; Levitan, R.D.; Carter, J.C.; Kaplan, A.S.; Kennedy, J.L. Evidence that "food addiction" is a valid phenotype of obesity. *Appetite* **2011**, *57*, 711–717. [CrossRef] [PubMed]
68. Gearhardt, A.N.; Davis, C.; Kuschner, R.; Brownell, K.D. The addiction potential of hyperpalatable foods. *Curr. Drug Abuse Rev.* **2011**, *4*, 140–145. [CrossRef] [PubMed]
69. Schulte, E.M.; Avena, N.M.; Gearhardt, A.N. Which Foods May Be Addictive? The Roles of Processing, Fat Content, and Glycemic Load. *PLoS ONE* **2015**, *10*, 117959. [CrossRef] [PubMed]
70. Wiss, D.A.; Criscitelli, K.; Gold, M.; Avena, N. Preclinical evidence for the addiction potential of highly palatable foods: Current developments related to maternal influence. *Appetite* **2017**, *115*, 19–27. [CrossRef] [PubMed]
71. Treasure, J.; Leslie, M.; Chami, R.; Fernández-Aranda, F. Are trans diagnostic models of eating disorders fit for purpose? A consideration of the evidence for food addiction. *Eur. Eat. Disord. Rev.* **2018**, *26*, 83–91. [CrossRef]
72. Berridge, K.C. "Liking" and "wanting" food rewards: Brain substrates and roles in eating disorders. *Physiol. Behav.* **2009**, *97*, 537–550. [CrossRef]
73. Kessler, R.M.; Hutson, P.H.; Herman, B.K.; Potenza, M.N. The neurobiological basis of binge-eating disorder. *Neurosci. Biobehav. Rev.* **2016**, *63*, 223–238. [CrossRef]
74. Gregorowski, C.; Seedat, S.; Jordaan, G. A clinical approach to the assessment and management of co-morbid eating disorders and substance use disorders. *BMC Psychiatry* **2013**, *13*, 289. [CrossRef]
75. Fernández-Aranda, F.; Steward, T.; Jimenez-Murcia, S. Substance-Related Disorders in Eating Disorders. In *Encyclopedia of Feeding and Eating Disorders*; Wade, T., Ed.; Springer: Singapore, 2016; pp. 1–5.
76. Innamorati, M.; Imperatori, C.; Balsamo, M.; Tamburello, S.; Belvederi Murri, M.; Contardi, A.; Tamburello, A.; Fabbricatore, M. Food cravings questionnaire-trait (FCQ-T) discriminates between obese and overweight patients with and without binge eating tendencies: The italian version of the FCQ-T. *J. Pers. Assess.* **2014**, *96*, 632–639. [CrossRef]
77. Moreno, S.; Rodríguez, S.; Fernandez, M.C.; Tamez, J.; Cepeda-Benito, A. Clinical validation of the trait and state versions of the food craving questionnaire. *Assessment* **2008**, *15*, 375–387. [CrossRef]
78. Ferrer-Garcia, M.; Pla-Sanjuanelo, J.; Dakanalis, A.; Vilalta-Abella, F.; Riva, G.; Fernandez-Aranda, F.; Sánchez, I.; Ribas-Sabaté, J.; Andreu-Gracia, A.; Escandón-Nagel, N.; et al. Eating behavior style predicts craving and anxiety experienced in food-related virtual environments by patients with eating disorders and healthy controls. *Appetite* **2017**, *117*, 284–293. [CrossRef] [PubMed]
79. Van den Eynde, F.; Koskina, A.; Syrad, H.; Guillaume, S.; Broadbent, H.; Campbell, I.C.; Schmidt, U. State and trait food craving in people with bulimic eating disorders. *Eat. Behav.* **2012**, *13*, 414–417. [CrossRef] [PubMed]
80. Meule, A.; Küppers, C.; Harms, L.; Friederich, H.C.; Schmidt, U.; Blechert, J.; Brockmeyer, T. Food cue-induced craving in individuals with bulimia nervosa and binge-eating disorder. *PLoS ONE* **2018**, *13*, e0204151. [CrossRef] [PubMed]
81. Giel, K.E.; Speer, E.; Schag, K.; Leehr, E.J.; Zipfel, S. Effects of a food-specific inhibition training in individuals with binge eating disorder—Findings from a randomized controlled proof-of-concept study. *Eat. Weight Disord.* **2017**, *22*, 345–351. [CrossRef]
82. Turton, R.; Nazar, B.P.; Burgess, E.E.; Lawrence, N.S.; Cardi, V.; Treasure, J.; Hirsch, C.R. To Go or Not to Go: A Proof of Concept Study Testing Food-Specific Inhibition Training for Women with Eating and Weight Disorders. *Eur. Eat. Disord. Rev.* **2018**, *26*, 11–21. [CrossRef]
83. Arias Horcajadas, F.; Sánchez Romero, S.; Gorgojo Martínez, J.J.; Almódovar Ruiz, F.; Fernández Rojo, S.; Llorente Martin, F. Clinical differences between morbid obese patients with and without binge eating. *Actas españolas Psiquiatr.* **2006**, *34*, 362–370.
84. Fandiño, J.; Moreira, R.O.; Preissler, C.; Gaya, C.W.; Papelbaum, M.; Coutinho, W.F.; Appolinario, J.C. Impact of binge eating disorder in the psychopathological profile of obese women. *Compr. Psychiatry* **2010**, *51*, 110–114. [CrossRef]
85. Fichter, M.M.; Quadflieg, N.; Brandl, B. Recurrent overeating: An empirical comparison of binge eating disorder, bulimia nervosa, and obesity. *Int. J. Eat. Disord.* **1993**, *14*, 1–16. [CrossRef]

86. Villarejo, C.; Jiménez-Murcia, S.; Álvarez-Moya, E.; Granero, R.; Penelo, E.; Treasure, J.; Vilarrasa, N.; Gil-Montserrat de Bernabé, M.; Casanueva, F.F.; Tinahones, F.J.; et al. Loss of Control over Eating: A Description of the Eating Disorder/Obesity Spectrum in Women. *Eur. Eat. Disord. Rev.* **2014**, *22*, 25–31. [CrossRef]
87. Álvarez-Moya, E.M.; Jiménez-Murcia, S.; Granero, R.; Vallejo, J.; Krug, I.; Bulik, C.M.; Fernández-Aranda, F. Comparison of personality risk factors in bulimia nervosa and pathological gambling. *Compr. Psychiatry* **2007**, *48*, 452–457. [CrossRef]
88. Fassino, S.; Daga, G.A.; Pierò, A.; Leombruni, P.; Rovera, G.G. Anger and personality in eating disorders. *J. Psychosom. Res.* **2001**, *51*, 757–764. [CrossRef]
89. Brockmeyer, T.; Hahn, C.; Reetz, C.; Schmidt, U.; Friederich, H.C. Approach bias and cue reactivity towards food in people with high versus low levels of food craving. *Appetite* **2015**, *95*, 197–202. [CrossRef] [PubMed]
90. Hetherington, M.M.; Macdiarmid, J.I. Pleasure and excess: Liking for and overconsumption of chocolate. *Physiol. Behav.* **1995**, *57*, 27–35. [CrossRef]
91. Sevinçer, G.M.; Konuk, N.; Bozkurt, S.; Coşkun, H. Food addiction and the outcome of bariatric surgery at 1-year: Prospective observational study. *Psychiatry Res.* **2016**, *244*, 159–164. [CrossRef]
92. Heriseanu, A.I.; Hay, P.; Corbit, L.; Touyz, S. Grazing in adults with obesity and eating disorders: A systematic review of associated clinical features and meta-analysis of prevalence. *Clin. Psychol. Rev.* **2017**, *58*, 16–32. [CrossRef]
93. White, M.A.; Grilo, C.M. Psychometric properties of the Food Craving Inventory among obese patients with binge eating disorder. *Eat. Behav.* **2005**, *6*, 239–245. [CrossRef]
94. Grilo, C.M. The assessment and treatment of binge eating disorder. *J. Pract. Psychiatry Behav. Health* **1998**, *4*, 191–201. [CrossRef]
95. Ng, L.; Davis, C. Cravings and food consumption in binge eating disorder. *Eat. Behav.* **2013**, *14*, 472–475. [CrossRef]
96. Davis, C.A.; Levitan, R.D.; Reid, C.; Carter, J.C.; Kaplan, A.S.; Patte, K.A.; King, N.; Curtis, C.; Kennedy, J.L. Dopamine for "wanting" and opioids for liking: A comparison of obese adults with and without binge eating. *Obesity* **2009**, *17*, 1220–1225. [CrossRef]
97. Eldredge, K.L.; Agras, W.S. Weight and shape overconcern and emotional eating in binge eating disorder. *Int. J. Eat. Disord.* **1996**, *19*, 73–82. [CrossRef]
98. Geliebter, A.; Aversa, A. Emotional eating in overweight, normal weight, and underweight individuals. *Eat. Behav.* **2003**, *3*, 341–347. [CrossRef]
99. Polivy, K.; Herman, C. Etiology of binge eating: Psychological mechanisms. In *Binge Eating: Nature, Assessment, and Treatment*; Fairburn, C., Wilson, G., Eds.; Guilford Press: New York, NY, USA, 1993; pp. 173–205.
100. Greeno, C.G.; Wing, R.R.; Shiffman, S. Binge antecedents in obese women with and without binge eating disorder. *J. Consult. Clin. Psychol.* **2000**, *68*, 95–102. [CrossRef] [PubMed]

© 2019 by the authors. Licensee MDPI, Basel, Switzerland. This article is an open access article distributed under the terms and conditions of the Creative Commons Attribution (CC BY) license (http://creativecommons.org/licenses/by/4.0/).

Article

Food Addiction Is Associated with Irrational Beliefs via Trait Anxiety and Emotional Eating

Laurence J. Nolan * and Steve M. Jenkins

Department of Psychology, Wagner College, 1 Campus Rd., Staten Island, NY 10301, USA
* Correspondence: LNolan@wagner.edu; Tel.: +1-718-390-3358

Received: 31 May 2019; Accepted: 22 July 2019; Published: 25 July 2019

Abstract: Irrational beliefs (IB) are believed, in cognitive behavioral therapies, to be a prime cause of psychopathologies including anxiety, depression, problem eating, and alcohol misuse. "Food addiction" (FA), which has been modeled on diagnostic criteria for substance use disorder, and emotional eating (EE) have both been implicated in the rise in overweight and obesity. Both FA and EE are associated with anxiety. Thus, in the present study, the hypothesis that IB is associated with FA and with EE was tested. Furthermore, possible mediation of these relationships by trait anxiety and depression (and EE for IB and FA) was examined. The responses of 239 adult participants to questionnaires measuring FA, IB, EE, depression, trait anxiety, and anthropometrics were recorded. The results revealed that IB was significantly positively correlated with FA and EE (and depression and trait anxiety). Furthermore, only EE mediated the effect of IB on FA and this was not moderated by BMI. Finally, trait anxiety (but not depression) mediated the effect of IB on EE. Exploratory analysis revealed a significant serial mediation such that IB predicted higher FA via elevated trait anxiety and emotional eating in that order. The results of this study suggest that IB may be a source of the anxiety that is associated with EE and FA and suggest that clinicians may find IB a target for treatment of those persons who report experiences of EE and FA. IB may play a role in food misuse that leads to elevated BMI.

Keywords: food addiction; irrational beliefs; emotional eating; anxiety; food misuse

1. Introduction

"Food addiction" (FA) has been suggested as a factor in the increased prevalence of overweight and obesity. Proponents of FA suggest that some energy-dense highly palatable foods (or specific additives to foods such as salt or refined sugar) generate addiction-like behaviors in those who consume them [1]. FA, as measured by the Yale Food Addiction Scale (YFAS), is associated with binge eating behavior (BED) [2–4], bulimia nervosa [5], night eating syndrome [6], and with elevated BMI even in the absence of BED [4,7,8]. The FA concept is not without controversy. Some critics prefer an alternate description that focuses on the behavior (i.e., "eating addiction") and suggest there is little evidence for an addicting substance in food. They instead suggest that overeating may be a form of habitual food "abuse" [9] or represent a possible food use disorder [10]. Others have suggested that there is not enough evidence yet to conclude that FA is a distinct entity that explains overeating [11]. Nonetheless, there has been significant interest in FA in the scientific community in recent years [12]. The YFAS, which was based on DSM IV substance dependence criteria, has now been updated to reflect the substance use disorder criteria described in the DSM-5 and dubbed the YFAS2 (which has since been provided in a shortened form [13]).

In cognitive behavioral therapies (CBT), irrational beliefs are believed to be a prime cause of psychopathologies including problem eating and addictive behavior. Ellis [14] and Beck [15] proposed that individuals often have habitual affect-eliciting thought patterns (referred to as irrational beliefs by

Ellis) that can lead to dysfunctional emotional and/or behavioral responses. These irrational beliefs originate from a core process of perfectionism [14] or absolutist thinking [16] and the idea that one should be extremely upset when things go wrong and that it is crucial to be successful and approved of by everyone [15]. This absolutist thinking inevitably leads to anxiety and, in turn, may lead to irrational coping strategies such as substance use and uncontrolled eating typified by emotional eating and FA.

In a meta-analysis of 100 independent samples, irrational beliefs were found to be moderately correlated with psychological distress [17]. More specifically, irrational beliefs were associated with trait anxiety [18,19]. While Rohsenow and Smith [18] did not find a connection between irrational beliefs and depression (as measured by Minnesota Multiphasic Personality Inventory), in daily reports of mood over several months, there was an association of irrational beliefs and reports of depression. Others reported that irrational beliefs were related to depression as measured by the Beck Depression Inventory in a sample of women [20]. Mayhew and Edeleman [21] found that irrational beliefs were predictive of poor coping strategies and low self-esteem. Irrational beliefs have been associated with addictive behaviors such as drug misuse [22–25] and problem gambling (see [26] for review) although, in the gambling studies, irrational beliefs are often assessed using different measures than they are in studies of depression and anxiety.

Several studies (mostly involving samples of undergraduate women without eating disorder diagnoses) have reported a link between irrational beliefs and problem eating. Ruderman [27] reported that irrational beliefs were associated with dietary restraint (the cognitive control of food consumption as measured by the Revised Restraint Scale or RRS), particularly the concern with dieting subscale. Studies examining the relationship between irrational beliefs and subclinical eating disorder symptoms are more common. Irrational beliefs predicted a number of bulimia symptoms in undergraduate women [20,28,29]. In addition, irrational beliefs were found to be predictive of drive for thinness, body dissatisfaction, ineffectiveness, and poor interceptive awareness as measured by the Eating Disorders Inventory [21]. More recently, Tomotaki et al. [30] reported that obsession with eating, dieting, and obese-phobia were predicted by irrational beliefs. Irrational beliefs were found to be higher in women with high body dissatisfaction when compared to a group diagnosed with eating disorders and a group with low body dissatisfaction [31].

While irrational beliefs have been associated with dietary restraint, it has not been examined in relation to emotional eating. Emotional eating is generally viewed as a response to negative emotion or distress [32,33] or ego-threat [34], and has been associated with overeating, binge eating, bulimia nervosa, and obesity (see [32]). There is a positive association between emotional eating and anxiety in persons with obesity (but not in persons with a BMI between 18 and 25) [35] and in a sample of children and adolescents [36]. Irrational beliefs and depression were positively correlated in a sample of women [37]. Thus, irrational beliefs may be associated with emotional eating, possibly as the source of anxiety and/or depression.

The research findings described above suggest that irrational beliefs could predict FA and emotional eating. If they do, it is likely that there would be mediating variables. FA is positively correlated to depression in persons with obesity. Furthermore, FA has been associated with depression in persons with obesity [2,3] and in students and the general population [6,38]. FA has also previously been associated with emotional eating [1,2] and with anxiety [39]. The present study was conducted to determine whether the presence of irrational beliefs predicts higher FA symptoms. Furthermore, if such a relationship exists, it may be mediated by depression, trait anxiety, and/or emotional eating, and may depend on BMI. Absolutist irrational beliefs are predicted to produce psychological distress via activation of anxiety. Maladaptive responses such as emotional or uncontrolled eating (e.g., FA) may occur in response to this anxiety. Thus, the following hypotheses (H) were tested. Irrational beliefs and FA are positively correlated (H1). Furthermore, the effect of irrational beliefs on FA is mediated by trait anxiety, depression, and/or emotional eating (H1a). It was also hypothesized that there would be a positive relationship between irrational beliefs and emotional eating (H2) and that the effect of irrational beliefs on emotional eating is mediated by trait anxiety and/or depression (H2a). Finally, it was hypothesized that there would be a moderation of these relationships by BMI; that the effect of irrational beliefs would depend on the value of BMI (H3). Moderation by gender was also hypothesized but not examined in the present study due to the relatively low number of men in the sample.

2. Materials and Methods

2.1. Participants

The participants were 239 adults; the sample included individuals who identified as women ($n = 176$), men ($n = 60$), or as having non-binary gender ($n = 3$). The sample was mostly composed of undergraduate students. The mean age of participants was 20.72 years (SEM = 0.42; range = 18–57) and their mean BMI was 24.07 kg/m^2 (SEM = 0.32; 5.4% had a BMI less than 18.5; 62.3% between 18.5 and 24.9; 20.9% were between 25 and 29.9; 11.3% ≥ 30). When asked in an open-ended question to describe their ethnic or racial background, participants described themselves as having Arab (1.3%), African (6.3%), Asian (4.2%), European (73.2%), South Asian (3.8%), or more than one primary (4.6%) ancestry. In addition, some of the participants (6.7%) also indicated Hispanic ancestry. In total, 17.3% of the participants met the criteria for FA. In total, 20.9% of the sample had a depression score of 50 or higher suggesting the presence of depression. See Table 1 for mean scores on questionnaires.

2.2. Measures

2.2.1. Food Addiction

FA was assessed by the Modified Yale Food Addiction Scale 2.0 (mYFAS2) which is designed to evaluate indications of "addiction" toward foods according to the DSM-5 criteria for substance use disorder but with fewer questions than the YFAS 2.0 [13]. The mYFAS2 has been validated against questionnaires that measure related constructs, and has a Kuder–Richardson's alpha of 0.86. In the present study, Kuder–Richardson's alpha was 0.77. The mYFAS2 has 13 items (11 for FA symptoms and 2 for distress) and is scored by counting the number of diagnostic criteria that are met. A person is considered to have FA when 3 or more of the criteria are met, and there is impairment or distress present. In analyses presented below, the mYFAS2 was entered as a continuous variable (number of symptoms).

2.2.2. Irrational Beliefs

Irrational beliefs were measured by the Shortened General Attitude and Belief Scale (SGABS) [40]. The SGABS is a 26-item scale that uses Likert-type ratings with responses ranging from 1 (strongly disagree) to 5 (strongly agree). The SGABS has one rational beliefs subscale and six irrationality subscales (need for achievement, need for approval, need for comfort, demand for fairness, self-downing, and other downing) that are summed to create a total irrationality score (higher scores indicate stronger irrational beliefs). Several instruments are available to measure irrational beliefs, some of which may be more sensitive to affect than to cognitions (for a review see [41]). The SGABS is based on the Ellis model of psychopathology [41] and its score has been shown to be less affected by mood than some previous measures [40]. The reliability for this sample was very good (Cronbach's $\alpha = 0.86$).

2.2.3. Eating Styles

Eating styles were measured by the Dutch Eating Behavior Questionnaire (DEBQ) which contains three subscales: restrained eating (DEBQr), emotional eating (DEBQe), and external eating (DEBQx) [42]. All 33 questions are rated on a 5-point Likert-type scale with "never" and "very often" as the anchors. The restraint (cognitive restraint of eating) and external eating (eating in the presence of external cues) scales each have 10 items while the DEBQe contains 13 items. Score for each subscale is the mean rating. In the present sample, the Cronbach's α for the DEBQe, DEBQr, and DEBQx were 0.94, 0.79, and 0.94 respectively.

2.2.4. Trait Anxiety

Trait anxiety was measured using the State-Trait Anxiety Inventory for Adults (STAI) [43]. The STAI differentiates between trait anxiety and state anxiety with 40 items (4-point Likert-type scale) regarding how participants generally feel (trait) and how they feel at this moment (state). Only the trait measure was used in the statistical analysis. In the present study, the Cronbach's α was 0.93.

2.2.5. Depression

Depression was assessed using the Self-report Depression Scale (SDS) [44]. The SDS score is the sum of responses to 20 questions to which the participant responds on a 4-point Likert-type scale. The total score can range from 20 to 80 with most depressed persons scoring between 50 and 69 [45]. The Cronbach's α for the SDS was 0.87 for this sample.

2.3. Procedure

The hypotheses were pre-registered with the Open Science Framework after data collection had commenced but prior to examination of the data (doi: 10.17605/OSF.IO/QWSRD). The procedure was approved by the Wagner College Human Experimentation Review Board (code F18–10).

All participants completed questionnaires using an online platform (Qualtrics, Provo, UT, USA). Questionnaires were presented in randomized order after informed consent was obtained. Questions regarding age, height and weight, gender, and ethnicity were presented after questionnaires. All participants were debriefed as to the purpose of the study after the survey was completed. In the debriefing, none of the participants appeared to be aware of the study's purpose. Students participated in groups at scheduled times in a computer laboratory as one way to complete a research requirement for an introductory psychology course ($n = 174$). Other participants were recruited via the university daily email bulletin and via a link (which took them to the same website on the university server as that used by the students) posted on Facebook. These participants were entered into a lottery to win one of four $25 gift cards if they wished.

2.4. Data Analysis

To ensure quality of data, records were screened for inappropriate responses to open-ended questions, lack of response variation (e.g., giving the same answer to all questions), and unusually short survey completion times.

Statistical analysis was performed using IBM SPSS (version 24). The variables were first correlated to establish whether there was a relationship between irrational beliefs and FA and whether each was correlated to trait anxiety, depression, eating styles, and BMI. Then, mediation multiple regression analysis was conducted using the PROCESS plug-in for SPSS (release 2.16.3) [46] using 5000 bootstrap samples. Variables were mean-centered and heteroscedasticity-consistent standard errors were used. Residuals were checked for stochasticity prior to the analysis. The unstandardized beta coefficients (B), confidence intervals (CI) and adjusted R^2 are reported. Planned analysis included testing whether the relationship between irrational beliefs and FA was mediated by emotional eating, trait anxiety, and/or depression and whether these would be moderated by BMI. That is, the causal hypothesis was that irrational beliefs produce an elevation in FA by increasing trait anxiety, depression, and/or emotional eating. Furthermore, the mediation of the effect of irrational beliefs on emotional eating by trait anxiety and/or depression was also examined (the hypothesis that irrational beliefs cause elevated emotional eating by increasing trait anxiety and/or depression). The results of these analyses led to an exploratory analysis of the indirect pathway between irrational beliefs, trait anxiety, depression, emotional eating, and FA in that order (serial mediation). This final analysis tested the hypothesis that irrational beliefs increase FA symptoms via a pathway from higher trait anxiety and/or depression to higher emotional eating.

3. Results

3.1. Confirmatory Analyses

3.1.1. Correlations

To examine the relationships among the questionnaire variables and test hypotheses H1 and H2, Pearson correlation coefficients were performed. The results indicated that irrational beliefs were positively correlated with all measures (most strongly with depression and trait anxiety; see Table 1). Correlation coefficients indicated that body mass index was significantly positively correlated only with dietary restraint and FA and not with irrational beliefs or other measures (see Table 2).

Table 1. Pearson correlation coefficients among and descriptive statistics for the questionnaire measures.

Measure	SGABS	DEBQe	DEBQx	DEBQr	SDS	mYFAS2	STAI
SGABS		0.298 [4]	0.166 [1]	0.242 [2]	0.498 [4]	0.282 [3]	0.601 [4]
DEBQe			0.530 [4]	0.388 [4]	0.338 [4]	0.485 [4]	0.422 [4]
DEBQx				0.219 [2]	0.039	0.409 [4]	0.057
DEBQr					0.182 [1]	0.328 [4]	0.240 [2]
SDS						0.342 [4]	0.800 [4]
mYFAS2							0.362 [4]
Mean	59.09	2.38	3.06	2.81	40.91	1.62	43.87
SEM	0.79	0.06	0.04	1.01	0.66	0.14	0.79

[1] $p \leq 0.01$, [2] $p \leq 0.001$, [3] $p < 0.0001$, [4] $p < 0.00001$. SGABS: Shortened General Attitude Belief Scale. DEBQe: Dutch Eating Behavior Questionnaire Emotional Eating. DEBQx: Dutch Eating Behavior Questionnaire External Eating. DEBQr: Dutch Eating Behavior Questionnaire Restrained Eating. SDS: Self-report Depression Scale. mYFAS2: Modified Food Addiction Scale 2.0. STAI: State-Trait Anxiety Inventory.

Table 2. Pearson correlation coefficients among the questionnaire measures and body mass index (BMI).

Measure	SGABS	DEBQe	DEBQx	DEBQr	SDS	mYFAS2	STAI
BMI	−0.020	0.082	0.037	0.161 [1]	−0.024	0.269 [2]	0.032

[1] $p < 0.05$, [2] $p < 0.0001$. SGABS: Shortened General Attitude Belief Scale. DEBQe: Dutch Eating Behavior Questionnaire Emotional Eating. DEBQx: Dutch Eating Behavior Questionnaire External Eating. DEBQr: Dutch Eating Behavior Questionnaire Restrained Eating. SDS: Self-report Depression Scale. mYFAS2: Modified Food Addiction Scale 2.0. STAI: State-Trait Anxiety Inventory.

3.1.2. Multiple Regression Mediation Analysis

Multiple Mediation of the Effect of Irrational Beliefs on FA

Multiple mediation analysis (PROCESS model 4) was used to determine whether the effect of irrational beliefs on FA was mediated by depression, trait anxiety, and/or emotional eating (H1a). BMI was included as a covariate because of its correlation with the criterion variable, FA. The results (see Figure 1) indicated that, although irrational beliefs significantly predicted elevated scores on depression, trait anxiety, and emotional eating measures, only emotional eating mediated the effect; there was a significant indirect effect of irrational beliefs on FA through emotional eating ($B = 0.02$; 95%CI: 0.010–0.034) but not though trait anxiety ($B = 0.00$; 95%CI: −0.021–0.021) or depression ($B = 0.02$; 95%CI: −0.001–0.035). In the mediation model, there was no significant direct effect of irrational beliefs on FA ($B = 0.02$, $t = 1.17$, $p = 0.242$; 95%CI: −0.010–0.040). The total effect was statistically significant ($B = 0.05$, $t = 4.20$, $p < 0.001$; 95%CI: 0.027–0.074).

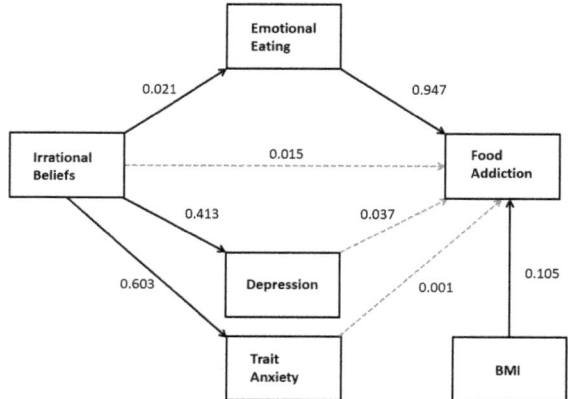

Figure 1. The relationship between irrational beliefs and food addiction (FA) is mediated by emotional eating but not depression or trait anxiety. Solid arrows indicate statistically significant regression coefficients. Nonsignificant BMI effects on mediators are not included for simplicity.

Moderated Mediation by BMI of the Effect of Irrational Beliefs on FA

In order to examine whether BMI moderated the direct and indirect (e.g., through emotional eating) effects of irrational beliefs on FA (H3), BMI was added to the model presented above as a moderator (a moderated mediation model, PROCESS model 15). The results of this analysis again showed emotional eating to be the only significant mediator of the effect of irrational beliefs on FA and indicated that there were no significant interactions between BMI and other effects on FA (see Figure 2). There was no significant conditional direct effect of irrational beliefs on FA at mean-1SD, mean, and mean+1SD values of BMI. In addition, there was a statistically significant conditional indirect effect of irrational beliefs on FA at mean-1SD BMI (95%CI: 0.008–0.037), mean BMI (95%CI: 0.009–0.034) and mean+1SD BMI (95%CI: 0.007–0.038); each coefficient was virtually identical ($B = 0.019$). The index for moderated mediation was not statistically significant (95%CI: −0.0017–0.0017). Finally, there was no moderation of the nonsignificant mediated effects of depression and trait anxiety on FA.

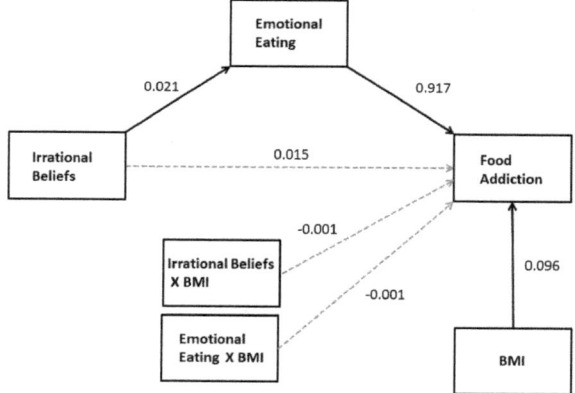

Figure 2. The statistical model for the moderated mediation of the effect of irrational beliefs on FA with BMI as moderator. Solid arrows indicate statistically significant regression coefficients. Nonsignificant mediator variables and BMI effects on mediators are not shown for simplicity.

Multiple Mediation of the Effect of Irrational Beliefs on Emotional Eating

Mediation analysis (PROCESS model 4) was used to determine whether the effect of irrational beliefs on emotional eating was mediated by depression and/or trait anxiety (H2a). The results indicated that, although irrational beliefs significantly predicted elevated scores on depression and trait anxiety, only the latter mediated the effect of irrational beliefs on FA (see Table 3). There was a significant indirect effect of irrational beliefs on emotional eating through trait anxiety ($B = 0.02$; 95%CI: 0.006–0.027) but not through depression ($B = 0.00$; 95%CI: −0.007–0.007; see Figure 3). In the mediation model, there was no significant direct effect of irrational beliefs on emotional eating ($B = 0.01$, $t = 0.83$, $p = 0.406$; 95%CI: −0.007–0.016). The total effect was statistically significant ($B = 0.02$, $t = 4.69$, $p < 0.001$, 95%CI: 0.012–0.030).

Table 3. Multiple mediation model predicting DEBQe from SGABS. Total effect: $R^2 = 0.181$, $F(3,235) = 15.98$, $p < 0.0001$.

Predictor	Coeff.	SE	p	Lower	Upper
Mediator: STAI $R^2 = 0.362$, $F(1,237) = 148.52$, $p < 0.0001$					
SGABS	0.602	0.049	<0.0001	0.505	0.670
Mediator: SDS $R^2 = 0.248$, $F(1,237) = 75.85$, $p < 0.0001$					
SGABS	0.413	0.048	<0.0001	0.320	0.507
Criterion: DEBQe $R^2 = 0.182$, $F(3,235) = 15.98$, $p < 0.0001$					
SGABS	0.005	0.006	0.406	−0.007	0.016
STAI	0.027	0.009	0.002	0.010	0.044
SDS	0.000	0.008	0.989	−0.017	0.016

STAI: State-Trait Anxiety Inventory. SDS: Self-report Depression Scale. DEBQe: Dutch Eating Behavior Questionnaire Emotional Eating. SGABS: Shortened General Attitude Belief Scale.

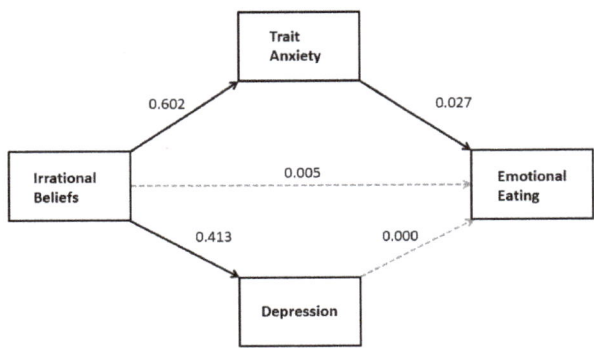

Figure 3. The relationship between irrational beliefs and emotional eating is mediated by trait anxiety but not depression. Solid arrows indicate statistically significant regression coefficients.

3.2. Exploratory Analyses: Serial Mediation of the Effect of Irrational Beliefs on FA

The results presented above indicate a strong relationship between irrational beliefs and depression and trait anxiety but neither predicted FA. Because irrational beliefs are associated with emotional eating via trait anxiety and anxiety and depressed mood have been associated with emotional eating, an additional serial mediation analysis was conducted. Based on the results of the planned correlational and mediation analyses presented above, the hypothesis that the indirect pathway between irrational beliefs and FA would include trait anxiety and depression was examined. Specifically, it was proposed that irrational beliefs would increase trait anxiety which, in turn, would increase depression. Depression and/or trait anxiety was expected to increase emotional eating which, in turn, would increase higher food addiction. BMI was included as a covariate in the model because of its correlation with FA. The results of this analysis (PROCESS model 6; see Table 4) largely supported this hypothesis with the exception that depression was not associated with elevated emotional eating. As depicted in Figure 4, there was a significant indirect effect of irrational beliefs on food addiction through trait anxiety and emotional eating ($B = 0.02$; 95%CI: 0.007–0.028). While trait anxiety was a significant predictor of higher depression score ($B = 0.65$, $t = 15.56$, $p < 0.001$; 95%CI: 0.568–0.732), the indirect path including depression was not statistically significant ($B = 0.00$; 95%CI: −0.006–0.007).

Table 4. Serial mediation analysis predicting mYFAS2 from SGABS. Total effect: $R^2 = 0.155$, $F(2,236) = 16.93$, $p < 0.0001$.

Predictor	Mediator: STAI $R^2 = 0.363$, $F(2,236) = 72.58$, $p < 0.0001$			95% Confidence Interval	
	Coeff.	SE	p	Lower	Upper
SGABS	0.603	0.050	<0.0001	0.504	0.702
BMI	0.109	0.171	0.525	−0.227	0.445
Mediator: SDS $R^2 = 0.643$, $F(3,235) = 180.88$, $p < 0.0001$					
SGABS	0.020	0.044	0.656	−0.067	0.106
STAI	0.653	0.042	<0.0001	0.570	0.735
BMI	−0.100	0.083	0.229	−0.263	0.063
Mediator: DEBQe $R^2 = 0.186$, $F(4,234) = 13.03$, $p < 0.0001$					
SGABS	0.005	0.006	0.387	−0.007	0.017
STAI	0.026	0.009	0.004	0.008	0.043
SDS	0.001	0.008	0.933	−0.016	0.017
BMI	0.013	0.120	0.299	−0.011	0.036
Criterion: mYFAS2 $R^2 = 0.334$, $F(5,233) = 14.51$, $p < 0.0001$					
SGABS	0.015	0.013	0.242	−0.010	0.040
STAI	0.001	0.018	0.964	−0.035	0.037
SDS	0.037	0.022	0.092	−0.006	0.079
DEBQe	0.947	0.194	<0.0001	0.565	1.328
BMI	0.105	0.026	.0001	0.054	0.157

STAI: State-Trait Anxiety Inventory. SGABS: Shortened General Attitude Belief Scale. SDS: Self-report Depression Scale. DEBQe: Dutch Eating Behavior Questionnaire Emotional Eating. mYFAS2: Modified Food Addiction Scale 2.0.

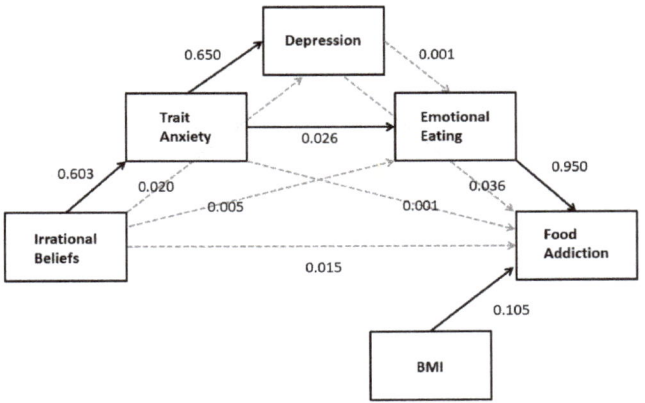

Figure 4. The relationship between irrational beliefs and FA is mediated by the serial pathway through trait anxiety and emotional eating. Solid arrows indicate statistically significant regression coefficients. Nonsignificant BMI effects on mediators are not included for simplicity.

4. Discussion

The results show that, as hypothesized, FA and emotional eating were each positively associated with irrational beliefs. The results of this study are consistent with cognitive behavioral theory and confirm previous findings using different measures of both irrational beliefs and psychopathology that irrational beliefs are associated with elevated trait anxiety and depression. While irrational beliefs did predict higher trait anxiety, depression, and emotional eating, only emotional eating mediated the effect of irrational beliefs on FA. This mediated effect was the same across values of BMI; thus, contrary to prediction, we were unable to show that BMI moderated the mediation of the effect of irrational beliefs on FA by emotional eating. The results also confirmed that the effect of irrational beliefs on emotional eating was mediated by trait anxiety. These findings suggested examination of a serial mediation which found that the indirect effect of irrational beliefs on FA also included trait anxiety. That is, the only significant pathway indicated that irrational beliefs increased trait anxiety which, in turn, increased emotional eating, which finally led to higher number of FA symptoms. The findings that trait anxiety and not depression mediated the associations between irrational beliefs and FA and emotional eating are consistent with the suggestion that anxiety is more important than depression of symptoms of problem eating [see 29]. Finally, we have confirmed that irrational beliefs were positively associated with restrained eating (using a measure other than the RRS) and have found that irrational beliefs were positively correlated with external eating (eating in the presence of food) both of which have been implicated in problem eating and control of body weight. The role of irrational beliefs in restraint and external eating warrants additional exploration as these relationships may be mediated by anxiety or depression.

The research method employed does not allow for causal relationships to be determined with certainty. However, mediation analysis depends upon a theory of causality in order to determine the order of placement of variables in statistical models. In the present analysis, the assumptions were that irrational beliefs are the cause of the psychopathologies and coping behaviors measured because, in CBT, irrational beliefs are considered prime sources of psychopathology. In traditional psychotherapy, it was often believed that activating events (i.e., negative occurrences) result in emotional consequences (i.e., psychological distress). However, according to Ellis [47], this theory is inaccurate or at best incomplete. In the research that led to the development of Rational Emotive Behavior Therapy (REBT), Ellis found that the emotional consequences of an activating event were primarily dictated by the belief a person holds about the activating event. For example, if an employee is reprimanded by an employer for a minor mistake, she or he could think "I am incompetent and will surely be fired from my job, and then no one will ever want to hire me!" The emotional consequence of such a belief would most likely be a significant level of anxiety. If that same person had the more rational thought "It is unfortunate that I made a mistake, but I am human so it will happen sometimes. It is highly unlikely that I will be fired if I make a minor mistake once in a while.", the emotional consequence would likely be one of mild concern or annoyance. Hence, one's emotional state is usually the result of how he or she interprets an event, rather than the event itself. Ellis [47] found that individuals who frequently interpret reality from a distorted, or irrational, perspective are likely to have anxiety or depressive disorders. He also found that when people are emotionally disturbed, they seek out ways to cope with the distress. Coping mechanisms can be adaptive (i.e., make appropriate changes, practice acceptance, exercise, etc.) or maladaptive (i.e., substance use, self-harm, uncontrolled eating). Indeed, some persons may eat to regulate mood and escape anxiety [48,49].

Irrational beliefs-based uncontrolled eating may not necessarily lead to weight gain; in the present study, irrational beliefs were not correlated with BMI. The research on the relationship between irrational beliefs and alcohol consumption indicates that irrational beliefs are associated with problems with alcohol use and not amount of alcohol used [22,23] or frequency of use or getting drunk [24]. Furthermore, perceived lack of control over alcohol use is correlated with irrational beliefs [24]. This is interesting in relation to FA as, unlike alcohol consumption, everyone needs to eat food but those with FA constitute a subset of people who have problem eating who often feel that they cannot control eating. Indeed, lack of control over eating is the most commonly reported FA symptom [50]. Irrational beliefs may lead to FA; FA is common in those with high BMI [7]. It is important to determine whether irrational beliefs are associated with energy consumption from food which may lead to higher BMI in some people. Recent findings suggest that psychological distress is associated with elevated BMI via higher FA and emotional eating [51]. Irrational beliefs may be a source of that psychological distress.

This study has several limitations. While mediation models often use causal wording (i.e., direct and indirect "effects"), the results are correlational and the direction of effect speculative. The sample is composed mostly of students who report having normal body weight. While students are of interest in the study of anxiety and other psychopathology due to high rates of anxiety, depression [52,53] and problem eating [54], their overrepresentation in the present study may limit the generalizability of the findings. Furthermore, while significant, the conditional effects are somewhat weak which may be due to the relatively low percentage of people with BMI greater than 25. In addition, while nearly a fifth of the sample meets the criterion for FA, the number of symptoms is rather low in the sample as a whole. Given the association of emotional eating and FA in persons with high BMI, the relationships reported here would be expected to be higher in those with high BMI.

5. Conclusions

In conclusion, these results suggest for the first time that irrational beliefs may underlie problem eating such as emotional eating and FA via trait anxiety. Additional research may consider whether anxiety sensitivity, the fear that somatic arousal leads to catastrophic consequences, more strongly mediates the relationship between irrational beliefs and FA than trait anxiety. Anxiety sensitivity is distinct from trait anxiety and has been related to substance misuse and emotional eating, particularly in those with high BMI [55]. The results of the present study also suggest that irrational beliefs may be an appropriate target for clinicians when treating problem eating; CBT is an effective approach in treatment of addictive behaviors and problem eating [56].

Author Contributions: Conceptualization, L.J.N. and S.M.J.; Formal analysis, L.J.N.; Investigation, L.J.N.; Methodology, L.J.N. and S.M.J.; Project administration, L.J.N.; Visualization, L.J.N.; Writing—original draft, L.J.N.; Writing—review and editing, L.J.N. and S.M.J.

Funding: This research received no external funding.

Conflicts of Interest: The authors declare no conflict of interest.

References

1. Gearhardt, A.N.; Corbin, W.R.; Brownell, K.D. Preliminary validation of the Yale food addiction scale. *Appetite* **2009**, *52*, 430–436. [CrossRef] [PubMed]
2. Davis, C.; Curtis, C.; Levitan, R.D.; Carter, J.C.; Kaplan, A.S.; Kennedy, J.L. Evidence that 'food addiction' is a valid phenotype of obesity. *Appetite* **2011**, *57*, 711–717. [CrossRef] [PubMed]
3. Gearhardt, A.N.; White, M.A.; Masheb, R.M.; Morgan, P.T.; Crosby, R.D.; Grilo, C.M. An examination of the food addiction construct in obese patients with binge eating disorder. *Int. J. Eat. Disord.* **2012**, *45*, 657–663. [CrossRef] [PubMed]
4. Gearhardt, A.N.; Boswell, R.G.; White, M.A. The association of "food addiction" with disordered eating and body mass index. *Eat. Behav.* **2014**, *15*, 427–433. [CrossRef] [PubMed]
5. De Vries, S.K.; Meule, A. Food addiction and bulimia nervosa: New data based on the Yale Food Addiction Scale 2.0. *Eur. Eat. Disord. Rev.* **2016**, *24*, 518–522. [CrossRef] [PubMed]
6. Nolan, L.J.; Geliebter, A. "Food addiction" is associated with night eating severity. *Appetite* **2016**, *98*, 89–94. [CrossRef] [PubMed]
7. Meule, A. How prevalent is "food addiction"? *Front. Psychiatry* **2011**, *2*, 61. [CrossRef] [PubMed]
8. Pedram, P.; Wadden, D.; Amini, P.; Gulliver, W.; Randell, E.; Cahill, F.; Vasdev, S.; Goodridge, A.; Carter, J.C.; Zhai, G.; et al. Food addiction: Its prevalence and significant association with obesity in the general population. *PLoS ONE* **2013**, *8*, e74832. [CrossRef]
9. Hebebrand, J.; Albayrak, Ö.; Adan, R.; Antel, J.; Dieguez, C.; de Jong, J.; Leng, G.; Menzies, J.; Mercer, J.G.; Murphy, M.; et al. "Eating addiction", rather than "food addiction", better captures addictive-like eating behavior. *Neurosci. Biobehav. Rev.* **2014**, *47*, 295–306. [CrossRef]
10. Nolan, L.J. Is it time to consider the "food use disorder"? *Appetite* **2017**, *115*, 16–18. [CrossRef]
11. Long, C.G.; Blundell, J.E.; Finlayson, G. A systematic review of the application and correlates of YFAS-diagnosed 'food addiction' in humans: Are eating-related 'addictions' a cause for concern or empty concepts? *Obes. Facts* **2015**, *8*, 386–401. [CrossRef] [PubMed]
12. Meule, A. Back by popular demand: A narrative review on the history of food addiction research. *Yale J. Biol. Med.* **2015**, *88*, 295–302. [PubMed]
13. Schulte, E.M.; Gearhardt, A.N. Development of the modified Yale food addiction scale version 2.0. *Eur. Eat. Disord. Rev.* **2017**, *25*, 302–308. [CrossRef] [PubMed]
14. Ellis, A. *Reason and Emotion in Psychotherapy*; Lyle Stuart: New York, NY, USA, 1962.
15. Beck, A.T. *Cognitive Therapy and the Emotional Disorders*; International Universities Press: New York, NY, USA, 1976.
16. Ellis, A. The essence of RET—1984. *J. Ration. Emot. Cogn. Behav. Ther.* **1984**, *2*, 19–25. [CrossRef]
17. Vîslă, A.; Flückiger, C.; Grosse Holtforth, M.; David, D. Irrational beliefs and psychological distress: A meta-analysis. *Psychother. Psychosom.* **2016**, *85*, 8–15. [CrossRef] [PubMed]
18. Rohsenow, D.J.; Smith, R.E. Irrational beliefs and predictors of negative affective states. *Motiv. Emot.* **1982**, *6*, 299–314. [CrossRef]
19. Deffenbacher, J.L.; Zwemer, W.A.; Whisman, M.A.; Hill, R.A.; Sloan, R.D. Irrational beliefs and anxiety. *Cognit. Ther. Res.* **1986**, *10*, 281–291. [CrossRef]
20. Wertheim, E.H.; Poulakis, Z. The relationship among the General Attitude and Belief Scale, other dysfunctional cognition measures, and depressive or bulimic tendencies. *J. Ration. Emot. Cogn. Behav. Ther.* **1993**, *10*, 219–233. [CrossRef]
21. Mayhew, R.; Edelmann, R.J. Self-esteem, irrational beliefs and coping strategies in relation to eating problems in a non-clinical population. *Person. Individ. Differ.* **1989**, *10*, 581–584. [CrossRef]
22. Camatta, C.D.; Nagoshi, C.T. Stress, depression, irrational beliefs, and alcohol use and problems in a college student sample. *Alcohol. Clin. Exp. Res.* **1995**, *19*, 142–146. [CrossRef]
23. Hutchinson, G.T.; Patock-Peckham, J.A.; Cheong, J.; Nagoshi, C.T. Irrational beliefs and behavioral misregulation in the role of alcohol abuse among college students. *J. Ration. Emot. Cogn. Behav.* **1998**, *16*, 61–74. [CrossRef]
24. Nagoshi, C.T. Perceived control of drinking and other predictors of alcohol use and problems in a college student sample. *Addict. Res.* **1999**, *7*, 291–306. [CrossRef]

25. Denoff, M.S. Irrational beliefs as predictors of adolescent drug abuse and running away. *J. Clin. Psychol.* **1987**, *43*, 412–423. [CrossRef]
26. Jacobsen, L.H.; Knudsen, A.K.; Krogh, E.; Pallesen, S.; Molde, H. An overview of cognitive mechanisms in pathological gambling. *Nordic Psychol.* **2007**, *59*, 347–361. [CrossRef]
27. Ruderman, A.J. Restraint and irrational cognitions. *Behav. Res. Ther.* **1985**, *23*, 557–561. [CrossRef]
28. Ruderman, A.J. Bulimia and irrational beliefs. *Behav. Res. Ther.* **1986**, *24*, 193–197. [CrossRef]
29. Lohr, J.M.; Parkinson, D.L. Irrational beliefs and bulimia symptoms. *J. Ration. Emot. Cogn. Behav. Ther.* **1989**, *4*, 253–262. [CrossRef]
30. Tomotake, M.; Okura, M.; Taniguchi, T.; Ishimoto, Y. Traits of irrational beliefs related to eating problems in Japanese college women. *J. Med. Investig.* **2002**, *49*, 51–55.
31. Möller, A.T.; Bothma, M.E. Body dissatisfaction and irrational beliefs. *Psychol. Rep.* **2001**, *88*, 423–430. [CrossRef]
32. Lindeman, M.; Stark, K. Emotional eating and eating disorder psychopathology. *Eat. Disord.* **2001**, *9*, 251–259. [CrossRef]
33. Van Strien, T.; Ouwens, M.A. Effects of distress, alexithymia and impulsivity on eating. *Eat. Behav.* **2007**, *8*, 251–257. [CrossRef] [PubMed]
34. Wallis, D.J.; Hetherington, M.M. Stress and eating: The effects of ego-threat and cognitive demand on food intake in restrained and emotional eaters. *Appetite* **2004**, *43*, 39–46. [CrossRef] [PubMed]
35. Schneider, K.L.; Appelhans, B.M.; Whited, M.C.; Oleski, J.; Pagoto, S.L. Trait anxiety, but not trait anger, predisposes obese individuals to emotional eating. *Appetite* **2010**, *55*, 701–706. [CrossRef] [PubMed]
36. Goossens, L.; Braet, C.; Van Vlierberghe, L.; Mels, S. Loss of control over eating in overweight youngsters: The role of anxiety, depression and emotional eating. *Eur. Eat. Disord. Rev.* **2009**, *17*, 68–78. [CrossRef] [PubMed]
37. Ouwens, M.A.; van Strien, T.; van Leeuwe, J.F. Possible pathways between depression, emotional and external eating. A structural equation model. *Appetite* **2009**, *53*, 245–248. [CrossRef]
38. Meadows, A.; Nolan, L.J.; Higgs, S. Self-perceived food addiction: Prevalence, predictors, and prognosis. *Appetite* **2017**, *114*, 282–298. [CrossRef] [PubMed]
39. Parylak, S.L.; Koob, G.F.; Zorrilla, E.P. The dark side of food addiction. *Physiol. Behav.* **2011**, *104*, 149–156. [CrossRef]
40. Lindner, H.; Kirkby, R.; Wertheim, E.; Birch, P. A brief assessment of irrational thinking: The Shortened General Attitudes and Belief Scale. *Cognit. Ther. Res.* **1999**, *23*, 651–663. [CrossRef]
41. Bridges, K.R.; Harnish, R.J. Role of irrational beliefs in depression and anxiety: A review. *Health* **2010**, *2*, 862–877. [CrossRef]
42. Van Strien, T.; Frijters, J.E.R.; Bergers, G.P.A.; Defares, P.B. The Dutch Eating Behavior Questionnaire (DEBQ) for assessment of restrained, emotional, and external eating behavior. *Int. J. Eat. Disord.* **1986**, *5*, 295–315. [CrossRef]
43. Spielberger, C.D.; Gorsuch, R.L.; Lushene, R.; Vagg, P.R.; Jacobs, G.A. *Manual for the State-Trait Anxiety Inventory*; Consulting Psychologists: Palo Alto, CA, USA, 1983.
44. Zung, W.W. A self-rating depression scale. *Arch. Gen. Psychiatry* **1965**, *12*, 63–70. [CrossRef] [PubMed]
45. Carroll, B.J.; Fielding, J.M.; Blashki, T.G. Depression rating scales: A critical review. *Arch. Gen. Psychiatry* **1973**, *28*, 361–366. [CrossRef]
46. Hayes, A.F. *Introduction to Mediation, Moderation, and Conditional Process Analysis: A Regression-Based Approach*; Guilford: New York, NY, USA, 2017.
47. Ellis, A. *Overcoming Destructive Beliefs, Feelings, and Behaviors: New Directions for Rational Emotive Behavior Therapy*; Prometheus Books: Amherst, NY, USA, 2001.
48. Greeno, C.G.; Wing, R.R. Stress-induced eating. *Psychol. Bull.* **1994**, *115*, 444–464. [CrossRef] [PubMed]
49. Heatherton, T.F.; Baumeister, R.F. Binge eating as escape from self-awareness. *Psychol. Bull.* **1991**, *110*, 86–108. [CrossRef] [PubMed]
50. Meule, A.; Gearhardt, A. Food addiction in the light of DSM-5. *Nutrients* **2014**, *6*, 3653–3671. [CrossRef] [PubMed]
51. Bourdier, L.; Orri, M.; Carre, A.; Gearhardt, A.N.; Romo, L.; Dantzer, C.; Berthoz, S. Are emotionally driven and addictive-like eating behaviors the missing links between psychological distress and greater body weight? *Appetite* **2018**, *120*, 536–546. [CrossRef] [PubMed]

52. Eisenberg, D.; Gollust, S.E.; Golberstein, E.; Hefner, J.L. Prevalence and correlates of depression, anxiety, and suicidality among university students. *Am. J. Orthopsychiatry* **2007**, *77*, 534–542. [CrossRef]
53. Beiter, R.; Nash, R.; McCrady, M.; Rhoades, D.; Linscomb, M.; Clarahan, M.; Sammut, S. The prevalence and correlates of depression, anxiety, and stress in a sample of college students. *J. Affect. Disord.* **2015**, *173*, 90–96. [CrossRef]
54. Zivin, K.; Eisenberg, D.; Gollust, S.E.; Golberstein, E. Persistence of mental health problems and needs in a college student population. *J. Affect. Disord.* **2009**, *117*, 180–185. [CrossRef]
55. Hearon, B.A.; Utschig, A.C.; Smits, J.A.; Moshier, S.J.; Otto, M.W. The role of anxiety sensitivity and eating expectancy in maladaptive eating behavior. *Cogn. Ther. Res.* **2013**, *37*, 923–933. [CrossRef]
56. Schulte, E.M.; Joyner, M.A.; Potenza, M.N.; Grilo, C.M.; Gearhardt, A.N. Current considerations regarding food addiction. *Curr. Psychiatry Rep.* **2015**, *17*, 563. [CrossRef] [PubMed]

© 2019 by the authors. Licensee MDPI, Basel, Switzerland. This article is an open access article distributed under the terms and conditions of the Creative Commons Attribution (CC BY) license (http://creativecommons.org/licenses/by/4.0/).

Review

Ethical, Stigma, and Policy Implications of Food Addiction: A Scoping Review

Stephanie E. Cassin [1,2,3], Daniel Z. Buchman [4,5,6], Samantha E. Leung [2,7], Karin Kantarovich [2,7], Aceel Hawa [8], Adrian Carter [9,10] and Sanjeev Sockalingam [2,3,7,8,*]

1. Department of Psychology, Ryerson University, 350 Victoria St., Toronto, ON M5B 2K3, Canada; stephanie.cassin@psych.ryerson.ca
2. Centre for Mental Health, University Health Network, Network - Toronto General Hospital, 200 Elizabeth Street, 8th Floor, Toronto, ON M5G 2C4, Canada; samantha.leung@uhn.ca (S.E.L.); kar.kantarovich@gmail.com (K.K.)
3. Department of Psychiatry, University of Toronto, 250 College Street, Toronto, ON M5T 1R8, Canada
4. University of Toronto Joint Centre of Bioethics, 155 College Street, Suite 754, Toronto, ON M5T 1P8, Canada; daniel.buchman@uhn.ca
5. Bioethics Program and Krembil Brain Institute, Toronto Western Hospital, University Health Network, 399 Bathurst Street, Toronto, ON M5T 1P8, Canada
6. Dalla Lana School of Public Health, University of Toronto, 155 College Street, 6th Floor, Toronto, ON M5T 1R8, Canada
7. Bariatric Surgery Program, University Health Network - Toronto Western Hospital, 399 Bathurst Street, East Wing – 4th Floor, Toronto, ON M5T 2S8, Canada
8. Department of Education, Centre for Addiction and Mental Health, 33 Russell Street, Toronto, ON M5S 2S1, Canada; aceel.hawa@gmail.com
9. School of Psychological Sciences, Monash University, Melbourne, VIC 3181, Australia; adrian.carter@monash.edu
10. UQ Centre for Clinical Research, University of Queensland, Herston, QLD 4029, Australia
* Correspondence: sanjeev.sockalingam@camh.ca; Tel.: +1-416-535-8501 (ext. x32178)

Received: 1 February 2019; Accepted: 20 March 2019; Published: 27 March 2019

Abstract: The concept of food addiction has generated much controversy. In comparison to research examining the construct of food addiction and its validity, relatively little research has examined the broader implications of food addiction. The purpose of the current scoping review was to examine the potential ethical, stigma, and health policy implications of food addiction. Major themes were identified in the literature, and extensive overlap was identified between several of the themes. Ethics sub-themes related primarily to individual responsibility and included: (i) personal control, will power, and choice; and (ii) blame and weight bias. Stigma sub-themes included: (i) the impact on self-stigma and stigma from others, (ii) the differential impact of substance use disorder versus behavioral addiction on stigma, and (iii) the additive stigma of addiction plus obesity and/or eating disorder. Policy implications were broadly derived from comparisons to the tobacco industry and focused on addictive foods as opposed to food addiction. This scoping review underscored the need for increased awareness of food addiction and the role of the food industry, empirical research to identify specific hyperpalatable food substances, and policy interventions that are not simply extrapolated from tobacco.

Keywords: ethics; food addiction; health policy; stigma

1. Introduction

The average weight of Americans increased dramatically in the 1980s, and this trend did not discriminate based on age, gender, or race [1]. Kessler notes that one remarkable change that coincided with the dramatic weight increase was the availability of "hyperpalatable" foods, specifically those engineered to combine optimal amounts of sugar, fat, and salt [2]. Some authors likened these hyperpalatable foods to addictive drugs, such as cocaine [3]. Though obesity only began garnering serious medical and public attention in recent years, the idea that obesity is a result of a food addiction was, in fact, first posited during the 1940s and 1950s [4,5]. However, this early conceptualization of food addiction resulted in much societal debate as it intensified weight stigma and contributed to ineffective policy changes around obesity and its consequences [5]. Since then, views on obesity and food addiction have been evolving, though the concept of food addiction is still highly debated today.

Food addiction is not an official diagnosis in the Diagnostic and Statistical Manual of Mental Disorders (DSM-5) [6]; however, health care professionals, researchers, patients, and the general public use the term, and a questionnaire (Yale Food Addiction Scale; YFAS) has been developed to diagnose its purported core features [7]. Many similarities have been noted between substance use disorders (e.g., alcohol, nicotine) and food addiction, including consuming more of the substance than intended or over a longer period of time, preoccupation with the substance, craving or strong urge to use the substance, and continued consumption despite knowledge of adverse effects [8].

Food addiction can have negative impacts on individuals living with obesity. Several studies have found that individuals with obesity who meet criteria for food addiction (as measured by the YFAS) demonstrate greater levels of eating disorder psychopathology, poorer general and health-related quality of life, greater depressive symptoms, and higher scores on impulsivity and self-control measures [9–11]. Together, this underscores the importance of studying the relationship between food addiction and obesity.

A food addiction model proposes that some individuals are addicted to certain foods and feel driven to engage in weight promoting eating behaviors, such as binge eating or compulsive overeating, when exposed to "addictive" food substances [7,12,13]. Foods with added fat and refined carbohydrates have been shown to be consumed in a more addictive manner [13,14] and craved more intensely [15] than less refined foods. This is thought to be, in part, a result of such hyperpalatable foods activating the mesolimbic reward-related pathway of the brain, which connects the ventral tegmental area of the midbrain to the nucleus accumbens in the ventral striatum via dopaminergic neurons [16,17]. Neuroimaging studies on individuals meeting the cut-off for food addiction according to the YFAS have been shown to have dysfunctional patterns of reward-related neural activation [12,18], and studies employing the use of genotyping have demonstrated higher multi-locus genetic profile scores in this group, which is associated with enhanced dopamine signaling [19].

Despite mounting evidence for the compulsive consumption of highly palatable foods, food addiction has continued to generate much controversy [20]. Discourses have centered around several issues, including diagnostic challenges of food addiction; arguments that food addiction is not distinct from binge eating disorder [21,22]; debates as to whether food addiction is akin to a behavioral addiction versus a substance use disorder [23]; and the notion that food (albeit not hyperpalatable foods), unlike other substances, is necessary for survival [24]. With respect to the diagnostic challenges, there is emerging evidence that food addiction is distinct from binge eating disorder [25] and may be best classified as a substance (food) use disorder as opposed to a behavioral (eating) addiction [8,26]. A recent systematic review examining the validity of food addiction concluded that the existing research supports food addiction as a unique construct [8]. Amongst the 35 articles (52 studies), the largest number of studies found evidence for brain reward dysfunction and impaired control, followed by studies on tolerance and withdrawal for foods with the greater addictive potential, specifically added sugars and fats.

In comparison to research examining the construct of food addiction and its validity, relatively little research has examined the potential ethical, stigma, and health policy implications of food addiction. Other addictions to substances, such as tobacco, cannabis, or alcohol, have seen an increase in the use of stigmatization as a policy tool in order to promote public health [27]. As a result, ethical issues have arisen surrounding public health communication and its effect on individuals and on society as a whole [28]. Bayer has argued that the use of stigma may reduce the prevalence of behaviors linked to disease and mortality [29]. People who identify as "fat" or "obese" have long received stigmatizing "fat-shaming" messages from healthcare providers and public health officials. Stigmatization of people with obesity often has the opposite effect and is ethically indivisible given the stress and emotional and dignitary harms it causes [30]. This further emphasizes the importance of developing a nuanced understanding of the potential impact of a food addiction model on stigma, particularly given that it could have divergent effects on externalized stigma towards others and internalized stigma towards the self.

A recent systematic review of nine empirical studies examining the link between food addiction and weight stigma concluded that limited evidence suggests that food addiction explanations may reduce self-blame and stigma from others, but these benefits may be offset by the adverse impact on self-efficacy and eating behaviors [31]. However, the authors also noted that the existing empirical research is sparse and inconsistent. To our knowledge, no scoping reviews have been conducted to examine the ethical and health policy implications of food addiction, or the implications on stigma more broadly beyond weight stigma. The purpose of this scoping review was to highlight what has been published in the food addiction literature and to address these current gaps to help inform current perceptions by identifying the potential ethical, stigma, and health policy implications of food addiction.

2. Materials and Methods

This review was conducted as per the Arksey and O'Malley methodological framework for scoping reviews [32]. As part of the development of our research questions and approach, a stakeholder summit was organized in collaboration with the Canadian Obesity Network (CON) involving a broad and diverse group of 18 stakeholders including local knowledge users (i.e., people with experience with obesity-related treatment and food addiction), representatives from national obesity organizations and patient advocacy groups, primary care physicians, and interprofessional research leaders nationally and internationally in food addiction and/or obesity management. Data from a pre-summit online questionnaire exploring perceptions related to the ethical and policy issues of food addiction in obesity informed summit discussions aimed at identifying key themes related to food addiction. The pre-summit online questionnaire was derived from previous studies on food addiction in obesity [33,34] and included open-ended questions regarding the potential implications of food addiction on stigma, ethics, and health policy. The following five stages were completed as part of the scoping review methodology.

Stage 1: Identifying the Research Questions

Based on the analysis of themes resulting from the stakeholder summit, this scoping review was conducted to address the following question: What are the potential ethical, stigma and policy implications of food addiction in obesity?

Inclusion/exclusion criteria

We included peer- and non-peer reviewed articles, editorials, commentaries, letters, and replies to authors that met the following criteria:

1. Were published in English;
2. Were human studies;
3. Established a link between obesity or binge eating and addiction or substance use;
4. Presented a conceptual or mechanistic model of food addiction (i.e., had to include terms such as "reward," "reward pathway," "compulsion," "substance use," or "pleasurable" somewhere in the article); and
5. Presented the issue of food addiction using a key concept (rather than quoting a prevalence).

We excluded book chapters, dissertations, and policy documents. Papers were also excluded if they only focused on obesity and did not relate to food addiction.

Stage 2: Identifying Relevant Studies

We searched for relevant articles published up until January 26, 2017, for each question using the following databases: Medline and PsycINFO from the OvidSP platform and CINAHL from the EBSCO platform. Each of the OvidSP databases was searched individually. The searches were conducted with assistance from a research librarian from an academic health sciences center. The searches were run with the limits of human studies and English language studies only. When appropriate, controlled vocabulary terms and/or text words were used in the subject component blocks. For example, obesity/bariatric terms included one block, food addiction terms included a second block, and ethics/stigma/policy terms included a third block. Lastly, we reviewed key special issues on food addiction in prominent eating disorder and obesity periodicals to identify additional articles that may have met our inclusion criteria and were missed by our initial search. A total of 410 articles were identified from these search strategies.

Stage 3: Selecting Studies

All references resulting from the literature search were reviewed and 46 duplicates were removed. Preliminary screening was conducted by titles and abstracts by two independent reviewers. Seventeen references were missing and were excluded from screening. Screened abstracts were separated into three categories: articles to include, unsure articles, and non-relevant articles to exclude. A secondary screening was then conducted by two additional independent reviewers on the articles that were deemed unsure from the first round of screening. This screening procedure resulted in 296 articles being excluded. Full text articles for the remaining 60 abstracts were then retrieved for data collection (see Figure 1 for PRISMA diagram).

Stage 4: Charting the Data

Data were extracted from all included full-text articles by one reviewer into a spreadsheet that included authors, year of publication, and article type and relevant excerpts, themes, and quotes. Each domain was then separated into individual spreadsheets with the relevant articles and their respective identifiers and variables for coding and analysis.

Stage 5: Collating, Summarizing, and Reporting the Results

Three independent reviewers completed this step. Each of the three main domains of ethics, stigma, and policy had two reviewers who independently reviewed the full text articles and coded the data. A thematic analysis approach was used to critically analyze the articles and generate open codes. All reviewers then came together to discuss, compare, and organize the relevant excerpts, themes, and quotes that were extracted for each of the three main domains. These iterative discussions allowed for a higher level of thematic analysis, generating broad categories and higher-level themes. Reviewers met nine times at 1- to 2-month intervals for a total of 18 h to review codes and to generate domains summarized below.

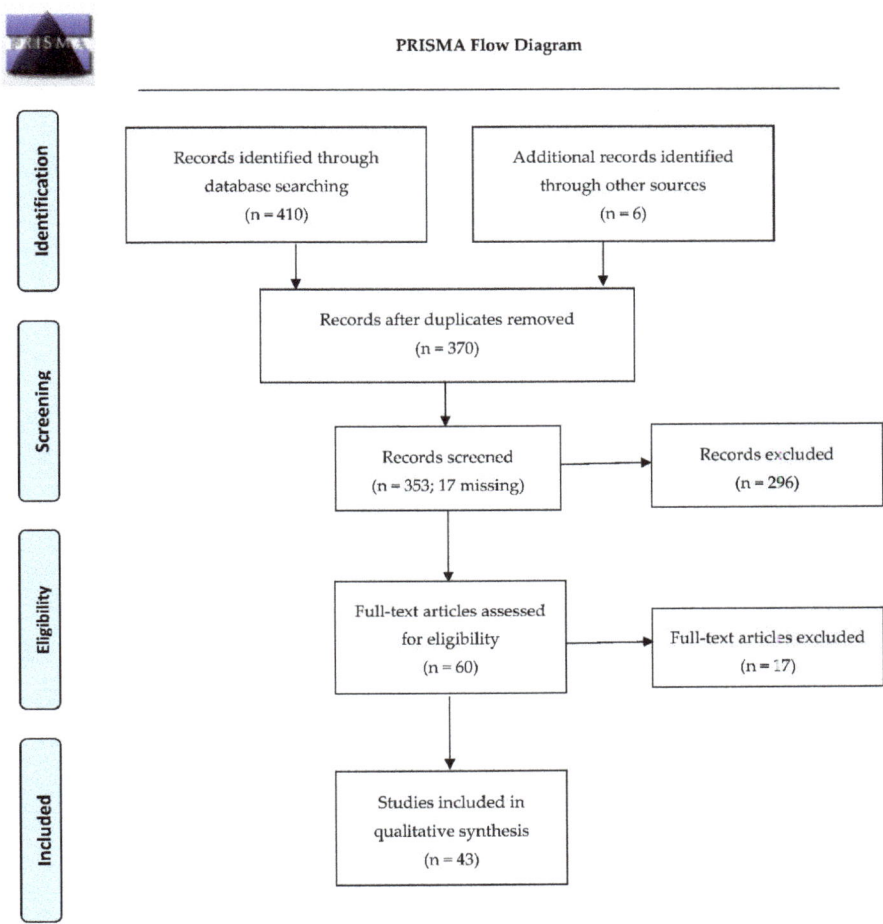

Figure 1. PRISMA flow diagram depicting the flow of information through the different phases of this scoping review. From: [35].

3. Results

Each of the three domains of ethics, stigma, and policy implications related to food addiction are described below. All articles included from the search strategy for each of the three domains are listed in Table S1. Of the 43 studies included in the analysis, 10 were reviews, 25 were original research articles, 5 were commentaries, 2 were opinions, and 1 was a brief. Table 1 provides a summary of themes and subthemes with sample quotes extracted from articles.

Table 1. Scoping review themes, subthemes, and representative quotes.

Themes and Sub-Themes	Authors	Sample Quote
ETHICS		
Personal control, will power, and choice	Foddy (2011) [36]	"The disease label implies a reduction in the autonomy of the addicted and obese. If their strong preferences for consuming food or drugs are merely the symptoms of disease, then these preferences should be viewed as irrelevant in assessing whether their behaviour is willful or not" (p. 86).
	Lee et al. (2014) [33]	"Equating obesity with food addiction could even justify the use of coercive treatments if obese individuals are seen to suffer from a form of addiction over which they have limited control" (pp. 5313–5314).
Blame and weight bias	Thibodeau et al. (2015) [37]	"The narrative of 'addiction' attributed relatively more blame to the behavior of an [obese] individual, evoking comparisons to alcoholism or drug abuse" (p. 29).
	Gearhardt et al. (2012) [38]	"[I]n addition to causing individual suffering weight bias creates personal and societal injustices for obese persons" (p. 409).
STIGMA		
Impact on self-stigma and stigma from others	Burmeister et al. (2013) [39]	"Food addiction symptoms were related to self-reports of internalized weight bias as well as a form of explicit weight bias in which participants indicated dislike of people who carry excess weight and fears related to becoming or remaining overweight. Given our culture's traditional views of addiction as a blameworthy illness, it may come as no surprise that obese food 'addicts' might tend to internalize these stigmatizing beliefs (p. 108)."
Differential impact of substance use disorder versus behavioral addiction	DePierre et al. (2013) [40]	"From the perspective of attribution theory, applying a food addict label to an obese individual could either ameliorate or exacerbate weight stigma. Attribution theory posits that the more a person is seen as responsible for his or her condition, the more people will blame and react to him or her negatively. Indeed, ascribing the cause of obesity to behavioral factors within personal control has been demonstrated to increase stigma, whereas people display fewer negative attitudes toward an individual whose overweight is attributed to biogenetic or physiological factors outside of personal control. Food addiction could be perceived as an external explanation for obesity, reducing blame and stigma. Alternatively, a food addict label could instead act as a behavioral causal attribution, leading obesity to be perceived simply as a result of overeating, potentially increasing weight bias (p. 11)."
	DePierre et al. (2014) [41]	"Food addiction was perceived as a problem of the mind more than either smoking or alcoholism, and received high endorsement as a behaviour resulting from personal unhappiness, indicating that food addiction is viewed as a mental or behavioural problem rather than a physical addiction. Such a perception may detract from beliefs that certain foods can be addictive, with food addiction instead being seen as rooted in an individual's psychological make-up. Additionally, given that mental illnesses elicit more stigma and blame than physical ailments, it is possible that food addiction may increase bias towards overweight/obese individuals with this disorder (p. 5)."
Additive stigma of addiction plus obesity and/or eating disorder	DePierre et al. (2013) [40]	"In the context of attribution theory, the food addict label may have increased blame toward obese individuals by attributing weight to eating behavior, where food addiction may be interpreted as a euphemism for overeating. Or perhaps, in line with Goffman's (1963) framework, categorizing obesity as the result of an addiction added to this 'abomination of the body' the stigma of a 'blemish of character'. Perceiving obesity as the result of a personal failing such as addiction may extend the domains in which it is stigmatized from the immediate social interaction to perceptions of competency for more solitary tasks (or, obesity may extend the otherwise concealable stigma of food addiction to social interactions), potentially explaining why an obese food addict was more negatively perceived than a food addict alone (p. 18)."

Table 1. Cont.

Themes and Sub-Themes	Authors	Sample Quote
POLICY		
Comparison to "Big Tobacco"	Allen et al. (2012) [42]	"It is also vital to consider the numerous elements unique to the obesity epidemic that may require novel policy strategies. Most evident is the fact that food is essential for survival, whereas tobacco use is viewed as a dispensable, or even recreational, activity."
Reducing access to addictive foods through regulation	Cullen et al. (2017) [43]	"[Restrictions] would probably help but if the person really wants a Coke every day, they would probably find the means to get it."
Food taxation	Pretlow RA. 2011 [44]	"Taxation of sugar-sweetened beverages, and possibly junk food and fast food, and restriction of such outlets to children, would seem warranted and even embraced by some children."
Limiting advertising for addictive foods	Gearhardt et al. (2012) [38]	"[The food industry] has pledged to remove its unhealthy products from schools and to market fewer unhealthy foods to children, in addition to a variety of other promises. Whether the public and elected officials find the industry trustworthy will determine in part how aggressive government will be in regulating industry practices."
Limited information on food addiction in obesity guidelines	Allen et al. (2012) [42]	"In these guidelines, nothing is said about the potential that an unhealthy diet may not be easily changed due to addictive aspects of hyperphagia. Furthermore, while they do recommend using interventions like taxation on high-energy food, they do not discuss more strict strategies like limiting the access of children to such food."

3.1. The Intersection of Ethics and Food Addiction

Our search yielded 20 articles related to ethics and food addiction. We identified individual responsibility as the primary theme of this section, with the following sub-themes: (i) personal control, will power, and choice; and (ii) blame and weight bias.

3.1.1. Personal Control, Will Power, and Choice

We found a tension in the literature between the traditional framings of addiction that emphasized personal control over food consumption, with medicalized, neuroscientific accounts of food addiction that emphasized reduced personal control. For example, Appelhans, Whited, Schneider, and Pagoto suggested, "[d]ietary lapses or failures should be conceptualized as the result of brain systems interacting with a toxic food environment, and not as a reflection of poor personal choices or lack of willpower" [45] (p. 1135). Some scholars argue that framing food addiction within a disease model may reduce individual blame [46] as the impairment in the person's autonomy and capacity for control reflects dysfunction in the brain's reward circuitry [36,44,47]. However, some scholars are skeptical that a food addiction label or a disease model will reduce responsibility or obviate perceptions that obesity is self-inflicted [5,48]. Others worry about the potential negative influence a brain disease model of food addiction might have on people with obesity. For example, Lee and colleagues suggest that a (brain) disease model could potentially justify coercive practices on a population that is thought to have reduced autonomy and limited control over their behavior [33].

A study by Lee and colleagues found that public support for a food addiction model of obesity was associated with the view that the individual had to overcome their condition through willpower and personal choice [34]. Other studies have reported that the public considers food addiction more of a personal choice than an addiction to other substances such as alcohol [41]. Green suggests that the free autonomous behavior to consume highly palatable foods could be deterred potentially by the implementation of sin taxes (or excise taxes) on said foods [49], but others, such as Franck, Grandi, and Eisenberg, question the evidence to support this recommendation [50]. Scholars have argued that the focus on individual-level factors, such as free will and personal choice, have failed to address the public health impact of high levels of highly palatable food consumption [51,52]. Some authors have argued that this narrow focus on individual-level factors "lets the food industry off the hook" for its contributions [53] (p. 2).

3.1.2. Blame and Weight Bias

Given the emphasis on personal responsibility in the literature, perspectives on who or what is to blame for food addiction are also far from stable. The literature reflects tensions between attributing blame for food addiction to the individual, the environment (including government and industry), or both, but seldom the alleged addictive properties of the food itself [5,41]. For instance, Thibodeau, Perko, and Flusberg found that participants who agreed with narratives that blamed the individual for their obesity (i.e., due to an addiction, a problem of individual behavior) were likely to support interventions that were penalizing, whereas support for narratives that blamed the environment were likely to support policy interventions that sought to protect people with obesity [37]. These tensions exist in a Western culture that values and legitimizes the over-consumption of foods and other material goods [51,54].

Weight bias is an ethically important concept in terms of its relevance to respect for persons, discrimination, and social justice, among others [38]. For example, entrenched harmful stereotypes that describe people with obesity as lazy and lacking in willpower underpins the moral view that they make poor choices with respect to food consumption and their health [38]. "Fat" people are considered "addict[ed] to inappropriate pleasures," which serves to further stigmatize and blame "fat" people [55] (p. 17).

3.2. Potential Implications of Food Addiction on Stigma

Our search yielded 19 articles related to stigma and food addiction. Identified themes included the following: (i) the impact on self-stigma and stigma from others, (ii) the differential impact of substance use disorder versus behavioral addiction on stigma, and (iii) the additive stigma of addiction plus obesity and/or eating disorder. As noted below, there was extensive overlap between several of the stigma and ethics themes.

3.2.1. The Impact on Self-Stigma and Stigma from Others

Related to the concepts of personal control, willpower, and choice, as well as blame and weight bias described above, several authors discussed potential implications of food addiction on self-stigma (internalized stigma) and stigma from others (externalized stigma). Similar to people who use substances, people with obesity receive messages from individuals and society that their obesity is a character flaw [56]. Empirical research using hypothetical case vignettes has found some evidence to suggest that a food addiction model may have a beneficial impact on stigma towards others who are overweight. Latner, Puhl, Murakami, and O'Brien randomly assigned participants to one of four conditions in which they read about either an addiction or non-addiction explanatory model of obesity, and subsequently read a vignette about a woman that was either obese or of normal weight [46]. Regardless of the participant's weight status, a food addiction explanatory model of obesity was found to result in less stigma, less blame, and less perceived psychopathology attributed to the woman in the vignette [46]. Moreover, a food addiction model resulted in less blame towards individuals with obesity in general, leading the authors to conclude that this explanatory model of obesity could have a beneficial impact on the pervasive prejudice against individuals with obesity.

Lee et al. conducted a large survey regarding public views of food addiction [34]. They found substantial support for the concept of food addiction, particularly among individuals with obesity [34]. They were less likely to agree that obesity is caused by overeating and they reported greater support for external causes of obesity. The authors speculated that a diagnosis of food addiction may reduce some of the guilt and self-blame associated with obesity. However, a correlational study by Burmeister, Hinman, Koball, Hoffman, and Carels found that weight-loss-seeking participants with a greater severity of food addiction, as measured by the YFAS, reported greater anti-fat attitudes (e.g., dislike of other people with excess weight), internalized weight bias (e.g., internalization of anti-fat attitudes), and body shame [39].

3.2.2. Differential Impact of Substance Use Disorder Versus Behavioral Addiction on Stigma

Despite evidence of shared neurobiological processes in substance use disorders and behavioral addictions, a false dichotomy permeates much of the discourse whereby substance use disorders are perceived as a neurobiological disease and behavioral addictions are perceived as a mental/psychological condition. Whether food addiction is considered a substance use disorder versus behavioral addiction may impact factors such as personal responsibility, autonomy, and blame attribution, which in turn have implications for internalized and externalized stigma. Stigma may decrease if food addiction is considered a substance use disorder because it provides an explanatory model for obesity "without invoking character flaws, such as lack of willpower" [45] (p. 1133), and shifts responsibility from the individual to the food substance and food industry. For example, Appelhans et al. state: "by emphasizing genetically-influenced neurobiological processes that confer vulnerability to overeating in a toxic food environment, the model enables dietetics practitioners to more effectively address obesity without promoting stigma" [45] (p. 1133). However, Rasmussen points out that the "neurochemical concept of obesity, as with drug addiction and eating disorders, can coexist with attribution of personal responsibility for the condition, and thus with blaming the impaired, addicted consumer rather than the supplier" (p. 217), and goes on to say that "it cannot be ruled out that attribution of addiction (as a chronic brain disease) will hurt more than it helps" [5] (p. 223).

Whereas opinions were mixed regarding whether considering food addiction as a substance use disorder would reduce internalized and externalized stigma, it was generally believed that considering food addiction as a behavioral addiction would either have little effect on stigma or actually increase stigma because it is perceived as a choice that is under personal control. DePierre, Puhl, and Luedicke conducted an empirical study to examine public perceptions of food addiction compared to nicotine and alcohol addiction. Food addiction was considered to be a disease to a greater extent than smoking and to be the product of personal choice to a greater extent than alcohol addiction, leading the authors to conclude that food addiction may be perceived as a behavioral addiction that is vulnerable to stigmatization rather than a substance addiction [41].

3.2.3. Additive Stigma of Addiction Plus Obesity/Eating Disorder

We found a tension in the literature regarding whether food addiction would reduce the stigma directed towards individuals with obesity or create an additional stigmatized identity. Several authors pointed out that substance-related disorders, such as nicotine, alcohol, and cocaine use disorders, are stigmatized to an even greater extent than obesity, a finding that may be due to the fact that everyone must eat food for survival, whereas a person must choose to seek out and use psychoactive substances of abuse [40,41,45]. It has also been suggested that food may be a less stigmatizing substance of abuse because it has less impact on others compared to nicotine (e.g., second-hand smoke) and alcohol (e.g., driving under the influence, impulsive behaviors) [40,41].

Several authors raised concerns that a diagnosis of food addiction could result in a double or additive stigma because the types of stigma associated with obesity and addiction differ [5,40,42,57]. For example, applying the types of stigma described by Goffman in this context, obesity could be considered a visible "abomination of the body," whereas addiction could be considered a "blemish of character" [40,58]. Foddy states: "I have said nothing about the myriad social dimensions in which we stigmatize and disadvantage addicts [sic] differently from the obese. It is hard to say exactly which group has the worse lot in this sense, since the stigma and discrimination takes different forms in each case" [36] (p. 87).

Rather than food addiction reducing stigma by providing an explanatory model for compulsive/binge eating and obesity that reduces blame and personal responsibility, several authors noted that food addiction may actually amplify the harms of stigma because obesity, eating disorders, and addictions are all stigmatized conditions, and neurobiological explanations do not necessarily modify attributions of personal responsibility. Allen et al. raised concerns about the "dietary obese being stigmatized as addicts [sic]" (p. 134) [42]. Rasmussen noted that "the stigmatizing attribution of addiction to people already stigmatized as obese may amplify the social harms that they suffer" (p. 217), "raising the risk that their combination might prove reinforcing" [5] (p. 223).

A few experimental studies have been conducted to examine the impact of food addiction on externalized stigma. Bannon, Hunter-Reel, Wilson, and Karlin randomly assigned participants to one of six conditions in which they read a vignette about a woman with obesity [59]. The independent variables were her binge eating status (present vs. absent) and the cause of her obesity (biological addiction vs. psychological vs. ambiguous). Unfortunately, the manipulation regarding the cause of her obesity was not successful, which prevented interpretation of the specific impact of food addiction on externalized stigma. However, the presence of binge eating increased externalized stigma such that participants rated individuals with obesity as less attractive and more blameworthy for their weight, and they desired greater social distance from them [59]. DePierre et al. directly examined the additive effect of food addiction on externalized stigma [40]. They found that labelling an individual as an "obese food addict" was more stigmatizing than either label on its own ("food addict" or "obese") and concluded that food addiction may increase the externalized stigma associated with obesity, but be less susceptible to stigma than other forms of addiction [40].

3.3. Potential Policy Implications of Food Addiction

Our search yielded 31 articles related to policy themes for addictive foods that were extrapolated to the construct of food addiction. Policy themes related to addictive foods were generally derived from comparisons to the tobacco industry. Articles identified specific policy interventions from "Big Tobacco" that could be applied to food addiction included the following: (i) reducing access to addictive foods, (ii) food taxation, and (iii) limiting advertising for addictive foods.

3.3.1. Comparisons to "Big Tobacco"

Many articles made parallels between the policy issues related to the food industry and the tobacco industry. The role of food taxation, reducing access to potentially addictive foods, litigation against the food industry, and limiting food advertising in addressing food addiction and obesity arise from this comparison to "Big Tobacco" and highlight a prominent theme amongst most articles focusing on policy issues of food addiction [38,52,53]. The association between the food industry and tobacco industry is contingent on a clear link between certain foods and their addictive potential. Authors argued that if certain foods are conceptualized as "a potentially disease-causing agent such as tobacco and alcohol then policy changes to encourage the intake of healthy food and decrease the intake of unhealthy foods are in order" [60] (p. 762). Furthermore, Schulte and colleagues argue that if the concept of food addiction is embraced by the public as it was for nicotine addiction, the shift in public perception of the substance (in this case, addictive foods) could generate enough support for change in public policies for obesity [61]. This argument has been supported by an online survey of 999 individuals from the U.S., including 52% of individuals who were overweight or obese, which showed that believing food is addictive increased support for obesity-related policies [62].

3.3.2. Reducing Access to Addictive Foods Through Regulation

In comparison to policies in the tobacco industry, the overall evidence for the effect of government regulations on limiting access to highly pleasurable food is still emerging and requires further exploration [63]. Current papers identified the role of government regulations on the food industry to support legislation, litigation, and regulation efforts to influence access to healthy as opposed to high fat or sugar foods [60,64]. Reports suggest that if food is deemed addictive, then governments need to establish clear policies to limit, restrict, or even ban addictive foods [52], such as limiting the number and location of fast food restaurants and soft drink vending machines in an area.

The support for policies focused on addictive food restriction for children and youth as opposed to adults. For example, in a survey of youth who were unable to lose weight using an online open-access website for obesity, participants felt that restriction of junk food and fast food outlets would be beneficial [36]. Additional articles supported early intervention efforts through increased regulation of pleasurable foods in children and youth settings such as schools, implementation of programs promoting healthy eating and food choices, and engagement of pediatricians and family physicians [65,66]. In contrast, a recent qualitative study with 23 individuals with obesity showed that individuals were skeptical about the effectiveness of restrictions on highly pleasurable foods, although they supported addictive food restrictions for children [43]. Nonetheless, reviews generally supported external control of addictive foods based on examples where cigarette smokers perceived external control as being influential or effective compared to individual motivating factors (e.g., willpower, personal choice) [67,68].

Despite this support for policy interventions focused on increased food regulation, concerns were expressed in articles around the lack of clarity around "addictive," or "good" and "bad," ingredients and foods [69]. The lack of clear evidence about foods categorized as "addictive" and the unique challenges with food being a necessary part of living (in contrast to alcohol) have complicated abstinence arguments and approaches thus far. Nonetheless, authors have used the comparison to the tobacco industry for informing food addiction policy interventions focused on restriction [56]. Empirical evidence has demonstrated that narratives where obesity was considered an addiction or a disorder led to increased public support for a protective approach to policy interventions for obesity; however, implications for obesity treatment are not fully understood [37].

Studies reiterated the importance of obesity experts being more explicit about food engineering and the need for greater public awareness of this issue beyond the traditional "obesogenic environment" discourses [44]. A greater focus on increasing public awareness about food engineering could lead to greater support for policy interventions focused on regulation and addictive food restriction. One article noted that a potential consequence of identifying and regulating specific "addictive" foods could be litigation and banning of the food industry for purposefully engineering foods to be more addictive [52].

3.3.3. Food Taxation

Themes related to food taxation also emerged predominantly from comparisons between the food and tobacco industries, as highlighted above [42,44]. The rationale stems from evidence for the taxation of tobacco products, which has estimated that a 10% increase in soft drink prices could result in approximately an 8–10% reduction in soft drink consumption by individuals [70,71]. Moreover, several countries, including France and Hungary, have implemented tax levies on high caloric density foods [41]. These national policy initiatives will be beneficial in evaluating the effects of these interventions longitudinally on mitigating food addiction.

Despite this early evidence, several factors complicate food taxation as a policy intervention to reduce access to caloric dense foods. In contrast to tobacco products, caloric dense foods are already cheaper than healthier food options [42]. Therefore, food taxation efforts would affect children and the poor disproportionately given the current low cost of caloric dense foods, although food taxes may potentially be used to offset costs for healthier foods in this population [42]. In addition, it is unclear what food ingredients are highly addictive and as a result, conclusive recommendations on specific ingredients warranting taxation remains a challenge [53].

3.3.4. Limiting Advertising for Addictive Foods

Studies have also articulated concerns about the early introduction of food advertising to young children and adolescents. Gearhardt and colleagues identified the need to limit food advertising to children to limit the potential long-term effects, such as food addiction [38]. Restriction of food marketing to youth has involved both government policies but also voluntary pledges by the food industry [38]. These initiatives have attempted to correct the default food industry messaging that has been inoculating children and youth for decades [71]. As a result, authors argue that voluntary self-regulatory efforts by the food industry are likely to be insufficient, and government protective policies are needed to counteract these trends in early childhood food marketing strategies by the food industry [38].

3.3.5. Limited Information on Food Addiction in Obesity Guidelines

Several papers focused on the limited discussion of food addiction within key obesity related guidelines and resources such as the World Health Organization (WHO) and Centre for Disease Control (CDC) [42]. Despite increasing public dialogue on food addiction, obesity guidelines have not included information on the controversies related to food addiction. This is complicated by the inability to differentiate specific addictive food ingredients and limited evidence for food addiction related interventions [72]. As a result, papers highlighted the need for an improved mechanistic understanding and the development and testing of new treatments for managing addictive foods.

Furthermore, there have been questions regarding the evidence for policy interventions for preventing and managing addictive foods in patients with obesity. Gostin and colleagues believe that the "lack of science" argument for food addiction policy interventions is a faulty argument given that no single intervention resulted in a substantial reduction in nicotine use and there is a need for a multi-level intervention approach [73]. Thus, further research is needed to clarify the construct of food addiction, risk factors and effectiveness of policy interventions to facilitate integration within obesity-related guidelines and resources.

4. Discussion

The current scoping review examined the potential ethical, stigma, and health policy implications of food addiction, identified the major themes in the literature, and provided insights into key challenges in these areas.

4.1. Potential Ethical Implications of Food Addiction

While various disease models of alcohol use disorders have circulated within the popular discourse since the early 19th century [74,75], more recently scholars have adopted the language of neuroscience to characterize addiction more generally as a chronic, relapsing brain disease [76,77]. We found that discussions in the literature lend themselves to moral debates over personal control, individual responsibility, and blame for the consequences of food consumption. Indeed, these moral framings are common in the obesity discourse [44,78]. However, the neurobiological theory of food addiction may place a disproportionate burden on the individual to treat their food addiction, which may provide less incentive for governments to scale up public health approaches, such as holding the food industry accountable for their food engineering practices as well as the social conditions that harm people who are defined as food addicted.

The focus of individual blame in the literature may be reflective of a culture of healthism, defined as the individual desire for health and well-being achieved primarily through lifestyle modification [79]. The moral assumptions underlying healthism is that individuals are responsible for making good choices as opposed to bad choices about their health. Our results suggest that healthism describes good choices as health-promoting and reflects self-control over eating while bad choices lead to food addiction or obesity. For instance, some dietary counselling recommendations include empowering the patient to take responsibility for their behaviors in the context of what they consider to be a toxic food environment [45]. Our results suggest that some authors were concerned about a culture that emphasizes healthism and individualism when the social conditions make the ability to make healthy eating choices near impossible [51]. Future research can build on the burgeoning literature in this area and provide deeper insights into the moral attitudes about food addiction held by diverse publics.

It will also be important for future research to explore if the empirical findings related to moral concepts align with the concerns raised in the conceptual philosophical and bioethics literatures.

4.2. Potential Stigma Implications of Food Addiction

Our review also examined the complex relationship between food addiction and stigma. Stigma is a multidimensional construct including many related concepts (e.g., labels, stereotypes, prejudice, discrimination), sources (e.g., self-stigma, public stigma, provider-based stigma [e.g., from healthcare providers], institutional stigma), characteristics (e.g., stigma of the physical body, stigma of character), and dimensions (e.g., social distance, perceptions of dangerousness) that can complicate analyses [80]. The limited empirical research conducted to date, much of which has experimentally manipulated the information provided in case vignettes, has generated mixed results regarding the impact of a food addiction model on externalized stigma. Whereas a food addiction explanation for obesity was found to reduce externalized stigma and blame towards a target in one study [45], another study found that labeling an individual as an "obese food addict" created an additive stigma that exceeded the stigma associated with either label on its own [40]. This finding is consistent with a recent qualitative study of individuals with obesity, which noted that some individuals who personally identified with the concept of food addiction felt reluctant to be described as an "addict" because they believed the label would increase self-stigma and stigma from others [43]. Interestingly, the one study identified through our scoping review that examined the association between food addiction and externalized stigma in a clinical sample found that patients with greater severity of food addiction reported greater anti-fat attitudes towards other individuals with excess weight [39]. The impact of food addiction on externalized stigma has been examined in student samples, community samples, and a clinical sample, and additional research is warranted to examine the impact of a food addiction model on externalized stigma among healthcare professionals.

In contrast to the mixed findings regarding the impact of a food addiction model on externalized stigma, the association between food addiction and internalized stigma has been fairly consistent. The aforementioned study conducted in a clinical sample found that patients with greater severity of food addiction reported greater internalized weight bias and body shame, as well as lower eating self-efficacy [39]. This finding is consistent with a recent study comparing three groups that varied with respect to food addiction (i.e., non-food-addicted, self-perceived food addicted, and food-addicted according to the YFAS) and found that internalized weight bias increased across the groups despite no differences in body mass index, whereas eating self-efficacy decreased across the groups [31]. Similarly, a recent systematic review examining the association between internalized weight bias and health-related variables reported that the published studies conducted to date have consistently found a positive association between food addiction and internalized weight bias [82]. Moreover, internalized weight bias was associated with a variety of negative mental and physical health outcomes including depression, anxiety, low self-esteem, poor body image, disordered eating, severity of obesity, reduced dietary adherence, and reduced motivation and self-efficacy to engage in health-promoting behaviors. Although much of the research conducted to date has been cross-sectional, which precludes conclusions regarding the direction of causality, if replicated in prospective research, such findings would lend support to concerns voiced by some authors that food addiction could have an adverse impact on internalized stigma and reduce the self-efficacy needed to improve eating behaviors [29]. Although the treatment implications of food addiction were beyond the scope of the current review, this important topic is recently receiving more attention in the literature [83,84] and the findings of the current review suggest that treatment approaches that reduce internalized weight bias and bolster self-efficacy would be advised.

Whether food addiction is best characterized as a behavioral addiction or a substance use disorder has been a topic of debate [26]. The results of the current review suggest that characterizing food addiction as a behavioral addiction likely has an adverse impact on externalized stigma compared to characterizing it as a substance use disorder. Despite evidence that behavioral addictions and substance use disorders have many shared features, including neurobiological mechanisms and genetic contributions [85,86], behavioral addictions are still commonly perceived as being more under personal control. The general public perceives food addiction as being less of a disease and more under personal control than alcoholism [41]. Increasing public awareness regarding the engineering of hyperpalatable foods and the validity of food addiction as a diagnostic construct that shares similarities with other substance use disorders [8] may be helpful as both a public health and stigma reduction intervention.

4.3. Potential Policy Implications of Food Addiction

Across reviewed articles, we found that the tobacco industry served as an example for many of the policy recommendations for managing food addiction in obesity. The challenges with these policy interventions were related to the lack of research on addictive food substances, specifically the current inability to classify foods more broadly in their addictive potential. Despite authors drawing parallels to "Big Tobacco," we identified a dearth of evidence for the effectiveness of food addiction as a driver for policy change in obesity. Therefore, many of the recommendations, such as food taxation, reducing advertising to children, and reducing access to addictive foods (especially in children and youth), were extrapolated from policy interventions observed in the tobacco industry as opposed to being specific to food addiction.

4.4. Limitations

Our results are subject to the following limitations associated with scoping review. Unlike a systematic review, we are unable to grade the quality of the studies or make any specific recommendations related to level of evidence. The articles included in this scoping review included commentaries and empirical research (e.g., experimental and correlational studies). Further, although our search used multiple methods to identify potential articles, it is possible that some studies were missed. This scoping review underscored the need for future research on the effectiveness of policy interventions to mitigate the effect of addictive foods on the obesity epidemic. The inclusion of food addiction within obesity guidelines was a notable omission in the literature that was identified in our review. It is possible that the absence of food addiction in obesity guidelines reflects the need for greater research on the role of food addiction on obesity management, including questions about its benefit for healthcare professionals and patients in obesity care.

5. Conclusions

This review is the first to explore the ethical, stigma, and policy issues related to food addiction using a scoping review methodology. The gaps in the evidence related to this scoping review underscore the need for additional research and increased clarity regarding the ethical and stigma related impact of a food addiction model. In addition, more rigorous studies are needed to move beyond policy interventions generated predominantly from past experiences in the tobacco industry.

Supplementary Materials: The following are available online at http://www.mdpi.com/2072-6643/11/4/710/s1, Table S1: Complete List of Articles from Search Strategy.

Author Contributions: S.E.C., D.Z.B., and S.S. contributed to the conception and design of the study. S.E.L., K.K., and A.H. conducted literature searches and provided summaries of previous research studies. All authors contributed equally to the analysis and interpretation of the data and the writing of the manuscript. All authors have approved the final manuscript.

Funding: This study was funded by the Canadian Institutes of Health Research (Grant No. 373261).

Acknowledgments: This project was funded by a Planning and Dissemination Grant from the Canadian Institutes of Health Research. We would like to thank Ashley Farrell for assisting with the scoping review methodology and literature search, as well as the individuals who participated in our food addiction retreat.

Conflicts of Interest: The authors declare no conflict of interest.

References

1. Kuczmarski, R.J.; Flegal, K.M.; Campbell, S.M.; Johnson, C.L. Increasing prevalence of overweight among US adults. The National Health and Nutrition Examination Surveys, 1960 to 1991. *JAMA* **1994**, *272*, 205–211. [PubMed]
2. Kessler, D.A. *The End of Overeating: Taking Control of the Insatiable American Appetite*; Rodale: New York, NY, USA, 2009.
3. Johnson, P.M.; Kenny, P.J. Dopamine D2 receptors in addiction-like reward dysfunction and compulsive eating in obese rats. *Nat. Neurosci.* **2010**, *13*, 635–641.
4. Parr, J.; Rasmussen, N. Making addicts of the fat: Obesity, psychiatry and the 'Fatties Anonymous' Model of Self-Help Weight Loss in the Post-War United States. In *Critical Perspectives on Addiction (Advances in Medical Sociology, Volume 14)*; Netherland, J., Ed.; Emerald Group Publishing Limited: Bingley, UK, 2012; pp. 181–200.
5. Rasmussen, N. Stigma and the addiction paradigm for obesity: Lessons from 1950s America. *Addiction* **2014**, *110*, 217–225.
6. American Psychiatric Association. *Diagnostic and Statistical Manual of Mental Disorders*, 5th ed.; American Psychiatric Association: Washington, DC, USA, 2013.
7. Gearhardt, A.N.; Corbin, W.R.; Brownell, K.D. Preliminary validation of the Yale food addiction Scale. *Appetite* **2009**, *52*, 430–436.
8. Gordon, E.L.; Ariel-Donges, A.H.; Bauman, V.; Merlo, L.J. What is the evidence for "food addiction"? A systematic review. *Nutrients* **2018**, *12*, 477.
9. Ivezaj, V.; White, M.A.; Grilo, C.M. Examining binge-eating disorder and food addiction in adults with overweight and obesity. *Obesity* **2016**, *24*, 2064–2069.
10. Chao, A.M.; Shaw, J.A.; Pearl, R.L.; Alamuddin, N.; Hopkins, C.M.; Bakizada, Z.M.; Berkowitz, R.I.; Wadden, T.A. Prevalence and psychosocial correlated of food addiction in persons with obesity seeking weight reduction. *Compr. Psychiatry* **2016**, *73*, 97–104. [PubMed]
11. Gearhardt, A.N.; White, M.A.; Masheb, R.M.; Morgan, P.T.; Crosby, R.D.; Grilo, C.M. An examination of the food addiction construct in obese patients with binge eating disorder. *Int. J. Eat. Disord.* **2012**, *45*, 657–663. [PubMed]
12. Gearhardt, A.N.; Davis, C.; Kuschner, R.; Brownell, K.D. The addiction potential of hyperpalatable foods. *Curr. Drug Abuse Rev.* **2011**, *4*, 140–145.
13. Schulte, E.M.; Avena, N.M.; Gearhardt, A.N. Which foods may be addictive? The roles of processing, fat content, and glycemic load. *PLoS ONE* **2015**, *10*, e0117959.
14. Curtis, C.; Davis, C. A qualitative study of binge eating and obesity from an addiction perspective. *Eat. Disord.* **2014**, *22*, 19–32.
15. Gihooly, C.H.; Das, S.K.; Golden, J.K.; McCrory, M.A.; Dallal, G.E.; Saltzman, E.; Kramer, F.M.; Roberts, S.B. Food cravings and energy regulation: The characteristics of craved foods and their relationship with eating behaviors and weight change during 6 months of dietary energy restriction. *Int. J. Obes.* **2007**, *31*, 1849–1858.
16. Smith, D.G.; Robbins, T.W. The neurobiological underpinnings of obesity and being eating: A rationale for adopting the food addiction model. *Biol. Psychiatry* **2013**, *73*, 804–810.
17. Volkow, N.D.; Wang, G.J.; Fowler, J.S.; Tomais, D.; Baler, R. Food and drug reward: Overlapping circuits in human obesity and addiction. *Curr. Top. Behav. Neurosci.* **2012**, *11*, 1–24. [PubMed]
18. Gearhardt, A.N.; Yokum, S.; Orr, P.; Stice, E.; Corbin, W.; Brownell, K.D. The neural correlates of 'food addiction'. *Arch. Gen. Psychiatry* **2011**, *68*, 808–816. [PubMed]
19. Davis, C.; Loxton, N.J.; Levitan, R.D.; Kaplan, A.S.; Carter, J.C.; Kennedy, J.L. 'Food addiction' and its association with a dopaminergic multilocus genetic profile. *Physiol. Behav.* **2013**, *118*, 63–69. [PubMed]
20. Fletcher, P.C.; Kenny, P.J. Food addiction: A valid concept? *Neuropsychopharmacology* **2018**, *43*, 2506–2513.
21. Burrows, T.; Skinner, J.; McKenna, R.; Rollo, M. Food addiction, binge eating disorder, and obesity: Is there a relationship? *Behav. Sci.* **2017**, *7*, 54.
22. Leigh, S.J.; Morris, M.J. The role of reward circuitry and food addiction in the obesity epidemic: An update. *Biol. Psychol.* **2018**, *131*, 31–42.
23. Muele, A.; Gearhardt, A.N. Food addiction in the light of DSM-5. *Nutrients* **2014**, *6*, 2652–2671.
24. Corwin, R.L.; Grigson, P.S. Symposium overview - food addiction: Fact or fiction? *J. Nutr.* **2009**, *139*, 617–619.

25. Davis, C. A commentary on the associations among 'food addiction', binge eating disorder, and obesity: Overlapping conditions with idiosyncratic clinical features. *Appetite* **2017**, *115*, 3–8. [PubMed]
26. Schulte, E.M.; Potenza, M.N.; Gearhardt, A.N. A commentary on the "eating addiction" versus "food addiction" perspectives on addictive-like food consumption. *Appetite* **2017**, *115*, 9–15.
27. Bell, K.; Salmon, A.; Bowers, M.; Bell, J.; McCullough, L. Smoking, stigma and tobacco 'denormalization': Further reflections on the use of stigma as a public health tool. *Soc. Sci. Med.* **2010**, *70*, 795–799.
28. Guttman, N.; Salmon, C.T. Guilt, fear, stigma and knowledge gaps: Ethical issues in public health communication interventions. *Bioethics* **2004**, *18*, 531–552. [PubMed]
29. Bayer, R. Stigma and the ethics of public health: Not can we but should we. *Soc. Sci. Med.* **2008**, *67*, 463–472.
30. Goldberg, D.S.; Puhl, R.M. Obesity stigma: A failed and ethically dubious strategy. *Hastings Cent. Rep.* **2013**, *43*, 5–6. [PubMed]
31. Reid, J.; O'Brien, K.S.; Puhl, R.; Hardman, C.A.; Carter, A. Food addiction and its potential links with weight stigma. *Curr. Addict. Rep.* **2018**, *5*, 192–201.
32. Arksey, H.; O'Malley, L. Scoping studies: Towards a methodological framework. *Int. J. Soc. Res. Methodol.* **2005**, *8*, 19–32.
33. Lee, N.M.; Hall, W.D.; Lucke, J.; Forlini, C.; Carter, A. Food Addiction and Its Impact on Weight-Based Stigma and the Treatment of Obese Individuals in the U.S. and Australia. *Nutrients* **2014**, *6*, 5312–5326.
34. Lee, N.M.; Lucke, J.; Hall, W.D.; Meurk, C.; Boyle, F.M.; Carter, A. Public Views on Food Addiction and Obesity: Implications for Policy and Treatment. *PLoS ONE* **2013**, *8*, e74836.
35. Moher, D.; Liberati, A.; Tetzlaff, J.; Altman, D.G. The PRISMA Group (2009). Preferred Reporting Items for Systematic Reviews and Meta-Analyses: The PRISMA Statement. *PLoS Med.* **2009**, *6*, e1000097. [CrossRef]
36. Foddy, B. Addicted to Food, Hungry for Drugs. *Neuroethics* **2011**, *4*, 79–89.
37. Thibodeau, P.H.; Perko, V.L.; Flusberg, S.J. The relationship between narrative classification of obesity and support for public policy interventions. *Soc. Sci. Med.* **2015**, *141*, 27–35. [PubMed]
38. Gearhardt, A.N.; Bragg, M.A.; Pearl, R.L.; Schvey, N.A.; Roberto, C.A.; Brownell, K.D. Obesity and public policy. *Annu. Rev. Clin. Psychol.* **2012**, *8*, 405–430. [PubMed]
39. Burmeister, J.M.; Hinman, N.; Koball, A.; Hoffmann, D.A.; Carels, R.A. Food addiction in adults seeking weight loss treatment. Implications for psychosocial health and weight loss. *Appetite* **2013**, *60*, 103–110. [PubMed]
40. DePierre, J.A.; Puhl, R.; Luedicke, J. A new stigmatized identity? Comparisons of a 'food addict' label with other stigmatized health conditions. *Basic Appl. Soc. Psych.* **2013**, *35*, 10–21.
41. DePierre, J.A.; Puhl, R.; Luedicke, J. Public perceptions of food addiction: A comparison with alcohol and tobacco. *J. Subst. Use* **2014**, *19*, 1–6.
42. Allen, P.J.; Batra, P.; Geiger, B.M.; Wommack, T.; Gilhooly, C.; Pothos, E.N. Rationale and consequences of reclassifying obesity as an addictive disorder: Neurobiology, food environment and social policy perspectives. *Physiol. Behav.* **2012**, *107*, 126–137.
43. Cullen, A.J.; Barnett, A.; Komesaroff, P.A.; Brown, W.; O'Brien, K.S.O.; Hall, W.; Carter, A. A qualitative study of overweight and obese Australians' views of food addiction. *Appetite* **2017**, *115*, 62–70. [PubMed]
44. Pretlow, R.A. Addiction to highly pleasurable food as a cause of the childhood obesity epidemic: A qualitative Internet study. *Eat. Disord.* **2011**, *19*, 295–307.
45. Appelhans, B.M.; Whited, M.C.; Schneider, K.L.; Pagoto, S.L. Time to abandon the notion of personal choice in dietary counseling for obesity? *J. Am. Diet. Assoc.* **2011**, *111*, 1130–1136. [PubMed]
46. Latner, J.D.; Puhl, R.M.; Murakami, J.M.; O'Brien, K.S. Food addiction as a causal model of obesity. Effects on stigma, blame, and perceived psychopathology. *Appetite* **2014**, *77*, 77–82. [PubMed]
47. Ho, A.L.; Sussman, E.S.; Pendharkar, A.V.; Azagury, D.E.; Bohon, C.; Halpern, C.H. Deep brain stimulation for obesity: Rationale and approach to trial design. *Neurosurg. Focus* **2015**, *38*, E8.
48. Ortiz, S.E.; Zimmerman, F.J.; Gilliam, F.D., Jr. Weighing in: The taste-engineering frame in obesity expert discourse. *Am. J. Public Health* **2015**, *105*, 554–559. [PubMed]
49. Green, R. The Ethics of Sin Taxes. *Public Health Nurs.* **2011**, *28*, 68–77. [PubMed]
50. Franck, C.; Grandi, S.M.; Eisenberg, M.J. Taxing Junk Food to Counter Obesity. *Am. J. Public Health* **2013**, *103*, 1949–1953.
51. Brownell, K.D.; Schwartz, M.B.; Puhl, R.M.; Henderson, K.E.; Harris, J.L. The need for bold action to prevent adolescent obesity. *J. Adolesc. Health* **2009**, *45*, S8–S17.

52. Pomeranz, J.L.; Teret, S.P.; Sugarman, S.D.; Rutkow, L.; Brownell, K.D. Innovative Legal Approaches to Address Obesity. *Milbank Q* **2009**, *87*, 185–213.
53. Hebebrand, J. Obesity prevention: Moving beyond the food addiction debate. *J. Neuroendocrinol.* **2015**, *27*, 737–738. [PubMed]
54. Blundell, J.E.; Finlayson, G. Food addiction not helpful: The hedonic component—Implicit wanting—Is important. *Addiction* **2011**, *106*, 1216–1218. [PubMed]
55. Mackenzie, R. Don't Let Them Eat Cake! A View from Across the Pond. *Am. J. Bioeth.* **2010**, *10*, 16–18. [PubMed]
56. deShazo, R.D.; Hall, J.E.; Skipworth, L.B. Obesity bias, medical technology, and the hormonal hypothesis: Should we stop demonizing fat people? *Am. J. Med.* **2015**, *128*, 456–460. [PubMed]
57. Rasmussen, N. Weight stigma, addiction, science, and the medication of fatness in mid-twentieth century America. *Sociol. Health Illn.* **2012**, *34*, 880–895.
58. Goffman, E. *Stigma: Notes on the Management of Spoiled Identity*; Simon & Schuster: New York, NY, USA, 1963.
59. Bannon, K.L.; Hunter-Reel, D.; Wilson, G.T.; Karlin, R.A. The effects of casual beliefs and binge eating on the stigmatization of obesity. *Int. J. Eat. Disord.* **2009**, *42*, 118–124.
60. Battle, E.K.; Brownell, K.D. Confronting a rising tide of eating disorders and obesity: Treatment vs. prevention and policy. *Addict. Behav.* **1996**, *21*, 755–765.
61. Schulte, E.M.; Tuttle, H.M.; Gearhardt, A.N. Belief in Food Addiction and Obesity-Related Policy Support. *PLoS ONE* **2016**, *11*, e0147557.
62. Moran, A.; Musicus, A.; Soo, J.; Gearhardt, A.N.; Gollust, S.E.; Roberto, C.A. Believing that certain foods are addictive is associated with support for obesity-related public policies. *Prev. Med.* **2016**, *90*, 39–46. [PubMed]
63. Alonso-Alonso, M.; Woods, S.C.; Pelchat, M.; Grigson, P.S.; Stice, E.; Faroogi, S.; Khoo, C.S.; Mattes, R.D.; Beauchamp, G.K. Food reward system: Current perspectives and future research needs. *Nutr. Rev.* **2015**, *73*, 296–307. [PubMed]
64. Gearhardt, A.N.; Corbin, W.R.; Brownell, K.D. Food addiction: An examination of the diagnostic criteria for dependence. *J. Addict. Med.* **2009**, *3*, 1–7.
65. Volkow, N.D.; Wise, R.A. How can drug addiction help us understand obesity? *Nat. Neurosci.* **2005**, *8*, 555–560. [PubMed]
66. Willette, A.L. Where have all the parents gone? Do efforts to regulate food advertising to curb childhood obesity pass constitutional muster? *J. Leg. Med.* **2007**, *28*, 561–577.
67. Morphett, K.; Partridge, B.; Gartner, C.; Carter, A.; Hall, W. Why Don't Smokers Want Help to Quit? A Qualitative Study of Smokers' Attitudes towards Assisted vs. Unassisted Quitting. *Int. J. Environ. Res. Public Health* **2015**, *12*, 6591–6607.
68. Uppal, N.; Shahab, L.; Britton, J.; Ratschen, E. The forgotten smoker: A qualitative study of attitudes towards smoking, quitting, and tobacco control policies among continuing smokers. *BMC Public Health* **2013**, *13*, 432.
69. Smith, T.G. All foods are habit-forming—What I want to know is which will kill me! *Addiction* **2011**, *106*, 1218–1220.
70. Andreyeva, T.; Long, M.W.; Brownell, K.D. The impact of food prices on consumption: A systematic review of research on the price elasticity of demand for food. *Am. J. Public Health* **2010**, *100*, 216–222.
71. Brownell, K.D.; Frieden, T.R. Ounces of prevention—The public policy case for taxes on sugared beverages. *N. Engl. J. Med.* **2009**, *360*, 1805–1808. [PubMed]
72. Gearhardt, A.N.; Grilo, C.M.; DiLeone, R.J.; Brownell, K.D.; Potenza, M.N. Can food be addictive? Public health and policy implications. *Addiction* **2011**, *106*, 1208–1212.
73. Gostin, L.O. Limiting what we can eat: A bridge too far? *Milbank Q.* **2014**, *92*, 173–176.
74. Rush, B. *An Inquiry into the Effects of Ardent Spirits Upon the Human Body and Mind: With an Account of the Means of Preventing, and of the Remedies for Curing Them*; James Loring: Boston, NY, USA, 1823.
75. Jellinek, E.M. *The Disease Concept of Alcoholism*; Hillhouse: New Haven, CT, USA, 1960.
76. Leshner, A.I. Addiction is a brain disease, and it matters. *Science* **1997**, *278*, 45–47. [PubMed]
77. Barnett, A.I.; Hall, W.; Fry, C.L.; Dilkes-Frayne, E.; Carter, A. Drug and alcohol treatment providers' views about the disease model of addiction and its impact on clinical practice: A systematic review. *Drug Alcohol Rev.* **2018**, *37*, 697–720.
78. Rich, E.; Evans, J. 'Fat ethics'—The obesity discourse and body politics. *Soc. Theory Health* **2005**, *3*, 341–358.

79. Crawford, R. Healthism and the medicalization of everyday life. *Int. J. Health Serv.* **1980**, *10*, 365–388. [PubMed]
80. Pescosolido, B.A.; Martin, J.K. The stigma complex. *Annu. Rev. Sociol.* **2015**, *41*, 87–116. [PubMed]
81. Meadows, A.; Nolan, L.J.; Higgs, S. Self-perceived food addiction: Prevalence, predictors, and prognosis. *Appetite* **2017**, *114*, 282–298. [PubMed]
82. Pearl, R.L.; Puhl, R.M. Weight bias internalization and health: A systematic review. *Obes. Rev.* **2018**, *19*, 1141–1163. [PubMed]
83. Schulte, E.M.; Joyner, M.A.; Schiestl, E.T.; Gearhardt, A.N. Future directions in "food addiction": Next steps and treatment implications. *Curr. Addict. Rep.* **2017**, *4*, 165–171.
84. Vella, S.C.; Pai, N.B. A narrative review of potential treatment strategies for food addiction. *Eat. Weight Disord.* **2017**, *22*, 387–393. [PubMed]
85. Grant, J.E.; Potenza, M.N.; Weinstein, A.; Gorelick, D.A. Introduction to behavioral addictions. *Am. J. Drug Alcohol Abuse* **2010**, *36*, 233–241.
86. Leeman, R.F.; Potenza, M.N. A targeted review of the neurobiology and genetics of behavioral additions: An emerging area of research. *Can. J. Psychiatry* **2014**, *58*, 260–273.

© 2019 by the authors. Licensee MDPI, Basel, Switzerland. This article is an open access article distributed under the terms and conditions of the Creative Commons Attribution (CC BY) license (http://creativecommons.org/licenses/by/4.0/).

Article

Obesity Stigma: Is the 'Food Addiction' Label Feeding the Problem?

Helen K. Ruddock [1,2], Michael Orwin [1,3], Emma J. Boyland [1], Elizabeth H. Evans [3] and Charlotte A. Hardman [1,*]

1. Department of Psychological Sciences, University of Liverpool, Liverpool L69 7ZX, UK
2. School of Psychology, University of Birmingham, Birmingham B15 2TT, UK
3. School of Psychology, Newcastle University, Newcastle NE1 7RU, UK
* Correspondence: charlotte.hardman@liverpool.ac.uk

Received: 25 July 2019; Accepted: 30 August 2019; Published: 4 September 2019

Abstract: Obesity is often attributed to an addiction to high-calorie foods. However, the effect of "food addiction" explanations on weight-related stigma remains unclear. In two online studies, participants ($n = 439$, $n = 523$, respectively, recruited from separate samples) read a vignette about a target female who was described as 'very overweight'. Participants were randomly allocated to one of three conditions which differed in the information provided in the vignette: (1) in the "medical condition", the target had been diagnosed with food addiction by her doctor; (2) in the "self-diagnosed condition", the target believed herself to be a food addict; (3) in the control condition, there was no reference to food addiction. Participants then completed questionnaires measuring target-specific stigma (i.e., stigma towards the female described in the vignette), general stigma towards obesity (both studies), addiction-like eating behavior and causal beliefs about addiction (Study 2 only). In Study 1, participants in the medical and self-diagnosed food addiction conditions demonstrated greater target-specific stigma relative to the control condition. In Study 2, participants in the medical condition had greater target-specific stigma than the control condition but only those with low levels of addiction-like eating behavior. There was no effect of condition on general weight-based stigma in either study. These findings suggest that the food addiction label may increase stigmatizing attitudes towards a person with obesity, particularly within individuals with low levels of addiction-like eating behavior.

Keywords: food addiction; obesity; stigma; eating behavior; attitudes

---

1. Introduction

According to recent statistics, more than one-third of the world's population is overweight or obesity. In the UK, these rates are even higher, with 64% of adults classed as having overweight or obesity [1]. Despite its prevalence, people with obesity frequently experience devaluation and discrimination (known as weight-related stigma) within educational, workplace, and healthcare settings [2]. Evidence also suggests that people may be more likely to face discrimination because of their weight than because of their ethnicity, gender, or sexual orientation [3]. Weight-related stigma has negative consequences for individuals' psychological and physical well-being [2,4,5] and may impede weight-loss by prompting maladaptive eating patterns and exercise avoidance [2].

Negative attitudes towards people with obesity can be exacerbated by beliefs about the *causes* of weight-gain. This is central to attribution theory, which suggests that people make judgements about the cause of a condition; in turn, these judgements determine their attitudes towards an individual [6,7]. For example, attributing obesity to factors that are within personal control (e.g., food choices) is thought to perpetuate obesity stigma [8]. Conversely, stigmatizing attitudes may be attenuated by the belief that weight-gain is caused by uncontrollable factors (e.g., genetics). In support of this, weight-related stigma was found to be most prevalent amongst individuals who believed that obesity was within personal control and caused by a lack of willpower, inactivity, and overeating [9,10]. Similar findings have been obtained from studies in which participants' causal beliefs about obesity were experimentally manipulated. Specifically, participants who read an article that stated that obesity is caused by overeating and a lack of exercise demonstrated more stigmatizing attitudes than participants in a 'no-prime' control condition or those who read a neutral article about research into memory skills [11,12]. Conversely, participants who were led to believe that obesity is caused by physiological factors (i.e., factors that are beyond personal control) demonstrated less weight-related stigma than those in a control condition [8,13].

One increasingly prevalent etiological theory is that obesity is caused by an addiction to high-calorie foods [14]. Proponents of this idea suggest that food and drugs have similar effects on the brain and argue that the clinical symptoms of substance abuse coincide with the behaviors and experiences of people who engage in compulsive overeating [15,16]. While this idea is widely debated throughout the scientific community (e.g., [17–19]), the concept of food addiction has been readily accepted by the general public [20]. Indeed, research suggests that the majority of people believe that obesity can be caused by food addiction [21], and up to half of people believe that they are themselves addicted to food [22–24]. In light of its popularity, it is important to establish how food addiction models of obesity might affect weight-related stigma.

A small number of studies have examined the effect of the food addiction label on obesity stigma. However, results to date have been inconsistent [25,26]. In one study [27], participants' attitudes towards a person with 'food addiction' were compared with attitudes towards persons with obesity, drug addiction, and disability. The study reported similarly high levels of stigma towards the "obese" and "food addict" labels and, when combined, these labels together elicited greater stigma than either label alone. These findings align with those obtained by Lee et al. [21] who found that, while the majority (72%) of survey respondents believed that obesity could be caused by a 'food addiction', more than half held the view that people with obesity are responsible for their condition (which would be expected to perpetuate obesity stigma). However, in contrast, Latner et al. [28] found that providing a food addiction explanation for obesity appeared to *reduce* weight-stigma. In this study, participants read one of two descriptions of a woman with obesity. In one condition (i.e., the 'food addiction' condition), the woman was described as fitting "the typical profile of someone who is addicted to food". In another condition (i.e., the 'non-addiction' condition), the woman was described as "someone who makes unhealthy food choices". The study found that participants in the food addiction condition displayed lower levels of stigma towards the woman, and towards people with obesity more generally, compared with those in the non-addiction condition.

Inconsistent findings in previous studies may be explained by differences in participants' causal beliefs about food addiction. Specifically, the effect of the "food addiction" label on obesity stigma may depend on the extent to which food addiction is perceived to be a legitimate medical condition. One qualitative study found that people with overweight and obesity were reluctant to label themselves as a food addict due to concerns that this would be viewed as an 'excuse' for overeating [29]. Indeed, providing excuses for weight gain may exacerbate negative attitudes towards those with obesity [30]. In contrast, attributing obesity to a medically diagnosed 'food addiction' may legitimize the condition and help to reduce weight-related stigma by removing personal responsibility from the individual [31,32].

To test these ideas, across two studies, we examined the effect of medically-diagnosed and self-diagnosed food addiction on weight-related stigma. Using a similar technique to Latner et al. [28], participants read one of three vignettes which described a woman with obesity. In the 'medical' condition, the vignette stated that the woman had been diagnosed with food addiction by her general practitioner (GP). In the 'self-diagnosed' condition, the vignette stated that the woman believed herself to be a food addict. There was no reference to food addiction in the control condition. Subsequent attitudes towards the woman (i.e., target-specific stigma) and obesity in general (i.e., general stigma) were then assessed. We hypothesized that weight-related stigma would be significantly lower in the medical condition, and higher in the self-diagnosed condition, relative to in the control condition. Based on previous findings [28], we predicted that the food addiction label would influence both target-specific and general weight-related stigma.

2. Study 1 Method

2.1. Participants

Female participants were invited to take part in a study into 'perceptions of employability among students'. Participants were recruited via social media advertisements and on internal webpages at the University of Liverpool, UK. Participants who were enrolled in the Psychology degree program at the University received course credits in exchange for taking part. A total of 440 participants completed the survey (533 participants started the study, but 93 did not complete all of the measures and so were excluded from analyses). To be eligible to take part, participants were required to be aged over 18 years old. The majority of participants were students (81%), and 90% of the sample were Caucasian. The mean age of participants was 21.2 y (SD = 7.1), and the mean self-reported body mass index (BMI) was 22.2 kg/m^2 (SD = 3.4). Participants with a self-reported BMI over 30 kg/m^2 (i.e., classified as having obesity) comprised 2.7% of the sample, 12.5% had a self-reported BMI between 25–29.9 kg/m^2 (i.e., 'overweight'), 76.8% had a self-reported BMI between 18.5–24.9 kg/m^2 (i.e., healthy weight), and 8.0% had a BMI below 18.5 kg/m^2 (i.e., 'underweight'). Participants provided informed consent prior to completing the study. Ethical approval was granted by the University of Liverpool's ethics committee (approval code: IPHS-1516-SMc-259-Generic RETH000619).

2.2. Procedure

The study was delivered via the online survey platform, Qualtrics (Qualtrics, Provo, UT, USA). Participants were asked to read an information sheet and, if they wished to continue with the study, were required to tick a consent box. On the first screen of the survey, a picture of a woman with obesity ("Paulina") was displayed, along with a short vignette which described her hobbies, family, and education (see online supplementary material). Paulina was also described as being 'very overweight'. Participants were randomly allocated to view one of three versions of the vignette: (1) In the 'medical' condition, the vignette stated that Paulina's "GP had recently diagnosed her as having a food addiction"; (2) in the 'self-diagnosed' condition, the vignette stated that Paulina "believes herself to be addicted to food"; (3) in the 'control' condition, there was no mention of food addiction. After reading the vignette, participants completed the measures in the following order: Modified Fat-Phobia Scale (M-FPS) (to assess target-specific stigma towards Paulina), employability questionnaire (included as part of the cover story), Anti-fat Attitudes (AFA; to assess general stigma towards people with obesity), and the Dutch Eating Behavior Questionnaire (DEBQ; to assess external, restrained, and emotional eating behavior). Participants were then asked to indicate their gender, age, ethnicity, occupation, and height and weight (which were used to calculate BMI). They then completed the item about self-perceived food addiction. After completing the study, participants read a debrief sheet which explained the true aim of the study.

2.3. Measures

2.3.1. Target Specific Stigma: Modified Fat-Phobia Scale (M-FPS)

The 14-item Fat Phobia Scale [33] was modified such that participants were asked to indicate their beliefs about a fictional individual named Paulina (Paulina was the name of the target female featured in the vignette. See Section 2.2). This scale consists of 14 pairs of antonyms which could be used to describe individuals with obesity (e.g., 'lazy' vs. 'industrious'). Higher scores on the M-FPS (i.e., indicative of more negative attitudes) have been positively associated with beliefs that obesity is within personal control [9]. Participants were required to indicate their perceptions of Paulina by selecting one of five points between each pair of words. A mean score was calculated for each participant. Higher scores on this measure indicated more negative attitudes towards Paulina. In the current sample, the internal reliability of the M-FPS was high (Cronbach's $\alpha = 0.834$).

2.3.2. General Stigma: Anti-fat Attitudes (AFA)

The AFA [8] consists of 13 items which assess stigmatizing attitudes toward individuals with obesity (e.g., "I dislike people who are overweight or obese"). Responses are provided on a 9-point scale ranging from 'Very strongly disagree' to 'Very strongly agree' (in Study 1, a 5-point Likert scale was used, but this was corrected to a 9-point scale in Study 2). Higher scores indicate stronger anti-fat attitudes. The scale comprises three subscales which assess dislike (i.e., obesity stigma), willpower (i.e., beliefs about weight controllability), and fear of fat (i.e., concerns about personal weight gain) (Cronbach's $\alpha = 0.796$).

2.3.3. Dutch Eating Behavior Scale (DEBQ)

The DEBQ [34] consists of 33 items which assess eating behavior. The scale comprises three subscales assessing Restrained Eating (DEBQ-R; 10 items), Emotional Eating (DEBQ-EM; 13 items), and External Eating (DEBQ-EX; 10-items). Previous research has demonstrated the ability of the DEBQ to predict restrictive eating tendencies [35], eating in response to external food-cues [36], and stress-induced eating [37]. Responses are recorded on a 5-point Likert-type scale ranging from 'Never' to 'Very often'. Higher scores indicate greater restrained, emotional, or external eating. The DEBQ was included to ensure that participants did not differ, between conditions, with regards to their eating behavior. The internal reliability for each of the subscales was high (DEBQ-R: Cronbach's $\alpha = 0.933$; DEBQ-EX: Cronbach's $\alpha = 0.869$; DEBQ-EM Cronbach's $\alpha = 0.932$).

2.3.4. Self-Perceived Food Addiction (SPFA)

To assess whether or not participants believed themselves to be a food addict, participants were presented with the statement "I believe myself to be a food addict" with response options "Yes" or "No". Similar measures have been used in previous research, and positive responses on this assessment have been associated with greater food reward, overeating [23,38], and fear of being stigmatized by others [22].

2.3.5. Employability Questions

For consistency with the study's cover story, seven items were included which assessed participants' beliefs about Paulina's employability (e.g., How likely would you be to employ Paulina?). Responses were recorded using Visual Analogue Scales (VAS) ranging from 0 (not at all) to 100 (extremely). Higher scores indicated more positive attitudes towards Paulina's employability. Analyses of the effect of condition on employability ratings are presented in the supplementary materials.

2.4. Data Analysis

A MANOVA was conducted to check whether participants differed between conditions on age, BMI, and DEBQ subscale scores. Chi-squared tests were conducted to check for any differences between the proportion of students/non-students and Caucasian/non-Caucasian participants allocated to each condition. To examine the effect of condition on target-specific and general stigma, two ANOVAs were conducted with the condition (i.e., control, medical, self-diagnosed) as the independent variable, and M-FPS (i.e., target specific stigma) and AFA (i.e., general stigma) scores as dependent variables. Where significant main effects were identified, these were followed up by inspecting pairwise comparisons.

We conducted exploratory analyses to examine whether self-reported BMI moderated the effect of condition on mean Modified Fat Phobia Scale (M-FPS) and Anti-Fat Attitudes (AFA) scores. To do this, we conducted two hierarchical multiple linear regression to examine the relative contributions of BMI (centered) and condition to mean M-FPS scores and AFA scores. All three conditions were dummy coded with the Control condition as the reference variable. To assign dummy codes, two dummy variables were created: D_1 (Medical) and D_2 (Self-diagnosed). Participants in the medical condition were assigned '1' to D_1, and '0' for D_2. Participants in the self-diagnosed condition were assigned '0' to D_1 and 1 to D_2. Participants in the control condition (i.e., the reference category) were assigned 0 to both D_1 and D_2. (see [39] for more information about dummy coding). Dummy-coded conditions were then entered into Step 1 of each regression model, along with BMI. The interaction terms (i.e., BMI × medical vs. control/self-diagnosed vs. control) were entered into Step 2 of the model.

Additional exploratory analyses were conducted to examine whether the effect of condition on target-specific and general stigma was moderated by participants' age or DEBQ subscales. Further details and results from these analyses are provided in the supplementary materials.

3. Results

3.1. Participant Characteristics

The MANOVA revealed that BMI differed significantly between conditions, $F(2,434) = 4.80$, $p = 0.009$, $\eta p^2 = 0.022$. This was due to a higher mean BMI in the medical condition relative to the self-diagnosed condition ($p = 0.002$). Participant characteristics as a function of condition are displayed in Table 1. Participants did not differ with regards to age or scores on DEBQ-subscales. Chi-squared tests (X^2) revealed no difference in the proportion of students/non-students and Caucasian/non-Caucasian participants in each condition.

Table 1. Participant characteristics as a function of condition.

Variable	Medical (N = 148)	Self-Diagnosed (N = 144)	Control (N = 146)	Between-Group Differences
Age (y)	21.09 (±6.44)	21.07 (±7.45)	21.38 (±7.32)	$F(2,435) = 0.09, p = 0.916$
BMI (kg/m^2)	22.60 (±3.22) *	21.60 (±2.95) *	22.04 (±2.93)	$F(2,432) = 3.64, p = 0.027$
DEBQ-Restraint	2.94 (±0.96)	2.72 (±0.90)	2.89 (±0.86)	$F(2,436) = 2.38, p = 0.094$
DEBQ-Emotion	2.97 (±0.90)	2.80 (±0.90)	2.84 (±0.86)	$F(2,436) = 1.43, p = 0.240$
DEBQ-External	3.34 (±0.69)	3.21 (±0.59)	3.35 (±0.71)	$F(2,436) = 1.95, p = 0.143$
Ethnicity (% Caucasian)	93.3	91.0	86.4	$X^2(2) = 4.12, p = 0.127$
Occupation (% students)	83.2	83.3	81.6	$X^2(2) = 0.186, p = 0.911$

Results are means (standard deviations) unless otherwise specified (* significant difference, $p < 0.05$).

3.2. Effect of Condition on Target-Specific and General Stigma

There was a main effect of condition on mean Modified Fat Phobia Scale (M-FPS) score (i.e., target-specific stigma), $F(2,437) = 9.07$, $p < 0.001$, $\eta p^2 = 0.040$. Pairwise comparisons revealed that, compared to those in the control condition, M-FPS scores were higher in the medical ($p < 0.001$) and self-diagnosed ($p = 0.001$) conditions (Figure 1) (Control condition: Mean = 3.47, SD = 0.47, range = 2.29–4.71; Self-diagnosed: Mean = 3.66, SD = 0.48, range = 2.71–4.93; Medical: Mean = 3.68, SD = 0.52, range = 1.00–5.00). There was no difference in mean M-FPS scores between those in the medical and self-diagnosed conditions ($p = 0.730$). There was no effect of condition on Anti-Fat Attitudes (AFA) total scores (i.e., general stigma), $F(2,437) = 0.754$, $p = 0.471$, (Control condition: Mean = 1.78, SD = 0.56, range = 0.31–3.46; Self-diagnosed: Mean = 1.71, SD = 0.56, range = 0.23–3.00; Medical: Mean = 1.72, SD = 0.56, range = 0.38–3.38).

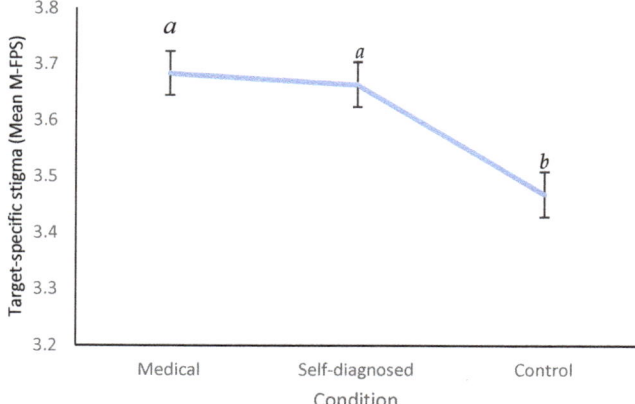

Figure 1. Mean Modified Fat Phobia Scale (M-FPS) scores (i.e., target specific stigma) as a function of condition. Different letters indicate significant differences. Higher scores indicate more negative attitudes towards Paulina (i.e., higher levels of target-specific stigma). Error bars denote standard error.

3.3. Moderating Effect of BMI

Hierarchical linear regression analyses were conducted to examine whether BMI moderated the effect of condition on target-specific (i.e., M-FPS scores) and general (AFA scores) stigma. Results from the exploratory analysis predicting M-FPS scores are provided in Table 2. In Step 1 and Step 2 of the model, M-FPS scores were significantly predicted by both condition (medical vs. control and self-diagnosed vs. control) and BMI; higher BMI was associated with lower M-FPS scores. However, M-FPS scores were not significantly predicted by the BMI × Condition interaction terms in Step 2 of the model.

Neither BMI nor condition predicted AFA scores in Step 1 of the model ($r^2 = 0.005$, $p = 0.510$), and the inclusion of interaction terms in Step 2 did significantly improve the fit of the model $r^2 = 0.015$, $p = 0.124$).

Table 2. Regression output with mean M-FPS (i.e., target-specific stigma) as the dependent variable.

Model	B	SE	t	p
Step 1				
BMI	−0.015 *	0.007	−2.119	0.035
Medical	0.230 **	0.056	4.109	<0.001
Self-diagnosed	0.189 **	0.056	3.360	0.001
Step 2				
BMI	−0.034 *	0.013	−2.547	0.011
Medical	0.223 **	0.056	3.992	<0.001
Self-diagnosed	0.190 **	0.056	3.360	0.001
BMI × Medical	0.031	0.017	1.816	0.070
BMI × Self-diagnosed	0.019	0.019	0.980	0.327

* $p < 0.05$, ** $p < 0.01$. Step 1: $r^2 = 0.051$, $p < 0.001$; Step 2: $r^2 = 0.058$, $p = 0.194$

4. Interim Discussion

Study 1 found that female participants who were exposed to medical and self-diagnosed food addiction vignettes exhibited more target-specific stigma towards a woman with obesity than those in the control condition. This is consistent with previous research in which the food addiction label was found to exacerbate stigmatizing attitudes towards an individual with obesity and 'food addiction' [27].

One possibility is that 'food addiction' stigma may be particularly high amongst those who perceive addiction to be within personal control [7]. This is supported by previous research in which perceiving addiction as a disease, rather than due to personal choice, was associated with reduced stigma towards people with addictive disorders [40,41]. Similarly, biogenetic explanations have been found to reduce stigma towards obesity, problematic eating, and substance abuse, relative to behavior-based explanations [10,31,42]. In Study 2, we examined whether the effect of food addiction condition on stigma would be moderated by the extent that addiction is viewed as a 'disease' relative to personal choice.

We also examined whether stigmatizing attitudes towards the target with food-addiction would be moderated by individuals' scores on a measure of addiction-like eating. Previous research has found that individuals with personal experience of addiction have less negative attitudes towards others with addiction [43]. Furthermore, social identity theory suggests that individuals view other 'in-group' members more favorably than out-group members [44]. Therefore, we predicted that the effect of condition on target-specific stigma would be attenuated in participants with greater levels of addiction-like eating behavior.

Finally, we examined whether the effect of condition on target-specific and general stigma would differ between males and females. Previous research has found that females demonstrate less obesity-related stigma and stigma towards the 'food addiction' label than males [27]. We, therefore, hypothesized that the exacerbating effect of the food addiction label on stigma would be most pronounced in males.

To summarize, Study 2 examined the following hypotheses: (1) The effect of condition on target-specific and general stigma would be attenuated in those with greater support for the disease model of addiction. (2) The effect of condition on stigma would be attenuated in those who score highly on a measure of addiction-like eating, relative to those who score lower on addiction-like eating. (3) The effect of condition on stigma would be attenuated in females, relative to males.

5. Study 2 Method

5.1. Participants

Male and female participants, aged over 18 years, were invited to take part in a study into 'employability perceptions'. A total of 523 (190 males; 314 females; 19 did not disclose their gender) participants completed the study. Six hundred and ten participants started the online survey, but 87 either did not complete it or were aged under 18 years old and were excluded from analyses. Participants were recruited from the University of Liverpool ($n = 333$) and Newcastle University ($n = 190$) in the UK. The mean age of participants was 27.1 (SD = 11.3) years, and the mean self-reported BMI was 23.6 kg/m^2 (SD = 4.1). Participants with self-reported BMI over 30 kg/m^2 (i.e., classified as having obesity) comprised 7.1% of the sample, 21.6% had a self-reported BMI between 25–29.9 kg/m^2 (i.e., 'overweight'), 64.4% had a self-reported BMI between 18.5–24.9 kg/m^2 (i.e., healthy weight), and 5.5% had a self-reported BMI below 18.5 kg/m^2 (i.e., 'underweight'). Just over half of the sample were university students ($n = 275$, 52.4%) and the majority were Caucasian ($n = 465$, 88.9%). Ethical approval was granted by the relevant ethics committee at each of the two sites (University of Liverpool approval code: IPHS-1516-SMc-259-Generic RETH000619; Newcastle University approval code 1485/4293).

5.2. Materials and Procedure

Study 2 used the same materials and procedure as Study 1 but with the following additional measures:

5.2.1. Addiction Belief Scale (ABS)

The ABS [39] was used to measure beliefs about addiction. Nine items assessed the belief that addiction is a disease (disease subscale, Cronbach's α = 0.590), and nine items assessed the belief that addiction is within personal control (free will subscale, Cronbach's α = 0.546). Items were rated on a 5-point Likert scale ranging from 'strongly disagree' to 'strongly agree'. Higher scores indicate greater support for the belief that addiction is akin to a disease (disease subscale), and a matter of personal choice (free will subscale).

5.2.2. Addiction-Like Eating Behaviour Scale (AEBS)

The AEBS [45] consists of 15 items which assess the presence of behaviors that are commonly associated with addiction-like eating (e.g., 'I continue to eat despite feeling full'). Responses are provided on 5-point Likert Scales ranging from 'Strongly disagree' to 'Strongly agree', and from 'Never' to 'Always'. The scale comprises two subscales: appetitive drive (9 items, Cronbach's α = 0.890) and low dietary control (6 items, Cronbach's α = 0.806). Higher scores indicate greater addiction-like eating behavior. Previous research suggests that this measure correlates positively with other measures of disinhibited eating (i.e., the Binge Eating Scale, [46]) and explains greater variance in BMI over and above other measures of 'food addiction' such as the Yale Food Addiction Scale [47].

5.2.3. Data Analysis

A MANOVA was conducted to check whether participants differed, between conditions, with regards to age, BMI, DEBQ subscales scores, and scores on the Addiction-like Eating Behaviour Scale (AEBS) and Addiction Belief Scale (ABS). Chi-squared tests were conducted to check for any differences between the proportion of students/non-students, Caucasian/non-Caucasian, and males/females allocated to each condition. As in Study 1, two univariate ANOVAs were conducted to examine the effect of condition on Anti-fat Attitudes (AFA; general stigma) and Modified-Fat Phobia Scale (M-FPS) scores (target-specific stigma). Gender was also included in the model as a between-subjects variable.

Hierarchical multiple linear regression analyses were conducted to examine whether any effects of condition on target-specific and general stigma were moderated by support for the 'disease' model of addiction (i.e., ABS-disease scores), and addiction-like eating behavior (i.e., AEBS scores). All three conditions were dummy coded with the Control condition as the reference variable. To assign dummy codes, two dummy variables were created: D_1 (Medical) and D_2 (Self-diagnosed). Participants in the medical condition were assigned '1' to D_1, and '0' for D_2. Participants in the self-diagnosed condition were assigned '0' to D_1 and 1 to D_2. Participants in the control condition (i.e., the reference category) were assigned 0 to both D_1 and D_2. (see [48] for more information about dummy coding). Dummy-coded conditions were then entered into Step 1 of each regression model, along with Addiction Belief Scale (disease subscale) or AEBS scores. The interaction terms (i.e., AEBS/Addiction Belief Scale (disease subscale) × medical vs. control/self-diagnosed vs. control) were entered into Step 2 of the model. Separate regression analyses were conducted to examine the ability of each interaction term to predict AFA scores (i.e., general stigma) and M-FPS scores (i.e., target-specific stigma). Addiction Belief Scale (disease subscale) and AEBS scores were centered prior to analyses.

6. Results

6.1. Participant Characteristics

Participants did not differ between conditions on any of the assessed characteristics (Table 3).

Table 3. Participant characteristics as a function of condition (Study 2).

Variable	Medical (N = 178)	Self-Diagnosed (N = 175)	Control (N = 170)	Between-Group Differences
Age (y)	26.6 (11.1)	26.9 (10.9)	27.8 (12.0)	$F(2,511) = 0.34, p = 0.711$
BMI (kg/m^2)	23.6 (4.5)	23.6 (4.2)	23.5 (3.7)	$F(2,511) = 0.03, p = 0.974$
DEBQ-Restraint	2.66 (0.91)	2.67 (.86)	2.76 (0.90)	$F(2,511) = 0.47, p = 0.626$
DEBQ-Emotion	2.67 (0.90)	2.64 (.98)	2.77 (0.99)	$F(2,511) = 1.16, p = 0.314$
DEBQ-External	3.29 (0.58)	3.26 (.57)	3.38 (0.55)	$F(2,511) = 2.44, p = 0.088$
AEBS	36.57 (9.65)	35.99 (9.87)	36.05 (8.70)	$F(2,511) = 0.33, p = 0.720$
ABS-disease	25.80 (3.75)	25.19 (3.92)	25.86 (4.41)	$F(2,511) = 1.45, p = 0.236$
ABS-Free Will	30.01 (3.29)	29.95 (3.72)	30.15 (4.04)	$F(2,511) = 0.14, p = 0.873$
Ethnicity (% Caucasian)	89%	89%	88%	$X^2(2) = 0.119, p=0.942$
Occupation (% students)	57%	49%	52%	$X^2(2) = 2.08, p = 0.354$
Gender (% male)	42%	31%	38%	$X^2(2) = 4.95, p = 0.084$

Abbreviations: AEBS, Addiction-like Eating Behavior Scale; ABS, Addiction Beliefs Scale; DEBQ, Dutch Eating Behaviour Scale.

6.2. Effect of Condition and Gender on Target Specific Stigma

In contrast to Study 1, there was no main effect of condition on target-specific stigma, $F(2,517) = 0.69$, $p = 0.501$, (Control condition: Mean = 3.56, SD = 0.48, range = 2.43–5.00; Self-diagnosed: Mean = 3.63, SD = 0.47, range = 2.36–4.64; Medical: Mean = 3.63, SD = 0.47, range = 2.57–4.93). Contrary to hypothesis 3, there was no gender × condition interaction for target-specific stigma, $F(2,517) = 1.18$, $p = 0.309$. However, there was a main effect of gender, $F(1,517) = 5.13$, $p = 0.024$, $\eta p^2 = 0.010$, such that males had significantly higher scores on the Modified Fat Phobia Scale (M-FPS) than females i.e., they showed higher levels of target-specific stigma (Males: M = 3.67, SE = 0.034; Females: M = 3.57, SE = 0.026).

6.3. Effect of Condition and Gender on General Stigma

As in Study 1, there was no effect of condition on Anti-fat Attitudes (AFA) scores (i.e., general stigma), $F(2,517) = 1.18$, $p = 0.308$, (Control: Mean = 4.34, SD = 1.00, range = 2.15–7.31; Self-diagnosed: Mean = 4.17, SD = 1.00, range = 1.54–7.15; Medical: Mean = 4.29, SD = 1.09, range = 1.31–7.85). Contrary to hypothesis 3, there was no gender × condition interaction, $F(2,517) = 0.02$, $p = 0.978$. There was also no main effect of gender on AFA scores, $F(1,517) = 0.02$, $p = 0.978$. For further analyses of gender differences on the AFA subscales, please see the supplementary materials (Figure S1).

6.4. Effect of Disease Beliefs on Stigma

Scores on the disease subscale of the Addiction Belief Scale (ABS) significantly predicted mean Modified-Fat Phobia Scale (M-FPS) scores in Step 1 and Step 2 of the model such that higher scores on the scale (i.e., greater belief that addiction is akin to a disease) were associated with greater target specific stigma (i.e., higher M-FPS scores) (Table 4). However, M-FPS scores were not significantly predicted by condition, and there was no condition × ABS-disease interaction, contrary to our hypothesis. Step 1: $r = 0.204$, $r^2 = 0.042$, $p < 0.001$; Step 2: $r = 0.204$, $r^2 = 0.042$, $p = 0.972$.

Table 4. Regression output for Addiction Belief Scale (ABS)-disease with M-FPS (target-specific stigma) as the dependent variable.

Model	B	SE	t	p
Step 1				
Medical	0.072	0.050	1.427	0.154
Self-diagnosed	0.091	0.051	1.797	0.073
ABS-disease	0.023 **	0.005	4.439	0.000
Step 2				
Medical	0.071	0.050	1.415	0.158
Self-diagnosed	0.090	0.051	1.781	0.076
ABS-disease	0.022 **	0.008	2.685	0.007
ABS-Disease × Medical	0.002	0.012	0.195	0.846
ABS-Disease × Self-diagnosed	0.000	0.012	−0.034	0.972

** $p < 0.01$. The control condition was used as the reference category against which medical and self-diagnosed conditions were compared. Abbreviations: ABS, Addiction Belief Scale. Step 1: $r^2 = 0.042$, $p < 0.001$; Step 2: $r^2 = 0.042$, $p = 0.972$.

Similarly, scores on the disease subscale of the ABS significantly predicted Anti Fat Attitude (AFA) scores (i.e., general stigma) in Step 1 and Step 2 of the model such that higher scores on the ABS-disease subscale predicted higher AFA scores (Table 5). Contrary to hypothesis 1, AFA scores were not significantly predicted by condition, and there was no interaction between condition and disease scores on AFA.

Table 5. Regression output for ABS-disease with Anti Fat Attitude (AFA; general stigma) as the dependent variable.

Model	B	SE	t	p
Step 1				
Medical	−0.056	0.109	−0.516	0.606
Self-diagnosed	−0.146	0.110	−1.331	0.184
ABS-disease	0.047 **	0.011	4.281	0.000
Step 2				
Medical	−0.053	0.109	−0.482	0.630
Self-diagnosed	−0.147	0.110	−1.337	0.182
ABS-disease	0.059 **	0.018	3.295	0.001
ABS-Disease × Medical	−0.016	0.027	−0.582	0.560
ABS-Disease × Self-diagnosed	−0.021	0.026	−0.791	0.429

** $p < 0.01$. The control condition was used as the reference category against which medical and self-diagnosed conditions were compared. Step 1: $r = 0.198$, $r^2 = 0.039$, $p < 0.001$; Step 2: $r = 0.201$, $r^2 = 0.040$, $p = 0.707$.

6.5. Addiction-Like Eating Behavior

Addiction-like Eating Behavior Scale (AEBS) scores and condition did not predict Modified Fat Phobia Scale (M-FPS) (target-specific stigma) scores in Step 1 of the model. However, the inclusion of the interaction terms in Step 2 significantly improved the fit of the model. Regression coefficients revealed a significant interaction between AEBS scores and medical (vs. control) condition on M-FPS scores (Table 6).

Table 6. Regression output for Addiction-like Eating Behavior Scale (AEBS) scores with M-FPS (target-specific stigma) as the dependent variable.

Model	B	SE	t	p
Step 1				
Medical	0.067	0.051	1.32	0.186
Self-diagnosed	0.071	0.051	1.38	0.168
AEBS	0.003	0.002	1.42	0.156
Step 2				
Medical	0.067	0.051	1.32	0.187
Self-diagnosed	0.071	0.051	1.39	0.165
AEBS	0.008	0.004	1.90	0.058
AEBS × Medical	−0.013 *	0.006	−2.35	0.019
AEBS × Self-diagnosed	0.000	0.006	−0.065	0.948

* $p < 0.05$. The control condition was used as the reference category against which medical and self-diagnosed conditions were compared. Abbreviations: AEBS, Addiction-like Eating Behavior Scale. Step 1: $r^2 = 0.009$, $p = 0.214$; Step 2: $r^2 = 0.023$, $p = 0.020$.

To further examine the interaction between AEBS scores and condition on M-FPS scores, we used the Johnson–Neyman technique [49] to identify the levels of addiction-like eating (i.e., AEBS scores) at which condition elicited a significant difference on M-FPS scores [50]. Using PROCESS (Version 3.1., [51]), the Medical (dummy-coded) condition was entered as the predictor variable, AEBS scores were entered as the moderator variable, and Self-diagnosed condition (dummy-coded) and the Self-diagnosed × AEBS interaction term were entered as covariates. Mean-FPS scores were entered as the dependent variable. This analysis showed that the Medical condition resulted in significantly greater M-FPS scores, relative to the Self-diagnosed and Control conditions (ps < 0.05), but only for those with low AEBS scores (centered AEBS score ≤ −2.81) (Figure 2). Findings are, therefore, consistent with our hypothesis that the effect of condition on stigma would be attenuated in those with higher levels of addiction-like eating behavior.

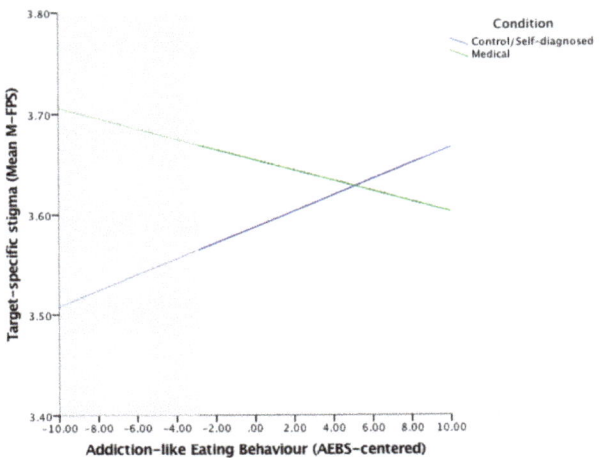

Figure 2. The effect of condition on M-FPS scores at different levels of addiction-like eating behavior (assessed using the AEBS). The shaded area represents the region of significance identified using the Johnson-Neyman technique.

The condition × AEBS scores model predicting (general stigma) AFA scores was not significant (Step 1: $r = 0.069$, $r^2 = 0.005$, $p = 0.484$; Step 2: $r = 0.084$, $r^2 = 0.007$, $p = 0.540$).

7. Discussion

Across two studies, we examined the effect of the food addiction label on stigmatizing attitudes towards an individual with obesity (i.e., target specific), and towards people with obesity more generally (i.e., general stigma). In Study 1, participants in both the medical and self-diagnosed food addiction conditions demonstrated greater target-specific stigma relative to the control condition. There was no effect of condition on general stigmatizing attitudes towards people with obesity. However, findings from Study 1 were not replicated in Study 2, in which we included both male and female participants. That is, we found no overall differences between the food addiction conditions and the control condition on target-specific stigma. The effect of condition on target-specific or general stigma was also not moderated by addiction disease beliefs (i.e., the extent to which addiction is perceived as a disease) or gender, in Study 2. However, there was a significant condition by addiction-like eating behavior interaction on target-specific stigma; participants who scored low on a measure of addiction-like eating demonstrated greater target-specific stigma in the Medical condition relative to Control and Self-diagnosed conditions. In contrast, target-specific stigma did not differ as a function of condition for those with high levels of addiction-like eating.

Findings from Study 1 are consistent with previous findings in which the food addiction label added to the stigma of obesity [27]. Higher levels of stigma towards the 'self-perceived' food-addicted target in the current study may reflect perceptions of food addiction as an 'excuse' for overeating. This is supported by qualitative evidence that individuals with overweight or obesity may be reluctant to label themselves as food addicts due to concerns that this would be perceived as an 'excuse' for their weight [29].

We predicted that the medical condition might legitimize the concept of food addiction and thereby reduce weight-related stigma (i.e., by removing personal responsibility from the individual). However, contrary to our hypothesis, in Study 1, we found that target-specific stigma was also higher in the medical condition compared to the control condition and did not differ from levels observed in the self-diagnosed condition. This finding is inconsistent with predictions from attribution theory [7] in which undesirable behaviors that are perceived as beyond personal control are thought to elicit less stigma than those that are perceived as controllable. One possibility is that food addiction explanations increase stigma by inadvertently emphasizing the behavioral aspect of obesity. That is, food addiction may imply a loss of control over eating, and previous studies have found that this may increase stigmatizing attitudes towards obesity [52]. Another possible explanation is that food addiction, unlike other biological causes of obesity, is believed to be within personal control and that medicalizing the term does not remove perceptions of personal responsibility. Indeed, Lee et al. [21] reported that almost three-quarters of people supported food addiction as a cause of obesity, and yet obesity was still viewed as a condition that individuals need to take responsibility for. Therefore, it may be the case that stigmatizing attitudes towards 'food addicted' individuals are dependent upon the extent that addiction is perceived as being outside of personal control and/or akin to a disease. In relation to this, Study 2 examined whether the effect of food addiction condition on stigma would be attenuated in those with greater support for the disease model of addiction (results discussed below).

Study 1 therefore suggests that the food addiction label exacerbated stigmatizing attitudes towards a woman with obesity, regardless of whether the food addiction was medically diagnosed or self-diagnosed. Notably, findings from Study 1 are inconsistent with those obtained in a previous study in which a 'food addiction' explanation for obesity elicited *lower* levels of target-specific and general stigma than a control explanation [28]. This inconsistency may be attributable to the control conditions used in ours and Latner et al.'s [28] study; in the current study, participants in the control condition were not provided with any explanation for the target's weight status. In contrast, participants in Latner et al.'s [28] study read that obesity is caused by repeatedly choosing to consume high-calorie foods. By emphasizing the role of personal choice, it is possible that the control condition used by Latner et al. [28] may have elicited greater stigma than a 'food addiction' explanation for obesity.

In Study 2, we found that greater support for the disease model of addiction was associated with greater target-specific and general stigma towards obesity. This finding was unexpected and is contrary to predictions derived from attribution theory. One possibility is that the perception of addiction as a 'disease' encourages the view that addicts are abnormal and perpetuates an 'us-them' distinction [53]. Holding disease views of addiction also suggests that the person's condition is irrevocable and permanent [54]. Another possibility is that causal beliefs about food addiction do not coincide with perceptions of other addictions. That is, individuals who support the 'disease' model for substance-based addictions may not necessarily attribute food addiction to a disease. Previous research supports this, indicating that addictions vary in the extent to which they are attributed to disease or personal choice. In particular, de Pierre et al. [40] found that food addiction was perceived as less of a disease and more within personal control compared with other addictions such as alcoholism. The measure of addiction beliefs (i.e., the ABS) used in the current study referred to addiction in general, and thus may not have reflected participants' beliefs about food addiction per se.

However, the moderating effect of addiction-like eating on target-specific stigma, observed in Study 2, suggest that medically diagnosed food addiction could exacerbate weight-related stigma but only for people with low levels of addiction-like eating tendencies. A possible explanation for this finding is that individuals with personal experience of problematic eating (i.e., high AEBS scores) may have identified more with the target in the vignette and thereby displayed less negative attitudes towards her food addiction (e.g., see [43,44]) as opposed to participants with low AEBS scores.

In Study 2, male participants demonstrated significantly higher target-specific stigma, relative to female participants. Males and females did not differ on a measure of general weight-related stigma. However, the lack of interaction between gender and condition is inconsistent with previous research [27] in which stigmatizing attitudes towards a 'food addicted' target were lower in females, relative to males. This null result may be explained by the fact that, in the current study, males had a significantly higher mean BMI than females (see Table S1). A previous study found that people with higher BMI hold less stigmatizing attitudes towards the 'food addict' label, relative to those with lower BMI [27]. Consistent with this, in Study 1, we found that higher BMI was associated with lower target-specific weight stigma. It is therefore possible that, in the current study, any moderating effect of gender on stigma may have been masked by the higher BMI of male, relative to female, participants. Future research should examine the moderating effect of gender on stigmatizing attitudes towards a food-addicted target in samples of males and females matched for BMI.

The inconsistent findings obtained across Studies 1 and 2 could not be attributable to the inclusion of males in Study 2 as the effect of condition on target-specific stigma was not moderated by gender. The sample tested in Study 2 comprised a larger proportion of older, non-students than the sample tested in Study 1. However, exploratory analyses revealed that the effect of condition on stigma was not moderated by student status or age (see online supplementary material). Differences between Studies 1 and 2 are, therefore, likely due to another (unknown) variable. Moreover, these findings suggest that the effects of the food addiction label on weight-related stigma may not be generalizable across populations.

There are several limitations to the current study that require consideration. Firstly, we note that the Addiction Belief Scale, used in Study 2, examined beliefs about the causes of addiction in general, and thus may not have captured individual differences in beliefs about the causes of food addiction. Future research could use an adapted version of the ABS (such as that used by de Pierre et al [40]) to test whether food addiction stigma is attenuated in individuals who have greater support for a disease model of food addiction. Secondly, we did not examine whether participants believed the food addiction explanation for obesity, nor did we check whether participants had guessed the study aims. It is, therefore, possible that the effect of the food addiction label on stigma, observed in Study 1, could be due to demand characteristics that were not present in Study 2. Thirdly, the use of a female target in the current study precludes the generalizability of our findings to males. Previous research suggests that females are more likely than males to be stigmatized due to their weight [55], and so attitudes towards the food addiction label may similarly differ as a function of the target's gender. Finally, it is important to consider that the findings may have been affected by the order in which the questionnaires were presented. In particular, the significant effect of condition on target-specific stigma (M-FPS) (in Study 1), and lack of effect of general stigma (AFA), may be due to the fact that participants completed the M-FPS immediately after reading the vignette, while general stigma (i.e., AFA scores) were assessed later in the study.

Future research should aim to clarify the effect of the food addiction label on weight-related stigma. This may be achieved by considering possible moderating effects of pre-existing beliefs about food addiction (e.g., the extent that it is a legitimate condition, whether it is controllable, etc.). There has been much debate in the scientific literature about whether addiction-like eating should be considered a substance-based 'food addiction' or a behavioral 'eating addiction' (e.g., [11]). Therefore, it will also be important to compare attitudes elicited by a 'food addiction' label, with attitudes towards an 'eating addiction' label. It would also be interesting to compare the effect on the stigma of medically-diagnosed food addiction, with other medical causes of weight gain (e.g., hypothyroidism). Doing so would provide insight into whether the potential exacerbating effect of medicalization on stigma is specific to the food addiction label or whether it extends to the medical model per se. It is also possible that emphasizing the non-behavioral aspect of food addiction (e.g., brain differences to food) may reduce any deleterious effect of a medical diagnosis on stigma. More broadly, the clinical implications of food addiction labels on weight-related stigma must now be considered. In particular, it is important to consider whether the food addiction label may affect people's approaches to treatment (e.g., seeking pharmacological solutions rather than psychotherapy). It is also possible that, by perpetuating weight-related stigma, the food addiction label could be detrimental to psychological well-being and undermine people's attempts to lose weight.

8. Conclusions

The results indicate that the food addiction label may exacerbate stigmatizing attitudes towards an individual with obesity. Furthermore, there is preliminary evidence that this effect may be most pronounced in people with pre-existing low levels of addiction-like eating behavior. Further research is needed to determine the longer-term effects of the food addiction label on weight stigma and clinical implications.

Supplementary Materials: The following are available online at http://www.mdpi.com/2072-6643/11/9/2100/s1, Figure S1: Scores on AFA-Willpower subscale as a function of condition and gender. Table S1: Participant characteristics as a function of gender.

Author Contributions: Conceptualization, E.J.B. and C.A.H.; Data curation, H.K.R. and M.O.; Formal analysis, H.K.R. and M.O.; Methodology, M.O., E.J.B. and C.A.H.; Supervision, E.H.E. and C.A.H.; Writing—original draft, H.K.R. and C.A.H.; Writing—review & editing, M.O., E.J.B., E.H.E. and C.A.H.

Funding: The work reported in this manuscript received no external funding.

Acknowledgments: The authors thank Kerry Boult, Richard Ensell, Helena Leech, and Belen Valle-Metaxas for assistance with data collection.

Conflicts of Interest: C.A.H. receives research funding from the American Beverage Association and speaker fees from the International Sweeteners Association for work outside of the submitted manuscript.

References

1. NHS Digital. Available online: Digital.nhs.uk/data-and-information/publications/statistical/sand-diet/statistics-on-obesity-physical-activity-and-diet-england-2019 (accessed on 27 June 2019).
2. Puhl, R.M.; Heuer, C.A. The stigma of obesity: A review and update. *Obesity* **2009**, *17*, 941–964. [CrossRef] [PubMed]
3. Brinsden, H.; Coltman-Patel, T.; Sievert, K. Weight Stigma in the Media. Available online: https://www.britishlivertrust.org.uk/wp-content/up (accessed on 27 June 2019).

4. Pearl, R.L.; Wadden, T.A.; Hopkins, C.M.; Shaw, J.A.; Hayes, M.R.; Bakizada, Z.M.; Alfaris, N.; Chao, A.M.; Pinkasavage, E.; Berkowitz, R.I.; et al. Association between weight bias internalization and metabolic syndrome among treatment-seeking individuals with obesity. *Obesity* **2017**, *25*, 317–322. [CrossRef] [PubMed]
5. Nolan, L.J.; Eshleman, A. Paved with good intentions: Paradoxical eating responses to weight stigma. *Appetite* **2016**, *102*, 15–24. [CrossRef] [PubMed]
6. Rush, L.L. Affective reactions to multiple social stigmas. *J. Soc. Psychol.* **1998**, *138*, 421–430. [CrossRef]
7. Weiner, B. An attributional theory of achievement motivation and emotion. *Psychol. Rev.* **1985**, *92*, 548–573. [CrossRef] [PubMed]
8. Crandall, C.S. Prejudice against fat people: Ideology and self-interest. *J. Personal. Soc. Psychol.* **1994**, *66*, 882–894. [CrossRef]
9. Pearl, R.L.; Lebowitz, M.S. Beyond personal responsibility: Effects of causal attributions for overweight and obesity on weight-related beliefs, stigma, and policy support. *Psychol. Health* **2014**, *29*, 1176–1191. [CrossRef]
10. Hilbert, A.; Rief, W.; Braehler, E. Stigmatizing attitudes toward obesity in a representative population-based sample. *Obesity* **2008**, *16*, 1529–1534. [CrossRef]
11. Teachman, B.A.; Gapinski, K.D.; Brownell, K.D.; Rawlins, M.; Jeyaram, S. Demonstrations of implicit anti-fat bias: The impact of providing causal information and evoking empathy. *Health Psychol.* **2003**, *22*, 68–78. [CrossRef]
12. Lewis, R.J.; Cash, T.F.; Jacobi, L.; Bubb-Lewis, C. Prejudice toward fat people: The development and validation of the antifat attitudes test. *Obes. Res.* **1997**, *5*, 297–307. [CrossRef]
13. Dejong, W. The stigma of obesity: The consequences of naive assumptions concerning the causes of physical deviance. *J. Health Soc. Behav.* **1980**, *21*, 75–87. [CrossRef] [PubMed]
14. Lerma-Cabrera, J.M.; Carvajal, F.; Lopez-Legarrea, P. Food addiction as a new piece of the obesity framework. *Nutr. J.* **2016**, *15*, 5. [CrossRef]
15. Schulte, E.M.; Avena, N.M.; Gearhardt, A.N. Which foods may be addictive? The roles of processing, fat content, and glycemic load. *PLoS ONE* **2015**, *10*, e0117959. [CrossRef]
16. Schulte, E.M.; Grilo, C.M.; Gearhardt, A.N. Shared and unique mechanisms underlying binge eating disorder and addictive disorders. *Clin. Psychol. Rev.* **2016**, *44*, 125–139. [CrossRef] [PubMed]
17. Rogers, P.J. Food and drug addictions: Similarities and differences. *Pharmacol. Biochem. Behav.* **2017**, *153*, 182–190. [CrossRef]
18. Ziauddeen, H.; Farooqi, I.S.; Fletcher, P.C. Obesity and the brain: How convincing is the addiction model? *Nat. Rev. Neurosci.* **2012**, *13*, 279–286. [CrossRef] [PubMed]
19. Hebebrand, J.; Albayrak, Ö.; Adan, R.; Antel, J.; Dieguez, C.; de Jong, J.; Leng, G.; Menzies, J.; Mercer, J.G.; Murphy, M.; et al. "Eating addiction", rather than "food addiction", better captures addictive-like eating behavior. *Neurosci. Biobehav. Rev.* **2014**, *47*, 295–306. [CrossRef] [PubMed]
20. Ruddock, H.K.; Hardman, C.A. Food addiction beliefs amongst the lay public: What are the consequences for eating behaviour? *Curr. Addict. Rep.* **2017**, *4*, 110–115. [CrossRef]
21. Lee, N.M.; Lucke, J.; Hall, W.D.; Meurk, C.; Boyle, F.M.; Carter, A. Public views on food addiction and obesity: Implications for policy and treatment. *PLoS ONE* **2013**, *8*, e74836. [CrossRef]
22. Meadows, A.; Nolan, L.J.; Higgs, S. Self-perceived food addiction: Prevalence, predictors, and prognosis. *Appetite* **2017**, *114*, 282–298. [CrossRef]
23. Ruddock, H.K.; Dickson, J.M.; Field, M.; Hardman, C.A. Eating to live or living to eat? Exploring the causal attributions of self-perceived food addiction. *Appetite* **2015**, *95*, 262–268. [CrossRef] [PubMed]
24. Hardman, C.A.; Rogers, P.J.; Dallas, R.; Scott, J.; Ruddock, H.; Robinson, E. "Food addiction is real": The effects of exposure to this message on self-diagnosed food addiction and eating behaviour. *Appetite* **2015**, *91*, 179–184. [CrossRef] [PubMed]
25. Reid, J.; Brien, K.S.O.; Puhl, R.; Hardman, C.A.; Carter, A. Food addiction and its potential links with weight stigma. *Curr. Addict. Rep.* **2018**, *5*, 192–201. [CrossRef]
26. Cassin, S.E.; Buchman, D.Z.; Leung, S.E.; Kantarovich, K.; Hawa, A.; Carter, A.; Sockalingam, S. Ethical, stigma, and policy implications of food addiction: A scoping review. *Nutrients* **2019**, *11*, 710. [CrossRef] [PubMed]
27. DePierre, J.A.; Puhl, R.M.; Luedicke, J. A new stigmatized identity? Comparisons of a "food addict" label with other stigmatized health conditions. *Basic Appl. Soc. Psychol.* **2013**, *35*, 10–21. [CrossRef]

28. Latner, J.D.; Puhl, R.M.; Murakami, J.M.; O'Brien, K.S. Food addiction as a causal model of obesity. Effects on stigma, blame, and perceived psychopathology. *Appetite* **2014**, *77*, 77–82. [CrossRef] [PubMed]
29. Cullen, A.J.; Barnett, A.; Komesaroff, P.A.; Brown, W.; O'Brien, K.S.; Hall, W.; Carter, A. A qualitative study of overweight and obese Australians' views of food addiction. *Appetite* **2017**, *115*, 62–70. [CrossRef]
30. Epstein, L.; Ogden, J. A qualitative study of GPs' views of treating obesity. *Br. J. Gen. Pract.* **2005**, *55*, 750–754. [PubMed]
31. Kvaale, E.P.; Haslam, N.; Gottdiener, W.H. The "side effects" of medicalization: A meta-analytic review of how biogenetic explanations affect stigma. *Clin. Psychol. Rev.* **2013**, *33*, 782–794. [CrossRef] [PubMed]
32. Schulte, E.M.; Tuttle, H.M.; Gearhardt, A.N. Belief in food addiction and obesity-related policy support. *PLoS ONE* **2016**, *11*, e0147557. [CrossRef]
33. Bacon, J.; Scheltema, K.; Robinson, B. Fat phobia scale revisited: The short form. *Int. J. Obes.* **2001**, *25*, 252. [CrossRef] [PubMed]
34. Van Strien, T.; Frijters, J.E.R.; van Staveren, W.A.; Defares, P.B.; Deurenberg, P. The predictive validity of the Dutch restrained eating scale. *Int. J. Eat. Disord.* **1986**, *5*, 747–755. [CrossRef]
35. Van Strien, T.; Bergers, G.P.A.; Defares, P.B. The Dutch eating behavior questionnaire (DEBQ) for assessment of restrained, emotional, and external eating behavior. *Int. J. Eat. Disord.* **1986**, *5*, 295–315. [CrossRef]
36. Van Strien, T.; Peter Herman, C.; Anschutz, D. The predictive validity of the DEBQ-external eating scale for eating in response to food commercials while watching television. *Int. J. Eat. Disord.* **2012**, *45*, 257–262. [CrossRef] [PubMed]
37. Oliver Georgina, L.; Wardle Jane, L.; Gibson, E.L. Stress and food choice: A laboratory study. *Psychosom. Med.* **2000**, *62*, 853–865. [CrossRef] [PubMed]
38. Ruddock, H.K.; Field, M.; Hardman, C.A. Exploring food reward and calorie intake in self-perceived food addicts. *Appetite* **2017**, *115*, 36–44. [CrossRef] [PubMed]
39. Schaler, J.A. The addiction belief scale. *Int. J. Addict.* **1995**, *30*, 117–134. [CrossRef]
40. DePierre, J.A.; Puhl, R.M.; Luedicke, J. Public perceptions of food addiction: A comparison with alcohol and tobacco. *J. Subst. Use* **2014**, *19*, 1–6. [CrossRef]
41. Cunningham, J.A.; Sobell, L.C.; Freedman, J.L.; Sobell, M.B. Beliefs about the causes of substance abuse: A comparison of three drugs. *J. Subst. Abus.* **1994**, *6*, 219–226. [CrossRef]
42. O'Brien, K.S.; Puhl, R.M.; Latner, J.D.; Mir, A.S.; Hunter, J.A. Reducing anti-fat prejudice in preservice health students: A randomized trial. *Obesity* **2010**, *18*, 2138–2144. [CrossRef]
43. Meurk, C.; Carter, A.; Partridge, B.; Lucke, J.; Hall, W. How is acceptance of the brain disease model of addiction related to Australians' attitudes towards addicted individuals and treatments for addiction? *BMC Psychiatry* **2014**, *14*, 373. [CrossRef] [PubMed]
44. Tajfel, H.; Turner, J. *The Social Identity Theory of Intergroup Behaviour. U: Worchel S. i Austin WG (ur.) Psychology of Intergroup Relations*; Nelson Hall: Chicago, IL, USA, 1986.
45. Ruddock, H.K.; Christiansen, P.; Halford, J.C.G.; Hardman, C.A. The development and validation of the addiction-like eating behaviour scale. *Int. J. Obes.* **2017**, *41*, 1710. [CrossRef] [PubMed]
46. Gormally, J.; Black, S.; Daston, S.; Rardin, D. The assessment of binge eating severity among obese persons. *Addict. Behav.* **1982**, *7*, 47–55. [CrossRef]
47. Gearhardt, A.N.; Corbin, W.R.; Brownell, K.D. Preliminary validation of the Yale food addiction scale. *Appetite* **2009**, *52*, 430–436. [CrossRef] [PubMed]
48. Alkharusi, H. Categorical variables in regression analysis: A comparison of dummy and effect coding. *Int. J. Educ.* **2012**, *4*, 202–210. [CrossRef]
49. Johnson, P.O.; Neyman, J. Tests of certain linear hypotheses and their application to some educational problems. *Stat. Res. Mem.* **1936**, *1*, 57–93.
50. Hayes, A.F.; Montoya, A.K. A tutorial on testing, visualizing, and probing an interaction involving a multicategorical variable in linear regression analysis. *Commun. Methods Meas.* **2017**, *11*, 1–30. [CrossRef]
51. Hayes, A.F. *Introduction to Mediation, Moderation, and Conditional Process Analysis*; The Guildford Press: New York, NY, USA, 2017.
52. Bannon, K.L.; Hunter-Reel, D.; Wilson, G.T.; Karlin, R.A. The effects of causal beliefs and binge eating on the stigmatization of obesity. *Int. J. Eat. Disord.* **2009**, *42*, 118–124. [CrossRef]
53. Phelan, J.C. Genetic bases of mental illness—A cure for stigma? *Trends Neurosci.* **2002**, *25*, 430–431. [CrossRef]

54. Phelan, J.C. Geneticization of deviant behavior and consequences for stigma: The case of mental illness. *J. Health Soc. Behav.* **2005**, *46*, 307–322. [CrossRef]
55. Puhl, R.M.; Andreyeva, T.; Brownell, K.D. Perceptions of weight discrimination: Prevalence and comparison to race and gender discrimination in America. *Int. J. Obes.* **2008**, *32*, 992–1000. [CrossRef] [PubMed]

© 2019 by the authors. Licensee MDPI, Basel, Switzerland. This article is an open access article distributed under the terms and conditions of the Creative Commons Attribution (CC BY) license (http://creativecommons.org/licenses/by/4.0/).

Article

The Effect of a Food Addiction Explanation Model for Weight Control and Obesity on Weight Stigma

Kerry S. O'Brien [1,*], Rebecca M. Puhl [2], Janet D. Latner [3], Dermot Lynott [4], Jessica D. Reid [1], Zarina Vakhitova [1], John A. Hunter [5], Damian Scarf [5], Ruth Jeanes [6], Ayoub Bouguettaya [1] and Adrian Carter [7]

1. School of Social Sciences, Faculty of Arts, Monash University, Melbourne 3800, Australia; jdorothea.reid@gmail.com (J.D.R.); zarina.vakhitova@monash.edu (Z.V.); ayoub.bouguettaya@monash.edu (A.B.)
2. Department of Human Development and Family Sciences, Rudd Center for Food Policy & Obesity, University of Connecticut, Storrs, CT 06269, USA; rebecca.puhl@uconn.edu
3. Department of Psychology, College of Social Sciences, University of Hawaii, Manoa, HI 96822, USA; jlatner@hawaii.edu
4. Department of Psychology, Faculty of Science and Technology University of Lancaster, Lancaster LA1 4YW, UK; d.lynott@lancaster.ac.uk
5. Division of Sciences, Department of Psychology, University of Otago, Dunedin 9016, New Zealand; jackie.hunter@otago.ac.nz (J.A.H.); damian@psy.otago.ac.nz (D.S.)
6. Curriculum & Pedagogy, Faculty of Education, Monash University, Melbourne 3800, Australia; ruth.jeanes@monash.edu
7. School of Psychology, Faculty of Medicine, Nursing, and Health Sciences, Monash University, Melbourne 3800, Australia; adrian.carter@monash.edu
* Correspondence: kerrykez@gmail.com

Received: 27 September 2019; Accepted: 7 January 2020; Published: 22 January 2020

Abstract: There is increasing scientific and public support for the notion that some foods may be addictive, and that poor weight control and obesity may, for some people, stem from having a food addiction. However, it remains unclear how a food addiction model (FAM) explanation for obesity and weight control will affect weight stigma. In two experiments ($N = 530$ and $N = 690$), we tested the effect of a food addiction explanation for obesity and weight control on weight stigma. In Experiment 1, participants who received a FAM explanation for weight control and obesity reported lower weight stigma scores (e.g., less dislike of 'fat people', and lower personal willpower blame) than those receiving an explanation emphasizing diet and exercise ($F_{(4,525)} = 7.675$, $p = 0.006$; and $F_{(4,525)} = 5.393$, $p = 0.021$, respectively). In Experiment 2, there was a significant group difference for the dislike of 'fat people' stigma measure ($F_{(5,684)} = 5.157$, $p = 0.006$), but not for personal willpower weight stigma ($F_{(5,684)} = 0.217$, $p = 0.81$). Participants receiving the diet and exercise explanation had greater dislike of 'fat people' than those in the FAM explanation and control group (p values < 0.05), with no difference between the FAM and control groups ($p > 0.05$). The FAM explanation for weight control and obesity did not increase weight stigma and resulted in lower stigma than the diet and exercise explanation that attributes obesity to personal control. The results highlight the importance of health messaging about the causes of obesity and the need for communications that do not exacerbate weight stigma.

Keywords: stigma; obesity; food addiction; weight bias; weight stigma; obesity prejudice reduction

1. Introduction

Research on the extent, nature, and impact of weight stigma (also termed weight bias, obesity stigma) suggests that weight stigma has increased over time in adults [1] and children [2] and is associated with a host of negative social and health outcomes [3]. For example, research shows that women perceived to be overweight or obese encounter discrimination in education, health, and employment settings [4–6]. Similarly, experiences of weight stigma are associated with poorer psychological and physical outcomes arising from stigma-related stress, including increased depression and anxiety [7], emotional and stress-related eating [8], and avoidance of health care settings [9]. As such, there is a need for research that seeks to understand factors that contribute to and reinforce weight stigma.

Attribution theories, and specifically, the attribution-value model [10] suggests that antipathy toward a specific group or target is maintained by beliefs about the controllability of specific group behaviours. In the case of overweight and obesity, studies show that weight stigma is increased by attributions about controllability of weight and obesity [11,12]. That is, because people are exposed to public health and media messages that weight is under personal control, people conclude that obesity must be due to an individual's personal failures, which in turn leads to greater weight stigma [13,14]. Dominant public health messages on the cause of overweight and obesity remain focused, if simplistically, on individual control of diet and physical activity [15]. This individualistic public health narrative is increasingly criticised [16], as it ignores research on the myriad of uncontrollable factors contributing to weight control and obesity, such as neurophysiology, environment, and the interplay with genetics/epigenetics. Experimental evidence suggests that changing people's attributions about the causes of obesity away from individual blame, and to more biologically and environmentally pre-determined factors, can help to reduce weight stigma [11]. Accordingly, correcting public misattributions about weight has the potential not only to improve knowledge about the complex causes of obesity, but also the potential to reduce weight stigma and discrimination [17].

Recent research posits that some people may have a neurobiological addiction to certain foods, particularly ultra-processed hyper-palatable foods. This addiction may, in part, contribute to people's food choices and consumption behaviour, and in turn obesity rates [18,19]. Termed *food addiction*, neuropsychological and behavioural research on the addictive properties of food identifies considerable overlap in the food and drug reward and addiction-related centres and pathways of the brain; as a result, food can be as rewarding and addictive as other addictive substances such as drugs, which share overlapping brain reward pathways [20,21]. Large-scale studies suggest that a significant proportion (15%) of the general population [22], and a greater proportion of those with obesity (up to 30%) meet criteria for a diagnosis of food addiction [23]. Furthermore, between 28% and 52% of the general population perceive themselves to be addicted to food [24].

The food addiction model (FAM) for weight control and obesity raises questions about whether a FAM explanation could be helpful or harmful in efforts to reduce weight stigma. While it could be argued that a FAM explanation might increase weight stigma as a result of labelling individuals with obesity as having an addiction, it is possible that a FAM explanation could instead reduce stigma towards people perceived to have obesity by reducing attributions of individual controllability of weight [25]. Research examining these questions is scarce [26]. Some research suggests that the addition of the addiction label to obesity is associated with increased vulnerability to stigmatization [27], and experimental work suggests that the FAM explanation may increase stigma associated with obesity [28]. In contrast, experimental research by Latner and colleagues [25] found that a FAM explanation for weight control and obesity resulted in less stigma and less blame for targets at both lower and higher body weights. These mixed findings indicate the need for additional research to establish whether a FAM explanation increases or decreases weight stigma relative to current public health messaging that suggests diet and exercise as the primary drivers of weight and obesity [29,30].

The present study aimed to examine whether a food addiction explanation for weight control would exacerbate or ameliorate weight stigma relative to the dominant public health messaging emphasizing personal control of diet and exercise. We conducted two experiments to assess the impact of a food addiction explanation for obesity on expressions of weight stigma. In line with previous research with the attribution value model [10] for weight stigma, we predicted that the FAM explanation would result in less weight stigma towards people perceived to be obese or "fat".

2. Methods (Experiment 1 and 2)

2.1. Participants

Table 1 details the demographic characteristics of participants in Experiment 1 and 2. For Experiment 1, a sample of $N = 652$ university (college) students was invited to participate in the experiment in return for course credit. Most students (86%; $N = 561$) agreed to participate, with $N = 530$ (94%) of those agreeing to participate providing data on the outcome variables. For Experiment 2, university (college) students ($N = 717$) were invited to participate in the experiment in return for course credit. Most (96%; $N = 696$) agreed to participate, with $N = 690$ providing data on the outcome variables.

Table 1. Participant characteristics for Experiment 1 and Experiment 2.

Parameter	Value
Experiment 1 ($N = 530$)	
Age (years)	$M = 19.7$, $SD = 1.8$ (range: 18–35)
Gender (female/male)	73.8% ($n = 391$)/26.2% ($n = 139$)
Body Mass Index (BMI)	$M = 22.5$, $SD = 4.1$
Percentage underweight (BMI ≤ 18.5 kg/m^2)	10.9% ($n = 57$)
Percentage normal weight (BMI = 18.5 < 25.0)	69.4% ($n = 367$)
Percentage overweight or obese (BMI > 25+)	20% ($n = 106$)
Experiment 2 ($N = 690$)	
Age (years)	$M = 19.7$, $SD = 2.51$ (range: 18–52)
Gender (female/male)	72.2% ($n = 498$)/27.7% ($n = 191$)
Body Mass Index (BMI)	$M = 22.4$, $SD = 4.1$
Percentage underweight (BMI ≤ 18.5 kg/m^2)	11.1% ($n = 77$)
Percentage normal weight (BMI = 18.5 < 25.0)	68.0% ($n = 469$)
Percentage overweight or obese (BMI > 25+)	20.9% ($n = 144$)

A priori sample size calculations indicated a required minimum sample of $N = 142$ for Experiment 1 and $N = 216$ for Experiment 2 (72 per group) to detect a small to moderate effect size ($d = 0.30$) between experimental groups with desired power at 0.80 and α set at 0.05 (two-sided). The present sample sizes were sufficient for experimental designs and planned analyses.

2.2. Design (Experiments 1 and 2)

Experiment 1 used a between-subjects experimental design to test the effect of a food addiction explanation ($N = 263$) for weight control and obesity vs. the diet and exercise explanation ($N = 267$) on weight stigma (prejudice towards "fat" people). The host university's Qualtrics research platform randomisation function with a 1:1 ratio was used for randomisation to conditions. Participants received either a simulated news article from The Guardian on the food addiction explanation for weight control and obesity (food addiction condition) or an identically formatted news article positing the dominant public health message that weight control and obesity are due to poor dieting and/or exercise behaviour. Single-item post-manipulation measures were taken for all variables.

Experiment 2 ($N = 690$) was identical to Experiment 1, but introduced a control group that received no newspaper article. The randomisation ratio was set at 2:1:1 with the control condition ($N = 346$) having two participants for every one participant allocated to the food addiction ($N = 167$) and diet and exercise conditions ($N = 175$).

2.3. Manipulation (Food Addiction vs. Diet and Exercise News Articles, vs. Control/No News Article)

Two newspaper articles were constructed for the experiments. The two articles appeared authentic and were structurally identical, using The Guardian newspaper format, with identical author, date/time of publication, word length, and text/photo placement. Both articles contained identical text reporting on research from The Lancet suggesting that mortality from obesity-related diseases was high and, for the first time, greater than mortality from starvation. However, the articles differed considerably in the text regarding explanations for weight control and obesity.

The *food addiction* article described the concept of food addiction and explained how foods can be addictive through the involvement of the pleasure/reward centres of the brain, and the release of dopamine when eating some foods, which in turn leads to cravings and a vicious cycle of addiction. The article named the originator of the term food addiction, and stated that approximately 20% of the population may have a food addiction, particularly to highly processed or convenience foods. The article also suggested that food addiction was a key factor in weight control and overweight and obesity.

The *diet and exercise* news article made no reference to food addictions or cravings, but instead focused on people's lack of physical exercise, sedentary lifestyles, and overconsumption of unhealthy foods. The article stated that these personal behaviours were the cause of obesity. The article cited research from experts stating that more self-control was needed when choosing what foods to eat, and it concluded by stating that diet and exercise programs are our best chance at reducing the obesity epidemic and that people need to get moving more.

2.4. Measures

Along with demographic characteristics including age, sex, height in centimetres, weight in kilograms, and ethnicity, we assessed participant's weight stigma (i.e., anti-fat prejudice, weight bias). We also included simple measures to assess whether the manipulations affected beliefs about the causes of obesity and weight gain and loss and a food addiction condition manipulation check. All participants received all of the measures summarized below.

2.4.1. Weight Stigma

To measure weight stigma we used the Anti-Fat Attitudes Test (AFAT), a psychometrically sound measure that has been widely used in the field to measure weight bias [31]. The AFAT is a 13-item scale comprised of three subscales assessing dislike of "fat people" which assesses antipathy towards people perceived to be "fat" (Dislike: 7 items, e.g., "I really don't like fat people much"), fear of becoming fat (Fear of Fat: 3 items, e.g., "I worry about becoming fat"), and belief that excess weight is due to a lack personal willpower (Willpower: 3 items, e.g., "Some people are fat because they have no willpower"). Participants indicate their agreement to items using a scale ranging from 0 = *very strongly disagree* to 9 = *very strongly agree*. The mean of the subscale items is used for analyses. Previous work has identified that the 'fear of fat' subscale functions as a measure of personal body image rather than weight stigma toward others per se, with one of the items lacking face validity, so this subscale was not analysed in the present study [32]. Cronbach's alpha's for the dislike and willpower subscales were good in the present sample: $\alpha = 0.87$ and $\alpha = 0.79$, respectively.

2.4.2. Belief in the Food Addiction Explanation

The food addiction support index (FASI) [24] was used to assess participant beliefs in, and support for, the food addiction explanation for eating, obesity, and weight gain, following exposure to the food addiction vs. diet and exercise news articles (manipulation). Participants responded to the five-item FASI (e.g., "Obesity should be treated as an addiction"), using a five-point scale ranging from 0 = *strongly disagree* to 4 = *strongly agree* with items summed to form a scale total from 0 to 20. Cronbach's alpha's in the present experiment was $\alpha = 0.83$.

2.4.3. Belief in Diet and Exercise for Weight Control

Two items from the dieting beliefs scale [33] that directly capture beliefs about exercise and dieting for the control of weight were used to assess the following exposure to the food addiction vs. diet and exercise news article (manipulation). Specifically, participants were asked to indicate their agreement using a six-point scale ranging from 1 = *not at all descriptive of my beliefs* to 6 = *very descriptive of my beliefs* to the statements "By restricting what one eats, one can lose weight" and "By increasing the amount that one exercises, one can lose weight". The two items were combined to form a score ranging from 1 to 12, with higher scores indicating greater belief that diet and exercise are responsible for weight control and obesity.

2.5. Food Addiction Manipulation Check

To assess whether participants attended to the information in the food addition article, we asked a short question assessing recall for a specific piece of information in the food addiction article. Specifically, we asked participants to identify via a multi-choice recognition response (four answer options) "Who first introduced the term Food Addiction?" Participant responses were coded as either incorrect = 0 or correct = 1.

2.6. Procedure

Upon entering the experiment via a web-link to the host university's Qualtrics research platform, participants in both experiments were provided with the title and description of the study, and then were asked to provide consent to participate. To limit bias in sampling and responding, the study used deception in the advertising and description of the experiment. Participants were told that the experiment was interested in how cognitive information processing styles affect public opinion on a range of political, health, and social issues and that researchers were interested in how people deal with being saturated by the wide range of media messages they receive via TV, computer, and mobile devices.

Participants first answered demographic questions before being presented with one of the two news articles (FAM condition vs. diet/exercise condition) or no article for those in the control group for Experiment 2. To enhance the authenticity of the experiment guise, a large set of distractor questions taken from measures assessing experiential and analytical thinking styles [34] were interspersed in the outcome and manipulation measures. These measures asked participants whether they enjoy the process of thinking deeply about issues, and they have been used successfully elsewhere [35]. Finally, participants were presented with the outcome and manipulation measures. The experiment took approximately 22 min on average to complete. Ethical approval for the study was sought and provided by the host university's Human Research Ethics Committee (Project ID: 8912).

2.7. Analysis

Chi-squared (X^2) and t-tests were used to assess whether randomisation resulted in balanced groups based on gender, age, and body mass index (BMI). Because weight stigma scores were not normally distributed in either experiment, we adopted a two-step transformation to normalise the data [36]. Subsequent normality checks showed the data to have no issues with skewness or kurtosis. ANCOVAs accounting for age, sex, and BMI as covariates were used to test for significant mean group differences on the dislike and willpower weight stigma measures. A Chi-squared test assessed accurate recognition of the food addiction originator in a probe question (manipulation check). ANOVA was also used to examine whether there were any differences on the manipulation measures (i.e., FASI, dieting/exercise beliefs, and food addiction article attention; i.e., who was the originator of food addiction term). We report adjusted means (M) and standard deviations (SD) along with F and p-values (significance was set at 0.05) for all primary outcomes.

3. Results (Experiments 1 and 2)

3.1. Preliminary Analysis Experiment 1

Preliminary analyses (X^2 and t-tests) assessing whether randomisation resulted in balanced groups on demographic characteristics showed there were no significant differences between groups for sex, age, or BMI scores (all p values > 0.27). A higher proportion of participants in the food addiction condition (61%) correctly recalled the name of the originator of the term food addiction than did those in the diet and exercise condition (39%, $X^2 = 23.663$, $p < 0.0005$).

3.2. Experiment 1 Food Addiction vs. Diet and Exercise

ANCOVA found significant group difference for the 'dislike' weight stigma measure: $F_{(4,525)} = 7.675$, $p = 0.006$. Participants exposed to the FAM explanation had significantly lower dislike of "fat people" (M = 1.82, SD = 1.46) scores than did the participants in the diet and exercise condition (M = 2.13, SD = 1.70). Similarly, participants in the FAM group endorsed significantly lower willpower stigma scores (M = 4.29, SD = 205) than participants in the diet and exercise explanation group (M = 4.68, SD = 1.97; $F_{(4,525)} = 5.393$, $p = 0.021$), indicating that those in the FAM condition were less likely to attribute excess weight to a lack of personal willpower.

We examined whether there were differences between groups in the posited beliefs about the causes of weight control and obesity (i.e., FASI, dieting/exercise beliefs). There were statistically significant group difference on FASI scores. Participants receiving the FAM explanation for obesity and weight control had greater belief in the FAM explanation for obesity and weight control (FASI) (M = 14.57, SD = 3.78) than participants receiving the traditional diet and exercise explanation (M = 13.66, SD = 3.80; $F_{(4,525)} = 8.823$, $p = 0.003$). There was not a statistically significant group difference for the dieting and exercise beliefs measure (FAM, M = 8.60, SD = 2.04; diet and exercise M = 8.50, SD = 2.03; $F_{(4,525)} = 0.100$, $p = 0.75$).

3.3. Preliminary Analysis Experiment 2

Preliminary analyses (X^2 and t-tests) assessing whether randomisation resulted in balanced groups on demographic characteristics showed there were no significant differences between groups for sex, age, or BMI scores (all p values > 0.46). A higher proportion of participants in the food addiction condition (58%) correctly recalled the name of the originator of the term food addiction than did those in the diet and exercise condition (42%; $X^2 = 15.047$, $p < 0.0005$).

3.4. Experiment 2 Food Addiction vs. Diet and Exercise vs. Control/No News Article

ANCOVA found significant group difference for the dislike of 'fat people' weight stigma scores ($F_{(5,684)} = 5.157$, $p = 0.006$). Post-hoc tests showed that participants exposed to FAM explanation had significantly lower dislike of 'fat people' (M = 1.85, SD = 1.57) than participants in the diet and exercise condition (M = 2.23, SD = 1.76, $p < 0.001$). Participants in the diet and exercise condition also endorsed higher weight stigma scores than those in the control condition (M = 1.70, SD = 1.54). There was no significant difference in dislike of 'fat people' scores between the FAM condition and the control group ($p = 0.56$). There was also no significant group difference for willpower weight stigma scores ($F_{(5,684)} = 0.217$, $p = 0.81$), with the FAM, diet and exercise, and control groups having similar mean scores (M = 4.71, SD = 2.05, and M = 4.81, SD = 1.86, M = 4.62, SD = 2.16, respectively).

We examined whether there were differences between groups in the posited beliefs about the causes of weight control and obesity (i.e., dieting/exercise beliefs, FASI). There was no significant group difference for the dieting/exercise beliefs measure ($F_{(5,682)} = 2.055$, $p = 0.13$; diet and exercise condition M = 4.62, SD = 0.95, control condition M = 4.38, SD = 1.06, FAM group M = 4.43, SD = 1.03). We found no statistically significant group difference in FASI scores ($F_{(5,680)} = 1.764$, $p = 0.17$; control M = 14.54, SD = 4.17, FAM M = 15.01, SD = 3.29, diet and exercise M = 13.99, SD = 3.90).

4. Discussion

We examined whether the FAM explanation for weight control and obesity versus the traditional public health messaging around control of diet and exercise would affect weight stigma. Relative to the dominant public health narrative that obesity stems from lack of control over diet and exercise, the FAM explanation resulted in lower weight stigma. In Experiment 1, both dislike of 'fat people' and perceptions that excess weight is a result of lack of willpower were lower in people presented with the FAM explanation. In a second experiment, we introduced a control group that was not exposed to information about obesity or causal models for obesity. Similar to Experiment 1, participants in Experiment 2 who received the FAM explanation for weight control and obesity displayed less weight stigma (dislike of "fat people") than participants in the diet and exercise condition. Importantly, there was no difference between the FAM and control groups in levels of weight stigma. Contrary to Experiment 1, in Experiment 2 there was no significant group difference with respect to perceptions that excess weight is caused by a lack of willpower (blame). The results of these two experiments are consistent with work by Latner and colleagues [25] who reported less stigma following exposure to a food addiction explanation. The results do not appear to support the notion that attaching an addiction label to people with obesity would exacerbate weight stigma. Accordingly, a simple interpretation of the results of both experiments is that the dominant public health messaging around the diet and exercise explanation for weight control and obesity exacerbates weight stigma, but the FAM explanation does not.

We found mixed support for the attribution value model of weight stigma [10]. Analysis of the FASI scores in Experiment 1 suggest that the difference in weight stigma scores between the FAM and diet and exercise conditions was due to changes in participants' attributions about the causes of obesity. However, we found no significant group difference on FASI scores in Experiment 2. Similarly, there was no group differences in participant beliefs about personal control of diet and exercise as a primary cause for obesity and weight control in Experiment 1 or 2. This finding is unexpected, as it was reasonable to predict a decrease in attributions of dieting and exercise as a causal explanation among participants who received the FAM explanation. It is possible that the FAM explanation resulted in less stigma because of a greater understanding of, and/or empathy for, those facing the challenges of weight management when one is addicted to food. As we did not assess constructs related to empathy, it will be important for future work to assess the relationship between perceptions of food addiction and levels of empathy towards people with obesity. It is apparent that those in the FAM condition did not dismiss the notion of diet and exercise as contributors to weight control and obesity. Participants may already have firmly established beliefs about personal control of weight given pervasive societal messages emphasizing this message. However, this possible pre-existing belief did not interfere with participants' abilities to receive and incorporate a different message about the contributors to obesity emphasized in the FAM perspective.

The perception that weight is determined by the individual's personal control of choices regarding dieting and exercise behaviours is widespread and accepted in society [37]. Belief in this dominant public health model for overweight and obesity is thought to be linked to weight stigma because it infers and/or attributes overweight and obesity to personal responsibility, lack of discipline, and laziness. The FAM explanation for obesity and weight control is garnering attention in several research fields [19,38,39] and is gaining traction in popular culture [40]. Indeed, studies suggest that around 15% of people meet criteria for a food addiction and anywhere from 28–52% of people believe they may be addicted to a food [22–24]. The present findings support suggestions that popular societal messages of blame and personal responsibility for weight may be partly responsible for the prevalence and rise of weight bias [1]. In contrast, an alternative explanation for obesity, the FAM explanation, may have a positive effect of reducing weight stigma. As our studies did not assess attitudes over time, it will be important for future work to examine whether changes in participants' causal attributions for obesity can be maintained following brief interventions or information about alternative contributors to obesity. It is likely that repeated exposure to FAM information and messages would be needed to sustain shifts in people's perspectives over time, especially in the face of continued and prominent societal messages emphasizing personal behaviours as the primary cause of obesity. Nevertheless, findings from both of our studies suggest that it is possible to shift weight stigma attitudes with a brief intervention emphasizing a FAM narrative for obesity.

Several limitations of this research should be noted. First, study participants were mostly young and female, with a majority who did not have overweight or obesity. Future studies should examine the FAM explanation in a more representative population sample, including those with diverse body sizes. Baseline data on weight stigma variables were not collected, which could otherwise have been included as potential covariates in the analyses. However, the decision not to collect baseline data on these variables was balanced against the importance of maintaining the study's deception, to avoid tipping off participants to the nature of the study and cuing them in the direction of socially desirable responding. Still, future studies could employ within-participant designs to examine potential changes in weight stigma across subjects upon exposure to the FAM model of weight. In addition, future research should explore the longer-term outcomes of public health messages, particularly those delivered in real-world settings.

5. Conclusions

The present study found lower weight stigma after exposure to a food addiction model of weight and greater weight stigma after exposure to a diet and exercise model of weight. While several studies have attempted to reduce weight stigma, with varying success, far more work is needed to address this prevalent and harmful societal problem. The improvement found in weight stigma following an addiction explanation, and the worsening of weight stigma following a diet and exercise explanation, has implications for public health messages about body weight and obesity. The results support the growing popularity of the food addiction model for eating and associated body weight and obesity. Increasing public understanding of the role of a food addiction explanation for eating behaviour and weight may help to alleviate weight stigma, including, potentially, internalised weight bias/self-stigma. At the same time, our research suggests that the current dominant public health message that largely attributes weight control and obesity to lack of personal control of, and responsibility for, diet and exercise needs to be changed as it appears to be supporting weight stigma [41]. Future research is needed to explore ways to modify current public health messaging so that it is not exacerbating weight stigma. Such messaging could describe the interplay between biological (e.g., FAM, genes) and/or environmental factors (e.g., food security, access to healthy and affordable foods, exercise-facilitating living and work environments) contributing to weight, as well as the importance of healthy eating and physical activity for all individuals, regardless of body size.

Author Contributions: Conceptualization: K.S.OB., A.C., J.D.R., J.D.L. data curation: K.S.OB., J.D.R., A.B.; formal analysis: K.S.OB., J.D.R.; project administration: K.S.OB., A.C., J.D.R.; supervision: K.S.OB., A.C.; writing—original draft: K.S.OB., J.D.L., Z.V., R.M.P., A.B.; writing—review and editing: K.S.OB., A.B., R.J., J.A.H., Z.V., D.S., D.L. All authors have read and agreed to the published version of the manuscript.

Funding: This research received no external funding.

Conflicts of Interest: The authors declare no conflict of interest.

References

1. Andreyeva, T.; Puhl, R.M.; Brownell, K.D. Changes in perceived weight discrimination among Americans, 1995–1996 through 2004–2006. *Obesity* **2008**, *16*, 1129–1134. [CrossRef]
2. Latner, J.D.; Stunkard, A.J. Getting worse: The stigmatization of obese children. *Obes. Res.* **2003**, *11*, 452–456. [CrossRef]
3. Puhl, R.M.; Heuer, C.A. The stigma of obesity: A review and update. *Obesity* **2009**, *17*, 941–964. [CrossRef]
4. Nutter, S.; Ireland, A.; Alberga, A.S.; Brun, I.; Lefebvre, D.; Hayden, K.A.; Russell-Mayhew, S. Weight bias in educational settings: A systematic review. *Curr. Obes. Rep.* **2019**, *8*, 185–200. [CrossRef] [PubMed]

5. Phelan, S.M.; Burgess, D.J.; Yeazel, M.W.; Hellerstedt, W.L.; Griffin, J.M.; van Ryn, M. Impact of weight bias and stigma on quality of care and outcomes for patients with obesity. *Obes. Rev.* **2015**, *16*, 319–326. [CrossRef] [PubMed]
6. O'Brien, K.S.; Latner, J.D.; Ebneter, D.; Hunter, J.A. Obesity discrimination: The role of physical appearance, personal ideology, and anti-fat prejudice. *Int. J. Obes.* **2013**, *37*, 455–460. [CrossRef] [PubMed]
7. Papadopoulos, S.; Brennan, L. Correlates of weight stigma in adults with overweight and obesity: A systematic literature review. *Obesity* **2015**, *23*, 1743–1760. [CrossRef]
8. O'Brien, K.S.; Latner, J.D.; Puhl, R.M.; Vartanian, L.R.; Giles, C.; Griva, K.; Carter, A. The relationship between weight stigma and eating behavior is explained by weight bias internalization and psychological distress. *Appetite* **2016**, *102*, 70–76. [CrossRef]
9. Mensinger, J.L.; Tylka, T.L.; Calamari, M.E. Mechanisms underlying weight status and healthcare avoidance in women: A study of weight stigma, body-related shame and guilt, and healthcare stress. *Body Image* **2019**, *25*, 139–147. [CrossRef]
10. Crandall, C.S.; D'Anello, S.; Sakalli, N.; Lazarus, E.; Wieczorkowska, G.; Feather, N.T. An attribution-value model of prejudice: Anti-fat attitudes in six nations. *Pers. Soc. Psychol. Bull.* **2001**, *27*, 30–37. [CrossRef]
11. O'Brien, K.S.; Puhl, R.M.; Latner, J.D.; Mir, A.S.; Hunter, J.A. Reducing anti-fat prejudice in preservice health students: A randomized trial. *Obesity* **2010**, *18*, 2138–2144. [CrossRef]
12. Pearl, R.L.; Lebowitz, M.S. Beyond personal responsibility: Effects of causal attributions for overweight and obesity on weight-related beliefs, stigma, and policy support. *Psychol. Health* **2014**, *29*, 1176–1191. [CrossRef]
13. Barry, C.L.; Brescoll, V.L.; Brownell, K.D.; Schlesinger, M. Obesity metaphors: How beliefs about the causes of obesity affect support for public policy. *Milbank Q.* **2009**, *87*, 7–47. [CrossRef] [PubMed]
14. Barry, C.L.; Jarlenski, M.; Grob, R.; Schlesinger, M.; Gollust, S.E. News media framing of childhood obesity in the United States from 2000 to 2009. *Pediatrics* **2011**, *128*, 132–145. [CrossRef] [PubMed]
15. World Health Organisation Fact Sheet. Available online: https://www.who.int/news-room/fact-sheets/detail/obesity-and-overweight#targetText=In%202016%2C%20more%20than%201.9,women)%20were%20obese%20in%202016 (accessed on 28 July 2019).
16. Kleinert, S.; Horton, R. Obesity needs to be put into a much wider context. *Lancet* **2019**, *393*, 724–726. [CrossRef]
17. Danielsdottir, S.; O'Brien, K.S.; Ciao, A. Anti-fat prejudice reduction: A review of published studies. *Obes. Facts* **2010**, *3*, 47–58. [CrossRef]
18. Gearhardt, A.N.; Davis, C.; Kuschner, R.; Brownell, K.D. The addiction potential of hyperpalatable foods. *Curr. Drug Res. Rev.* **2011**, *4*, 140–145. [CrossRef]
19. Adams, R.C.; Sedgmond, J.; Maizey, L.; Chambers, C.D.; Lawrence, N.S. Food addiction: Implications for the diagnosis and treatment of overeating. *Nutrients* **2019**, *11*, 2086. [CrossRef]
20. Alonso-Alonso, M.; Woods, S.C.; Pelchat, M.; Grigson, P.S.; Stice, E.; Farooqi, S.; Khoo, C.S.; Mattes, R.D.; Beauchamp, G.K. Food reward system: Current perspectives and future research needs. *Nutr. Rev.* **2015**, *73*, 296–307. [CrossRef]
21. Volkow, N.D.; Wise, R.A.; Baler, R. The dopamine motive system: Implications for drug and food addiction. *Nat. Rev. Neurosci.* **2017**, *18*, 741–752. [CrossRef]
22. Gearhardt, A.N.; Corbin, W.R.; Brownell, K.D. Development of the yale food addiction scale version 2.0. *Psychol. Addict. Behav.* **2016**, *30*, 113–121. [CrossRef] [PubMed]
23. Meadows, A.; Nolan, L.J.; Higgs, S. Self-perceived food addiction: Prevalence, predictors, and prognosis. *Appetite* **2017**, *114*, 282–298. [CrossRef] [PubMed]
24. Latner, J.D.; Puhl, R.M.; Murakami, J.M.; O'Brien, K.S. Food addiction as a causal model of obesity. Effects on stigma, blame, and perceived psychopathology. *Appetite* **2014**, *77*, 79–84. [CrossRef] [PubMed]
25. Reid, J.; O'Brien, K.S.; Puhl, R.; Hardman, C.A.; Carter, A. Food addiction and its potential links with weight stigma. *Curr. Addict. Rep.* **2018**, *5*, 192–201. [CrossRef]
26. DePierre, J.D.; Puhl, R.M.; Luedicke, J. Public perceptions of food addiction: A comparison with alcohol and tobacco. *J. Subst. Use* **2013**, *19*, 1–6. [CrossRef]
27. DePierre, J.A.; Puhl, R.M.; Luedicke, J. A new stigmatized identity? Comparisons of a 'Food Addict' label with other stigmatized health conditions. *Basic Appl. Soc. Psych.* **2013**, *35*, 10–21. [CrossRef]
28. Eller, D. Are You Addicted to Food? *Prevention magazine.* Available online: http://www.prevention.com/ (accessed on 30 November 2011).

29. Micco, N. Addicted to Food? 5 Tips to Control Your Cravings. *The Huffington Post*. Available online: http://www.huffingtonpost.com/eatingwell/could-you-be-addicted-to-_b_828270.html (accessed on 27 February 2011).
30. Crandall, C.S. Prejudice against fat people: Ideology and selfinterest. *J. Pers. Soc. Psychol.* **1994**, *66*, 882–894. [CrossRef]
31. O'Brien, K.S.; Hunter, J.A.; Banks, M. Implicit anti-fat bias in physical educators: Physical attributes, ideology, and socialisation. *Int. J. Obes.* **2007**, *31*, 308–314. [CrossRef]
32. Lee, N.M.; Lucke, J.; Hall, W.D.; Meurk, C.; Boyle, F.M.; Carter, A. Public views on food addiction and obesity: Implications for policy and treatment. *PLoS ONE* **2013**, *8*, e74836. [CrossRef]
33. Stotland, S.; Zuroff, D.C. A new measure of weight locus of control: The dieting beliefs scale. *J. Pers. Assess.* **1990**, *54*, 191–203. [CrossRef]
34. Epstein, S.; Pacini, R.; Denes-Raj, V.; Heier, H. Individual differences in intuitive experiential and analytical-rational thinking styles. *J. Pers. Soc. Psychol.* **1996**, *71*, 390–405. [CrossRef] [PubMed]
35. O'Brien, K.S.; Latner, J.D.; Halberstadt, J.; Hunter, J.A.; Anderson, J.; Caputi, P. Do antifat attitudes predict antifat behaviors? *Obesity* **2008**, *16*, S87–S92. [CrossRef] [PubMed]
36. Templeton, G.F. A two-step approach for transforming continuous variables to normal: Implications and recommendations for IS research. *Commun. Assoc. Inf. Syst.* **2011**, *28*, 41–58. [CrossRef]
37. Brownell, K.D.; Kersh, R.; Ludwig, D.S.; Post, R.C.; Puhl, R.M.; Schwartz, M.B.; Willett, W.C. Personal responsibility and obesity: A constructive approach to a controversial issue. *Health Aff.* **2010**, *29*, 379–387. [CrossRef]
38. Davis, C.; Curtis, C.; Levitan, R.D.; Carter, J.C.; Kaplan, A.S.; Kennedy, J.L. Evidence that 'food addiction'is a valid phenotype of obesity. *Appetite* **2011**, *57*, 711–717. [CrossRef]
39. Schulte, E.M.; Joyner, M.A.; Potenza, M.N.; Grilo, C.M.; Gearhardt, A.N. Current considerations regarding food addiction. *Curr. Psychiatry Rep.* **2015**, *17*, 19. [CrossRef]
40. Almendrala, A. Food Addiction vs. Eating Addiction: Why A Single Word Makes All The Difference [Internet]. *HuffPost Australia*. 2014. Available online: https://www.huffingtonpost.com.au/2014/09/23/food-addiction-eating-addiction_n_5844712.html?guccounter=1 (accessed on 23 November 2015).
41. Puhl, R.M.; Latner, J.D.; O'Brien, K.; Luedicke, J.; Forhan, M.; Danielsdottir, S. Cross-national perspectives about weight-based bullying in youth: Nature, extent and remedies. *Pediatr. Obes.* **2015**, *11*, 241–250. [CrossRef]

© 2020 by the authors. Licensee MDPI, Basel, Switzerland. This article is an open access article distributed under the terms and conditions of the Creative Commons Attribution (CC BY) license (http://creativecommons.org/licenses/by/4.0/).

MDPI
St. Alban-Anlage 66
4052 Basel
Switzerland
Tel. +41 61 683 77 34
Fax +41 61 302 89 18
www.mdpi.com

Nutrients Editorial Office
E-mail: nutrients@mdpi.com
www.mdpi.com/journal/nutrients